Electronics and Communications for Scientists and Engineers

Electronics and Communications for Scientists and Engineers

Martin Plonus

Department of Electronic and Computer Engineering
Northwestern University
Evanston, Illinois

San Diego   San Francisco   New York   Boston   London   Sydney   Tokyo

Front cover photograph: Index Stock Imagery, Inc.

This book is printed on acid-free paper.

Copyright © 2001 by HARCOURT/ACADEMIC PRESS

All Rights Reserved.
No part of this publication may be reproduced or transmitted in any form or by any means, electronic or mechanical, including photocopy, recording, or any information storage and retrieval system, without permission in writing from the publisher.

Requests for permission to make copies of any part of the work should be mailed to: Permissions Department, Harcourt, Inc., 6277 Sea Harbor Drive, Orlando Florida 32887-6777

Academic Press
A Harcourt Science and Technology Company
525 B Street, Suite 1900, San Diego, California 92101-4495, U.S.A.
http://www.academicpress.com

Academic Press
Harcourt Place, 32 Jamestown Road, London NW1 7BY, UK
http://www.academicpress.com

Harcourt/Academic Press
A Harcourt Science and Technology Company
200 Wheeler Road, Burlington, Massachusetts 01803
http://www.harcourt-ap.com

Library of Congress Catalog Card Number: 2001086295

International Standard Book Number: 0-12-533084-7

PRINTED IN THE UNITED STATES OF AMERICA
01 02 03 04 05 06 IPC 9 8 7 6 5 4 3 2 1

To my wife, Martina, and our children, Sabine, Jacqueline, Marcus, and Michelle.

CONTENTS

PREFACE

Focus of the Book

Although the audience for this book is the same as that for broad-based electrical engineering texts, this book differs in length, structure, and emphasis. Whereas the traditional texts for nonelectrical engineering cover circuits and electronics and then treat electrical machinery, we believe that it is more important for today's students to be knowledgeable in digital technology than in electrical machinery. After developing circuits and analog electronics in the early chapters of this text, we continue with digital electronics and conclude with chapters on the digital computer and on digital communications—chapters that are normally not included in books for non-EEs.

The text is intended for students who need to understand modern electronics and communication. Much of the material in the book is developed from the first principles, so previous courses on circuits, for example, are not required; only freshman math and physics courses, and the elementary treatment of circuits that freshman physics provides, are expected. The emphasis throughout the book is on applications and on understanding the underlying principles. For example, Chapter 8 is presented from the perspective of a user who needs to understand the various subsystems of a computer, including the relationship of hardware and software such as operating systems and application programs. Expertise in designing computers is thus left to more advanced courses. Similarly, Chapter 9 on digital communication is sufficiently detailed to present the information sampling and pulse code modulation necessary for an understanding of such diverse subjects as digital signal processing, the audio CD, and the Internet. More advanced topics are left to specialized communication texts.

Presenting and teaching circuits, electronics, and digital communications from a single textbook can be an advantage if nonmajors are limited to a single EE course, which seems to be the trend at many schools.

Motivation for the Book

Electrical engineering began in the power industry, rapidly progressed to electronics and communications, and then entered the computer age in the 1960s. Today, electrical and electronic devices, analog and digital, form the backbone of such diverse fields as computer engineering, biomedical engineering, and optical engineering, as well as financial markets and the Internet. For example, the electronics in a modern aircraft constitute about 50% of the total cost.

This text is an outgrowth of lecture notes for a one-term course titled Applications of Electronic Devices that is offered, on an elective basis, to non-electrical-engineering students. It provides a sufficiently deep understanding of this subject for students to interact intelligently with other engineers. The goal is not so much to teach design as to present basic material in sufficient depth so that students can appreciate and understand the application chapters on operational amplifiers, the digital computer, and digital communication networks. A suitable textbook for such a course did not exist. Typical electronics texts omit circuits and communications and are too detailed. On the other hand, texts on electrical engineering for non-EEs are very broad, with material on machinery and power engineering that is not relevant to electronics and communication. In addition, the breadth of these texts, when used in a one-term course, often forces the omission of certain sections, making the flow of the presentation choppy. Finally, encyclopedic books that are useful as references for designing circuits are much too advanced for nonmajors. What is needed is a text brief enough for a one-term course that begins with chapters on AC and DC circuits, then progresses to analog and digital electronics, and concludes with application chapters on contemporary subjects such as digital computers and digital communication networks—demonstrating the importance as well as the underlying basis of electronics in modern technology. These views were used as guidelines for writing this text.

Organization of the Book

The book has three basic parts: circuits, electronics, and communications. Because electronics is basically the combination of circuit elements R, L, and C and active elements such as a transistor, we begin the book with a study of circuits. DC circuits are presented first because they are simpler but still permit the development of general principles such as Thevenin's theorem, maximum power transfer, and "matching." Resistors, defined by Ohm's law, are shown to be energy conversion elements, and capacitors and inductors are energy storage elements. The distinction between ideal and practical sources is stressed before loop equations are introduced as a method for solving for currents and voltages anywhere in a circuit. AC circuits are considered in Chapter 2, where we first learn that in a circuit, currents and voltages can change significantly with changes in the frequency of the applied source. Resonance, band-pass action, and bandwidth are a consequence. Average power, effective values of AC or of any periodic waveform, transformers, and impedance matching complete the chapter. These two chapters provide the basic understanding of DC and AC circuits, of transient analyses, and of frequency response and in that sense serve as a foundation for the remainder of the book.

In Chapter 3 we add a new element, a diode, to a circuit. Omitting lengthy theory, we simply define a diode as a fast on–off switch which in combination with RLC elements makes possible clippers, clampers, voltage regulators, SCRs, etc. However, we emphasize its use in power supplies that change AC to DC. As DC powers most electronic equipment, a power supply is an important component in computers, TVs, etc. A simple power supply consisting of a rectifier and capacitor filter is designed. This simple design nevertheless gives the student an appreciation of the concept, even though modern power supplies can be quite complicated circuits.

In Chapter 4 we begin the study of electronics with the underlying physics of the *pn* junction, which can explain diode and transistor action for students who are baffled by these seemingly mystical devices. Equally baffling is the transistor's ability to amplify, which we approach by first considering a graphical analysis of an amplifier circuit. The notion of a load line imposed by the external circuit to a transistor and drawn on the transistor characteristic graphs seems to be acceptable to the student and is then easily extended to explain amplifier action. The load line and Q-point also help to explain DC biasing, which is needed for proper operation of an amplifier. Only then is the student comfortable with the mathematical models for small-signal amplifiers. After frequency response, square wave testing, and power amplifiers, we are ready to consider a complete system. As an example we dissect an AM radio receiver and see how the parts serve the system as a whole, noting that the electronics of most AM receivers these days come as integrated chips allowing no division into parts. Chapters 3, 4, and 5 cover analog electronics, and large parts of these chapters could be omitted if the choice is made to deemphasize analog and devote more class time to digital electronics.

Operational amplifiers are the subject of Chapter 6. This chapter can stand alone because it is to a large extent independent of the previous three chapters on analog electronics. After presenting the standard inverting op amp circuit, which is characterized by moderate but stable gain obtained by applying large amounts of negative feedback to the op amp, we consider a wide variety of practical op amp devices from summers, comparators, integrators, differential amplifiers, filters, A/D, and D/A converters. A final example of the analog computer is given primarily because it applies to control, teaches us a tad more about differential equations, and shows how a mechanical system can be effectively modeled and solved by electrical circuits.

The final three chapters consider the subject of digital electronics. The last chapter, even though on digital communication, is nonetheless rooted in electronics. Our objective for these chapters is to give the student a deeper understanding of the digital computer and the Internet, cornerstones of the digital revolution. Gates, combinatorial and sequential logic, flip-flops, and the microprocessor (Experiment 9), all building blocks for more complex systems, are considered in Chapter 7. We move to the digital computer in Chapter 8 and to communication networks in Chapter 9. These chapters are not so much intended to teach design skills as they are for the nonmajor to acquire a thorough understanding of the subject matter for a workable interaction with experts. In that sense, the chapter on the digital computer concentrates on those topics with which the user interacts such as programming languages, RAM and ROM memory, the CPU, and the operating system. Similarly, in Chapter 9 we cover the sampling process, Nyquist criterion, information rates, multiplexing, and pulse code modulation, all of which are necessary for an understanding of digital signal processing and digital communication networks such as the Internet.

Acknowledgments

I thank Dr. Carl J. Baumgaertner of Harvey Mudd College, Dr. Shawn Blanton of Carnegie Mellon University, Dr. Gary Erickson of Idaho State University, and Dr. Can E. Korman of The George Washington University for reviewing the manuscript. I also thank my colleagues at Northwestern University, Professors Larry Henschen, Mike Honig, Zeno Rekasius, and Alan Sahakian, for reviewing portions of the manuscript.

Martin Plonus

CHAPTER 1

Circuit Fundamentals

1.1 INTRODUCTION

Electronics deals with voltage and current interaction in a network of resistances R, inductances L, capacitances C, and active elements such as transistors. The purpose is usually to amplify signals or to produce signals of a desired waveform, typically at low levels of power. A study of electronics therefore should begin with the passive elements R, L, and C—a study usually referred to as *circuit theory*.

It should then be followed by the basics of transistors, which generally act as amplifiers or on–off switches. We can then proceed to electronic circuit design in which passive and active elements are combined to form elementary circuits such as a power supply, amplifier, oscillator, A/D converter, etc. In turn we can combine these elementary circuits to create useful devices such as radios, TVs, computers, etc.

The study of electronic circuits will essentially follow this path: DC circuit analysis, AC circuit analysis, basic solid-state theory, junction diodes, transistors, elementary amplifiers and op amps, small-signal amplifier circuits, and digital electronics, which are then used as building blocks to introduce digital communications and the Internet.

1.2 DIMENSIONS AND UNITS

In this book the mksa (meter-kilogram-second-ampere) system of units, now a subsystem of the SI units, is used. A dimensional analysis should always be the first step in checking the correctness of an equation.[1] A surprising number of errors can be detected at an early stage simply by checking that both sides of an equation balance

---

[1] A *dimension* defines a physical characteristic. A *unit* is a standard by which the dimension is expressed numerically. For example, a second is a unit in terms of which the dimension time is expressed. One should not confuse the name of a physical quantity with its units of measure. For example, power should not be expressed as work per second, but as work per unit time.

dimensionally in terms of the four basic dimensions. For example, Newton's second law gives the force F in newtons (N) as

$$F = ma \quad \text{mass(length)/(time)}^2$$

An increment of work dW in *joules* (J) is given by

$$dW = F\, dl \quad \text{mass(length)}^2/\text{(time)}^2$$

where dl is an incremental distance in *meters* (m). *Power*[2] in *watts* (W), which is the time rate of doing work in joules per second, is

$$P = \frac{dW}{dt} \quad \text{mass(length)}^2/\text{(time)}^3$$

The next quantity that we want to consider is the electric current, $I = dQ/dt$, measured in *amperes* (A), which is the rate of flow of electric charge Q, measured in *coulombs* (C). The smallest naturally occurring charge e is that possessed by an electron and is equal to $-1.6 \cdot 10^{-19}$ C. A coulomb, which is a rather large charge, can be defined as the charge on $6.28 \cdot 10^{18}$ electrons, or as the charge transferred by a current of 1 A. Any charged object is a collection of elementary particles, usually electrons. The possible values of total charge Q of such an object are given by

$$Q = \pm ne \quad \text{where } n = 0, 1, 2, \ldots$$

Electric charge is quantized and appears in positive and negative integral multiples of the charge of the electron. The discreteness of electric charge is not evident simply because most charged objects have a charge that is much larger than e.

1.3   BASIC CONCEPTS

1.3.1   Electric Field

Coulomb's law states that a force F exists between two charges Q_1 and Q_2. This force is given by $F = kQ_1Q_2/r^2$, where k is a proportionality constant and r is the distance between the charges. Hence each charge is in a force field[3] of the other one as each charge is visualized as producing a force field about itself. We can now define an *electric field* E as a force field per unit charge.

$$E = \frac{F}{Q} \tag{1.1}$$

---

[2]Note that in this book we also use W as the symbol for energy. This should not lead to any confusion as it should be self-evident which meaning is intended.

[3]There are various force fields. For example, if force acts on a *mass* it is referred to as a gravitational field, if it acts on an *electric charge* it is an electric field, and if it acts on a *current-carrying wire* it is a magnetic field.

For example, the electric field which acts on charge Q_1 can be stated as $E = F/Q_1 = kQ_2/r^2$. Hence for those that are more comfortable with mechanical concepts, one can think of an electric field as a force that acts on a charge.

1.3.2 Voltage

Voltage, or *potential difference*, can be introduced similarly by considering work. If we look at a small amount of work as a small motion in a force field, and replace the force by $F = QE$, we obtain $dW = QE \cdot dl$. We can now define a *voltage*, measured in *volts* (V), as work per unit charge, i.e.,

$$dV = \frac{dW}{Q} = E \, dl \tag{1.2}$$

Hence a small voltage corresponds to a small displacement of a charge in an electric force field.

It is useful to consider the inverse operation of (1.2), which expresses work as an integration in a force field, that is, $V = \int E \cdot dl$. The inverse operation gives the force as the gradient of the work, or

$$E = -\frac{dV}{dl} \quad \text{in volts/meter (V/m)} \tag{1.3}$$

For example, to create a fairly uniform electric field one can connect, say, a 12 V battery to two parallel metallic plates as shown in Fig. 1.1.

If the plates are separated by 10 cm, an electric field of $E = 12$ V/0.1 m = 120 V/m is created inside the space. If an electron were placed inside that space, it would experience a force of

$$-1.6 \cdot 10^{-19} \cdot 120 = -1.9 \cdot 10^{-17} \text{ N}$$

and would begin to move toward the positive plate. As another example of the usefulness of the gradient expression for the electric field (1.3), let us consider a receiving antenna which is in the field of a radio transmitter that radiates an electric field. If the electric field at the receiving antenna is 1 mV/m, then a 1-m-long antenna would develop a voltage of 1 mV. If a transmission line connects the antenna to a radio receiver, this voltage would then be available to the receiver for amplification.

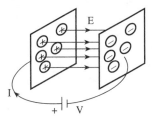

FIGURE 1.1   A uniform electric field E is created inside the space between two parallel plates.

1.3.3   Current

Current, measured in amperes (A), is the time rate at which charges Q pass a given reference point (analogous to counting cars that go by a given point on the road and dividing by the time interval of the count). Thus

$$I = \frac{dQ}{dt} \tag{1.4}$$

As Benjamin Franklin assumed charge to be positive, the direction of current (ever since his day) is given by the direction that positive charges would flow when subjected to an electric field. Hence in Fig. 1.1 the current I that flows in the wire connecting the battery to the plates has the direction of positive charge flow. On the other hand we now know that current in a conducting wire is due to moving electrons which carry a negative charge. The direction of I and electron flow are therefore opposite. Similarly, if the space between the plates is occupied by positive charges, flow would be from the left to the right plate (the direction of I), but if occupied by electrons, flow would be from the right to the left plate.[4]

1.3.4   Power

If we take the expression for power, which is the rate of doing work, and multiply it by dQ/dQ we obtain

$$P = \frac{dW}{dt} = \frac{dW}{dt}\frac{dQ}{dQ} = \frac{dW}{dQ}\frac{dQ}{dt} = VI \tag{1.5}$$

Hence *power is voltage multiplied by current*. When combined with Ohm's law, this is one of the most useful expressions in electronics.

1.3.5   Ohm's Law

Thus far we have a picture of current as a flow of charges in response to an applied electric field. Current in a conducting medium though, such as copper or aluminum, is fundamentally different. A piece of copper is neutral, overall and at every point. So, if copper has no free charge, how does a copper wire conduct a current? A conducting medium such as copper is characterized by an atomic structure in which the atoms have only one, weakly attached electron in the outer shell. Hence, even a small force, such as a small electric field, created by a small voltage across the copper wire will make the electrons move. While such motion within the wire takes place—that is, a current flows in the wire—charge neutrality throughout the wire is always preserved (charge does not accumulate in the conductor). Hence, when a current flows in a segment of copper wire,

---

[4]Note that a battery connected to two plates as shown in Fig. 1.1 produces a current only during a brief interval after connection. More on this when we consider capacitors. Current could also flow if the space between the plates is filled with free charges.

the copper does not make any net electrons available—electrons simply leave one end of the segment while at the other end the same number enters.

There is now a subtle difference between electrons moving in vacuum and electrons moving in copper or in any solid conductor. It appears that electrons in both media are free in a sense, except that an electron in copper is free to move only until it collides with one of the many copper atoms that occupy the metal. It is then slowed down and must again be accelerated by the electric field until the next collision. Hence the progression of electrons in a conducting medium is one of colliding with many atoms while moving slowly through the metal. The current through such a material is therefore determined by the resistance to electron flow due to collisions and by the voltage across the material which provides the energy for successive accelerations after collisions. *Resistivity* ρ is a material property that relates to the average interval between collisions. Incorporating the geometry of a conductor, as shown in Fig. 1.2., the resistance R of a bar ℓ meters long and A square-meters in cross-section is given by

$$R = \rho \frac{\ell}{A} \tag{1.6}$$

Hence the resistance increases with length but decreases with cross-sectional area. The unit of resistance is the *ohm*, denoted by the Greek letter Ω. The reciprocal of resistance, called *conductance G*, is also used; the unit of conductance is the *siemens* (S). A typical value of resistivity for a good conductor such as copper is $\rho = 1.7 \cdot 10^{-8}$ ohm-meters (Ω-m), whereas for a good insulator such as glass it is 10^{12} Ω-m. This is a very large difference and implies that even though copper can carry large amounts of current, for all practical purposes the ability of glass to carry even tiny amounts of current is negligible.

A current in a conductor is maintained by placing a voltage across it. A larger voltage provides more energy for the electrons, and hence more current flows in the conductor. Resistance is the constant of proportionality between current and voltage, that is,

$$V = RI \tag{1.7}$$

This is a fundamental relationship and is known as *Ohm's law*. It states that *whenever a conductor carries a current, a voltage must exist across the conductor.* Figure 1.2 shows

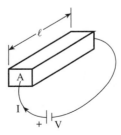

FIGURE 1.2 A resistor is formed when a material of resistivity ρ is shaped into a bar of length ℓ and cross-sectional area A.

a simple circuit to which (1.7) applies. The connecting wires between the voltage source and the bar are assumed to have zero resistance. Hence the voltage V is directly applied to the bar resistor.

1.3.6   Joule's Heating Law

During current flow in a metal, the repeated collisions of the electrons with the lattice atoms transfer energy to the atoms with the result that the temperature of the metal increases. A resistor can therefore be considered as an energy-transforming device: it converts electrical energy into heat. There are numerous everyday examples of this. An electric heater, a hair dryer, electric stove, etc., all contain resistors (usually tungsten wire) that give off heat when an electric current passes. The rate of energy conversion by a resistor can be expressed by substituting (1.7) in (1.5), which gives

$$P = VI = I^2 R = \frac{V^2}{R} \qquad (1.8)$$

The expression $P = I^2 R$ is known as *Joule's law* and P is power (equal to rate of change of energy) and is measured in *watts* (W). If we integrate this expression we obtain the thermal energy W dissipated in a resistor over a time interval T,

$$W = I^2 R T \qquad (1.9)$$

where it was assumed that current and resistance remain constant over the time interval. This is known as *Joule's heating law*, where the units of W are in joules (J).

1.3.7   Kirchoff's Laws

A circuit is an interconnection of *passive* (resistors, capacitors, inductors) and *active* (sources, transistors) elements. The elements are connected by wires or leads with negligible resistance. The circuit can be a simple one with one closed path, for example, a battery connected to a resistor (Fig. 1.2), or the circuit can be more elaborate with many closed paths. Figure 1.3*a* shows a simple, one-loop closed path circuit in which a battery forces a current to flow through two resistors (represented by "zig-zag" symbols) connected in series. We now observe, and we can state it as a rule, that current around the loop is continuous. That is, in each of the three, two-terminal elements the current entering one terminal is equal to the current leaving the other terminal at all times. In addition, the polarity convention can be used to differentiate between *sources* and *sinks*. For example, the current in the battery flows into the negative terminal and comes out at the positive terminal—this defines a source and voltage V_B is called a *voltage rise*. Resistors, which absorb energy, are called *sinks*. In a sink, the current enters the positive terminal and the voltage across a sink is called a *voltage drop*.[5] It should

---

[5]At times a battery can also be a sink. For example, connecting a 12 V battery to a 10 V battery (plus to plus and minus to minus) results in a one-loop circuit where current flows from the negative to positive

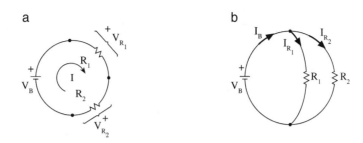

FIGURE 1.3   (a) A battery is connected to two resistors in series. (b) A battery with two resistors in parallel.

now be intuitively obvious that in a circuit, voltage rises should equal voltage drops. *Kirchhoff's voltage law* states this precisely: *the algebraic sum of the voltages around any closed path in a circuit is zero.* Mathematically this is stated as

$$\sum V_n = 0 \tag{1.10}$$

Applying (1.10) to the circuit of Fig. 1.3a, we obtain $V_B = V_{R_1} + V_{R_2}$.

A *node* is defined as a point at which two or more elements have a common connection, as for example in Fig. 1.3b, where two resistors in parallel are connected to a battery to form a two-node, two-loop circuit. *Kirchhoff's current law* states that at a node, the algebraic sum of the currents entering any node is zero. Mathematically this is stated as

$$\sum I_n = 0 \tag{1.11}$$

Applying (1.11) to the top node in the circuit of Fig. 1.3b, we obtain

$$I_B = I_{R_1} + I_{R_2}.$$

Hence, at any node, not all currents can flow into the node—at least one current must flow out of the node. In other words, a node is not a "parking lot" for charges: *the same number of charges that flow into a node must flow out.* This is not unlike a traffic junction, in which the number of cars that flow in must flow out.

1.4   CIRCUIT ELEMENTS

1.4.1   Resistors

A resistor in a simple circuit is shown in Fig. 1.2. We represent a resistor by its standard zig-zag symbol, as illustrated in Fig. 1.4. Low-resistance resistors which often have

terminal of the 12 V battery (a source), and flows from the positive to negative terminal of the 10 V battery (a sink). Of course what is happening here is that the 10 V battery is being charged by the 12 V battery.

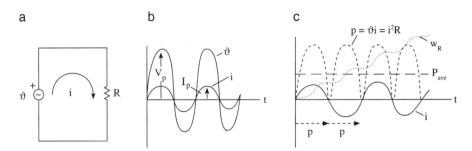

FIGURE 1.4   (a) A resistor with a voltage v applied. (b) A sinusoidal voltage causes an in-phase sinusoidal current in R. (c) Instantaneous power p in a resistor is pulsating but is always positive. The dashed arrows suggest that power is always flowing from source to resistor. Energy w_R continues to increase with time.

to dissipate power in the multiple watts range are usually wire-wound, whereas the more common, higher-resistance resistors are of carbon material in the shape of small cylinders or thin-film strips. Carbon is a nonmetallic material that can have a high resistivity. These resistors can have values up into the megaohm (MΩ) range and have small power ratings—typically $\frac{1}{4}$, $\frac{1}{2}$, and 1 W.

Ohm's law $v = Ri$ states that, for a constant R, voltage and current in a resistor are in phase. It is assumed that R remains a constant for a large range of voltages, currents, and temperatures. The in-phase property is best shown when a sinusoidal voltage is applied to the resistor and the resulting current is sketched, as in Fig. 1.4b.[6]

The current is proportional (in phase) to the applied voltage. In Fig. 1.4c we sketch the instantaneous power $p = vi = i^2 R$ and note that even though current reverses periodically, p is always positive, implying that power or energy always flows from the source to the resistor, where it is converted into heat and dissipated to the surroundings. The average power delivered to the resistor is obtained by integrating the instantaneous power over a period T of the sinusoid, that is,

$$P_{\text{ave}} = \frac{1}{T} \int_0^T i^2 R \, dt = \frac{V_p^2}{2R} = \frac{R I_p^2}{2} \tag{1.12}$$

where $I_p = V_p/R$. We can now observe that had the resistor been connected to a DC battery of voltage V, the power delivered to R would have been constant with a value $P = VI = I^2 R = V^2/R$. Hence if we equate DC power V^2/R to the corresponding average AC power (1.12), we conclude that

$$V = \frac{V_p}{\sqrt{2}} = 0.707 V_p \tag{1.13}$$

---

[6]From now on, the instantaneous values of currents or voltages that vary with time will be represented by lowercase letters, whereas uppercase letters will be used for constants such as DC voltage. For example, for the sinusoidal case, $v = v(t) = V_p \sin t$, where V_p is the peak or maximum value of the sine wave. In Fig. 1.4a, the symbol for a sinusoidal source is used, whereas in Figs. 1.1 and 1.2 the symbol for a DC source, i.e., a battery, was used.

This is called the *effective value* of an AC voltage; that is, a sinusoidal voltage of peak value V_p is as effective in delivering power to a resistor as a DC voltage of value $V_p/\sqrt{2}$. Effective or *rms* values will be considered in more detail in the following chapter.

To demonstrate that a resistor continues to absorb energy from a connected source, we can evaluate the energy supplied to a resistor, that is,

$$w_R = \int_0^t p\, dt' = R \int_0^t i^2 dt' = \frac{V_p^2}{R} \int_0^t \sin^2 t'\, dt' = \frac{V_p^2}{2R} \left[t - \frac{\sin 2t}{2} \right] \qquad (1.14)$$

Sketching the last expression of the above equation in Fig. 1.4c shows that w continues to increase, wiggling about the average term $V_p^2 t/2R$; this term is equal to Joule's heating law (1.9), and when differentiated with respect to time t also accounts for the average power given by (1.12).

1.4.2 Capacitors

A capacitor is a mechanical configuration that accumulates charge q when a voltage v is applied and holds that charge when the voltage is removed. The proportionality constant between charge and voltage is the capacitance C, that is,

$$q = Cv \qquad (1.15)$$

Most capacitors have a geometry that consists of two conducting parallel plates separated by a small gap. The C of such a structure is given by $C = \varepsilon A/\ell$, where ε is the *permittivity* of the medium between the plates, A is the area, and ℓ is the separation of the plates. Figure 1.1 shows such a parallel-plate capacitor (note that the large gap that is shown would result in a small capacitance; in practice, capacitors have a small gap, typically less than 1 mm).

The unit for capacitance is the *farad* (F), which is a rather large capacitance. Most common capacitors have values in the range of microfarads ($\mu F = 10^{-6}F$), or even picofarads ($pF = 10^{-12}F$), with the majority of practical capacitors ranging between 0.001 and 10 F. To obtain larger capacitances, we can either increase the area A, decrease the spacing ℓ, or use a dielectric medium with larger permittivity ε. For example, mica and paper have *dielectric constants*[7] of 6 and 2, respectively. Therefore, a parallel-plate capacitor of Fig. 1.1 with mica filling the space between the plates would have a capacitance six times that of a free-space capacitor. Most tubular capacitors are made of two aluminum foil strips, separated by an insulating dielectric medium such as paper or plastic and rolled into log form. It is tempting to keep reducing the spacing between the plates to achieve high capacitance. However, there is a limit, dictated by the dielectric breakdown strength of the insulating material between the plates. When this is exceeded, a spark will jump between the plates, usually ruining the capacitor by leaving

---

[7]A *dielectric constant* is defined as relative permittivity $\varepsilon_r = \varepsilon/\varepsilon_0$, where $\varepsilon_0 = 8.85 \cdot 10^{-12}$ F/m is the *permittivity of free space*.

a conducting track within the insulating material where the spark passed. Hence, knowing the breakdown electric field strength of a dielectric material (for air it is $3 \cdot 10^4$ V/cm, for paper $2 \cdot 10^5$ V/cm, and for mica $6 \cdot 10^6$ V/m) and using (1.3), which gives the electric field when the voltage and plate separation are specified, we can calculate the voltage which is safe to apply (that which will not cause arcing) to a capacitor of a given plate separation. A practical capacitor, therefore, has stamped on it not just the capacitance but also the voltage. For example, a stamp of 50 V_{DC} means, do not exceed 50 V of DC across the capacitor.

To determine how a current passes through a capacitor, we use (1.15), $q = Cv$, differentiate both sides of the equation with respect to time, and note that $i = dq/dt$; this results in

$$i = C\frac{dv}{dt} \tag{1.16}$$

for the capacitor current, where we have used lowercase letters $q, i,$ and v to denote that charge, current, and voltage can be time-changing and capacitance C is a constant. This expression shows that a constant voltage across a capacitor produces no current through the capacitor ($dv/dt = 0$). Of course, during the charging phase of a capacitor, the voltage changes and current flows.[8] If we now apply a sinusoidal voltage to the simple capacitor circuit of Fig. 1.5a, we see that the resultant current leads the applied voltage

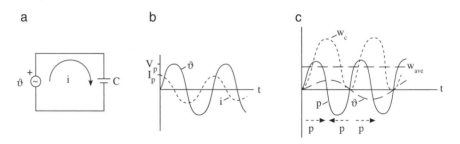

a   b   c

FIGURE 1.5   (a) A capacitor (depicted by the double line) with voltage v applied. (b) Sinusoidal voltage and current in C. (c) Instantaneous power and energy as well as average energy are sketched. (*Note:* amplitudes of p and w_C are not to scale.)

---

[8]During a brief time interval after the capacitor is connected to a battery, a charging current flows through the capacitor, that is, (1.16) gives a finite value for i because the capacitor voltage changes from zero for an initially uncharged capacitor to the battery voltage during charging. During the charging interval, dv/dt is not zero, therefore. Going back to our parallel-plate capacitor of Fig. 1.1, we infer that the charging current moves electrons from left to right through the battery, depositing electrons on the right plate, and leaving the left plate deficient of the same amount of electrons. The electrons on the charged plates do not come from the battery, but from the metallic plates, which have an abundance of free electrons. The battery merely provides the energy to move the charges from one plate to the other. Charging of a capacitor is considered in detail in Section 1.8.

by 90°, or v lags i by 90°, as shown in Fig. 1.5b. This is easily seen by use of (1.16): if $v = V_p \sin t$, then

$$i = V_p C \cos t = I_p \cos t = I_p \sin(t + \pi/2)$$

The angle of $\pi/2$ is also referred to as a 90° *degree phase shift*.

The instantaneous power in C is given by

$$p = vi = Cv\frac{dv}{dt} = \frac{CV_p^2}{2} \sin 2t \tag{1.17}$$

where $\sin 2t = 2 \sin t \cos t$ was used. Equation (1.17) is sketched in Fig. 1.5c. The positive and negative values of p imply that power flows back and forth, first from source to capacitor, and then from capacitor to source with average power $P_{\text{ave}} = 0$. The back and forth surging of power, at twice the frequency of the applied voltage, is alluded to by the dashed arrows for p. It thus appears that the capacitor, unlike a resistor, does not consume any energy from the source, but merely stores energy for a quarter-period, and then during the next quarter-period gives that energy back to the source. C is thus fundamentally different from R because R dissipates electrical energy as it converts it to heat. C, on the other hand, only stores electrical energy (in the charge that is deposited on the plates). To learn more about capacitance let us consider the energy stored in C, which is

$$w_C = \int p\, dt = \frac{1}{2}Cv^2 = \frac{CV_p^2}{2} \sin^2 t = \frac{CV_p^2}{4}(1 - \cos 2t) \tag{1.18}$$

In general, the energy stored in a capacitor is given by the $Cv^2/2$ term. For the specific case of applied voltage which is sinusoidal, the energy is represented by the last expression in (1.18). When a sketch of this expression is added to Fig. 1.5c, we see that the average energy, $CV_p^2/4$, does not increase with time. That is, the energy only pulsates as it builds up and decreases again to zero. If one compares this to the corresponding sketch for a resistor, Fig. 1.4c, one sees that for an energy-converting device, which R is, energy steadily increases with time as R continues to absorb energy from the source and convert it to heat.

EXAMPLE 1.1 An initially uncharged 1μF capacitor has a current, shown in Fig. 1.6, flowing through it. Determine and plot the voltage across the capacitor produced by this current.

Integrating the expression $i = C\, dv/dt$, we obtain for the voltage

$$v = \frac{1}{C} \int_{-\infty}^{t} i\, dt = \frac{1}{C} \int_{0}^{t} i\, dt + V_0$$

where V_0 is the initial voltage on the capacitor due to an initial charge. For $0 < t < 3$ ms, the current represented by the straight line is $i = 0.01 - 5t$, and since $V_0 = 0$ we obtain

$$v = 10^4(1 - 250t)t$$

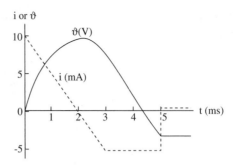

FIGURE 1.6 The dotted line is the capacitor current. The resultant voltage is represented by the solid line.

which is an equation of a parabola. At $t = 2, 3$ ms, voltage is $v = 10, 7.5$ V. For $3 < t < 5$ ms, $i = -5$ mA, which yields

$$v = \frac{1}{C} \int_3^t i \, dt + V_0 = -5(t - 3) + 7.5$$

and which sketches as the straight line. For $t > 5$ ms, $i = 0$ and the voltage remains a constant, $v = -2.5$ V.   ■

We can now summarize the characteristics of capacitors:

- Only a voltage that changes with time will produce a current through a capacitor. A capacitor is therefore an open circuit for direct current (DC).

- Since energy cannot change instantaneously (it is a continuous function of time), and since the energy stored in a capacitor is expressed in terms of voltage as $\frac{1}{2}Cv^2$, we conclude that voltage across a capacitor cannot change instantaneously (unless we want to entertain infinite currents, which is not practical). Capacitance has therefore smoothing properties for voltage, which has many important applications such as in filter design.

- A finite amount of energy can be stored, but because no mechanism for energy dissipation exists in an ideal capacitor none can be dissipated. For sinusoidal time variations this is easily seen since the 90° phase difference between current and voltage results in Expression (1.17), which gives $P_{ave} = 0$.

1.4.3 Inductors

The last of the common circuit elements is the inductor. Like a capacitor, it is an energy-storage device, and like a capacitor which stores the energy in its electric field between the plates, an inductor stores it in its *magnetic field*, which surrounds the inductor. As this is a book on electronics, we will not pursue the field interpretation of energy storage, but instead use the voltage and current which create these fields in capacitors and inductors, respectively. Thus we can say that a capacitor stores energy in the charges which are created when a voltage is applied to a capacitor, giving rise to the energy expression in terms of voltage, which from (1.18) is $w_C = \frac{1}{2}Cv^2$. Similarly, since current causes a magnetic field, we can derive the expression $w_L = \frac{1}{2}Li^2$, which gives the energy stored in an inductor, where i is the current flowing through the inductor and L is the inductance of the inductor (note the duality of these two expressions: C is to L as v is to i).

To derive the above formula, we begin with the notion that inductance L, like capacitance C, is a property of a physical arrangement of conductors. Even though any arrangement possesses some inductance, there are optimum arrangements that produce a large inductance in a small space, such as coils, which consist of many turns of fine wire, usually wound in many layers, very much like a spool of yarn. The definition of inductance rests on the concept of flux linkage. This is not a very precise concept unless one is willing to introduce a complicated topological description. For our purposes, it is sufficient to state that flux linkage Φ is equal to the magnetic field that exists in a coil multiplied by the number of turns of the coil. Inductance L is then defined as $L = \Phi/i$ (which is analogous to capacitance $C = q/v$), where i is the current in the coil that gives rise to the magnetic field of the coil. Recalling *Faraday's law*, $v = d\Phi/dt$, which gives the induced voltage in a coil when the coil is in a time-changing magnetic field, we obtain

$$v = L\frac{di}{dt} \tag{1.19}$$

which is the defining equation for voltage and current in an inductance L. Again, as in the case of C, we assume that L remains constant over a large range of voltages and currents.

The unit for inductance is the *Henry* (H). Inductors for filter applications in power supplies are usually wire-wound solenoids on an iron core with inductances in the range from 1 to 10 H. Inductors found in high-frequency circuits are air-core solenoids with values in the milli-Henry (mH) range.

As in the case of capacitance, assuming sinusoidal variations will quickly show the characteristics of inductance. If (as shown in Fig. 1.7a) a current source which produces a current $i = I_p \sin t$ is connected to an inductor L, then using (1.19), voltage across the inductor will be $v = LI_p \cos t$, which is sketched in Fig. 1.7b. Hence, for sinusoidal variation, voltage leads current by $90°$, or i lags v by the same amount in L. The instantaneous power is

$$p = vi = Li\frac{di}{dt} = \frac{LI_p^2}{2}\sin 2t \tag{1.20}$$

The positive and negative values of p imply that power[9] flows back and forth between the source and the inductor. Hence, like the capacitor, an inductor accepts energy from

---

[9]Strictly speaking, it is the *energy* that flows back and forth and power is the time rate of change of energy. However, when describing flow, the terms power and energy are used interchangeably in the popular literature.

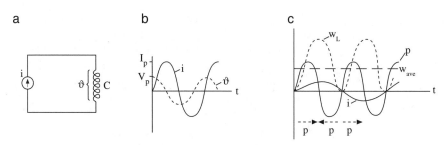

FIGURE 1.7   (a) An inductor (depicted by a spiral symbol) with current i applied. (b) Sinusoidal voltage and current in L. (c) Sketches of instantaneous power and energy, and average energy. (*Note*: amplitudes of p and w_L are not to scale.)

the source for a quarter-period and returns the energy back to the source over the next quarter-period. This is nicely illustrated when we consider energy, which is

$$w_L = \int p\, dt = \frac{1}{2} L i^2 = \frac{L I_p^2}{2} \sin^2 t \qquad (1.21)$$

In general, the energy stored in an inductor is given by the $Li^2/2$ term. For the sinusoidal case, the last term shows that the energy increases as the inductor accepts energy from the source and decreases again to zero as the inductor returns the stored energy to the source. This is illustrated in Fig. 1.7c. An example of large-scale energy storage is a new technology which will enable industry to store cheap, off-peak electrical energy in superconducting coils for use during periods of high demand. A large, steady current is built up in a coil during off-peak periods, representing $Li^2/2$ of energy that is available for later use.

EXAMPLE 1.2   A 1 H inductor has an initial 1 A current flowing through it, i.e., $i(t = 0) = I_0 = 1$ A. If the voltage across L is as shown in Fig. 1.8, determine the current through the inductor.

Integrating the expression $v = L\, di/dt$, we obtain for the current

$$i = \frac{1}{L} \int_{-\infty}^{t} v\, dt = \frac{1}{L} \int_{0}^{t} v\, dt + I_0$$

For $0 < t < 2$ s, we have $i = I_0 = 1$ A, because $v = 0$ as shown in the figure. For $2\text{ s} < t < 3$ s,

$$i = \int_{2}^{t} (-1)\, dt + I_0 = 3 - t$$

which gives the downward-sloping straight line for the current. For $t = 3$ s, $i = 0$, and for $t > 3$ s, the current remains at zero, i.e., $i = 0$, as the voltage for $t > 3$ s is given as $v = 0$.

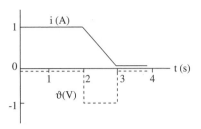

FIGURE 1.8   Current and voltage waveforms in a 1 H inductor.

This example demonstrates that even though the voltage makes finite jumps, the current changes continuously in an inductor.   ■

The characteristics of inductors can be summarized as follows:

- Only a current that changes with time can produce a voltage across an inductor. An inductor is therefore a short circuit for DC. Very high voltages can be produced across an inductor when the current through L is suddenly interrupted (an arc can form at the point of interruption, if the interruption is too sudden).

- As the energy (which cannot change instantaneously) stored in an inductor is given by $W_L = \frac{1}{2}LI^2$, we conclude that current through an inductor also cannot change instantaneously—unless we want to consider infinite voltages, which is not practical. Inductance has therefore smoothing properties for current. An inductor, for example, inserted in a circuit that carries a fluctuating current will smooth the fluctuations.

- A finite amount of energy can be stored, but because no mechanism for energy dissipation exists in an ideal inductor none can be dissipated.

1.4.4   Batteries

Joule's law states that a resistor carrying a current generates heat. The electrical energy is frequently supplied to the resistor by a battery, which in turn obtains its energy from chemical reactions within the battery. Hence, heat generation by R involves two transformations: from chemical to electrical to heat. The symbol for a battery is shown in Fig. 1.1 and in Fig. 1.9a, with the longer bar denoting the positive polarity of the battery terminals. Batteries are important sources of electrical energy when a constant voltage is desired.

Before we analyze practical batteries, let us first characterize ideal batteries or ideal voltage sources. An ideal battery is defined as one that maintains a constant voltage, say, V_B, across its terminals, whether a current is flowing or not. Hence, voltage V_B of an ideal battery is completely independent of the current, as shown in Fig. 1.9b. Such a source is also referred to as an *independent source* (a source connected in a circuit is said

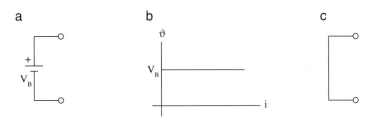

FIGURE 1.9   (a) An ideal battery. (b) The output characteristics of an ideal battery. (c) The internal resistance of an ideal battery is that of a short circuit.

to be independent if its value can be assigned arbitrarily[10]). Since an ideal battery will maintain a voltage V_B across its terminals even when short-circuited,[11] we conclude that such a source can deliver, in theory, infinite power (since $P = V^2/R$, as $R \to 0$, $P \to \infty$). Hence, the name *ideal source*. We also observe that the slope of the v-i curve in Fig. 1.9b is zero. Applying Ohm's law, $R = V/I$, to such a horizontal v-i line implies zero resistance. We therefore conclude that the internal resistance of an ideal source is zero. This explains why an ideal battery causes infinite current when short-circuited. Ignoring difficulties that infinities create, we learn that when looking into the terminals of an ideal battery, we see a short circuit (we are now using common circuits language). Saying it another way, if we somehow could turn a dial and decrease the voltage V_B of the ideal battery to zero, we would be left with a short circuit as shown in Fig. 1.9c.

It is common to represent voltage sources in circuit schematics by ideal sources, which is fine as long as there are no paths in the schematic that short such sources (if there are, then the schematic is faulty and does not represent an actual circuit anyhow). Practical sources, on the other hand, always have finite internal resistance, as shown in Fig. 1.10a, which limits the current to non-infinite values should the battery be short-circuited. Of course R_i is not a real resistor inside the battery, but is an abstraction of the chemistry of a real battery and accounts for the decrease of the terminal voltage when the load current increases. The internal voltage V_B is also referred to as the *electromotive force* (emf) of the battery. From our previous discussion, we easily deduce that powerful batteries are characterized by low internal resistance (0.005 Ω for a fully charged car battery), and smaller, less powerful batteries by larger internal resistance (0.15 Ω for an alkaline flashlight battery, size "C").

Another characteristic of practical batteries is their increasing internal resistance with discharge. For example, Fig. 1.10b shows the terminal voltages versus hours of continuous use for two types. The mercury cell maintains its voltage at a substantially constant

---

[10]There are special kind of sources in which the source voltage depends on a current or voltage elsewhere in the circuit. Such sources will be termed dependent sources or controlled sources.

[11]A short circuit is a zero-resistance path (current can flow but voltage across the path is zero). For example, a piece of copper wire can be considered as a short-circuit element. The opposite of a short is an open circuit, which is an infinite resistance path (voltage can exist across the path but current is zero). These two elements are modeled by the two positions of an on–off switch.

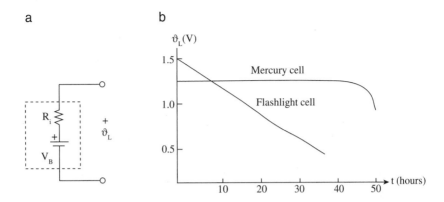

FIGURE 1.10   (a) A practical battery with *emf* V_B and internal resistance R_i. (b) Discharge characteristics of two types of batteries.

level of 1.35 V over its lifetime (but drops sharply when the battery is exhausted) in comparison to ordinary flashlight cells which start out at 1.55 V but decrease continually with use. Other types (lithium, 3.7 V, very long shelf life of over 10 years; nickel-cadmium, 1.25 V, sealed but rechargeable; lead-acid, 2 V, powerful and rechargeable, used as car batteries when connected in series as three-cell 6 V or six-cell 12 V units) fall somewhere between the two curves. The rate of decrease of available voltage as the battery discharges is determined by the chemical reaction within the battery. While battery chemistry is beyond the scope of this book, what is of interest to us is that the decreasing chemical activity during discharge can be associated with an increase of internal battery resistance. Hence, a fully charged battery can be viewed as possessing a low internal resistance, which gradually increases with battery use and becomes very large for a discharged battery.

Figure 1.11*a* shows a circuit in which a practical battery is connected to a load, represented by R_L, and delivers power to the load. R_L can be the equivalent resistance of a radio, a TV set, or any other electrical apparatus or machinery which is to be powered by the battery. The power available to the load is given by $i^2 R_L$. However, since the battery has an internal resistance, energy will also be dissipated within the battery. The internal loss is given by $i^2 R_i$ and will show up as internal heat. It is therefore dangerous to short a powerful battery, as all of the available energy of the battery will then be rapidly converted to internal heat and, unless the shorting element melts rapidly, a dangerous explosion is possible.

Let us now assume, for the time being, that R_i is constant but the load R_L is variable (represented by the arrow across R_L in Fig. 1.11*a*) and analyze the circuit as the burden on the battery is increased. Using Kirchhoff's voltage law (1.10), we obtain for the circuit

$$V_B = i R_i + i R_L \tag{1.22}$$

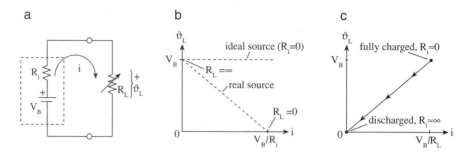

FIGURE 1.11 (a) A practical battery with a variable load connected. (b) Characteristics of a source with increasing load. (c) Characteristics of a source being depleted.

The voltage across the load resistor, $v_L = iR_L$, which is also the available voltage across the external battery terminals, is given from (1.22) as

$$v_L = V_B - iR_i \tag{1.23}$$

This is an equation of a straight line with constant slope of $-R_i$ and is plotted in Fig. 1.11b. The available voltage is therefore the emf of the battery minus the internal voltage drop of the battery. The current that flows in the series circuit is obtained from (1.22) as

$$i = \frac{V_B}{R_i + R_L} \tag{1.24}$$

As the load resistance R_L decreases, the burden on the battery increases. As shown in Fig. 1.11b, this is accompanied by a decrease in the available voltage v_L, usually an undesirable result. Eliminating i from (1.23) and (1.24) to give

$$v_L = V_B \frac{R_L}{R_i + R_L} \tag{1.25}$$

shows the decrease of v_L from V_B as R_L is decreased. Thus, when there is no load on the battery (R_L is very large), the available voltage is maximum at $v_L \approx V_B$, but for a large load ($R_L \approx 0$), the available voltage drops to $v_L \approx 0$. Utility companies, for example, have difficulty maintaining constant voltage during summer when the demand for electricity increases mostly because of energy-hungry air conditioning equipment.[12] Lower-than-normal voltage conditions (popularly referred to as brownouts) put an abnormal strain on customers' electrical equipment which leads to overheating and

---

[12]The circuit of Fig. 1.11b is a general representation of power delivery at a constant voltage. It applies to a flashlight battery delivering power to a bulb, a solar cell powering a calculator, a car battery starting an automobile, or a power utility delivering power to homes. All these systems have an internal emf and an internal resistance, irrespective of AC or DC power produced.

eventually to failure.[13] An obvious solution to brownouts is to decrease the internal resistance R_i of the generating equipment as this would decrease the slope of the curve in Fig. 1.11b by moving the intersection point V_B/R_i to the right, thus bringing the curve closer to that of an ideal source of Fig. 1.9b. Of course, low-R_i equipment means larger and more expensive generators.

To obtain Fig. 1.11b we have assumed that the internal resistance R_i remains constant as the load resistance R_L changes. Let us now consider the case when load R_L remains constant but R_i changes. An example of this is a battery being discharged by a turned-on flashlight which is left on until the battery is depleted. Figure 1.11c gives the v-i curve for battery discharge with the arrows indicating the progression of discharge. We see that the fully charged battery, starting out with a small internal resistance ($R_i \approx 0$), can deliver a current $i \approx V_B/R_L$ and a voltage $v_L \approx V_B$. After discharge, ($R_i \approx \infty$), the current (1.24) and terminal voltage (1.25) are both zero.

In summary, one can say that the reason that current goes to zero as the battery is discharged is not that the emf, whose magnitude is given by V_B, goes to zero, but that the internal resistance R_i changes to a very large value. A discharged battery can be assumed to still have its emf intact but with an internal resistance which has become very large. R_i is therefore a variable depending on the state of the charge and the age (shelf life) of the battery.

To measure the emf of the battery, we remove the load, i.e. we open-circuit the battery and as the current i vanishes we obtain from (1.23) that $v_L = V_B$; the voltage appearing across the battery terminals on an open circuit is the battery's emf. To measure the emf, even of an almost completely discharged battery, one can connect a high-resistance voltmeter (of $10^7\ \Omega$ or larger) across the battery terminals. Such a voltmeter approximates an open-circuit load and requires only the tiniest trickle of charge flow to give a reading. If the input resistance of the meter is much larger than R_i, the reading will be a measure of the V_B of the battery.

To measure the R_i of a battery, one can short-circuit the battery, for only a very brief time, by connecting an ammeter across the battery and reading the short-circuit current. (As this is a dangerous procedure, it should be done only with less powerful batteries, such as flashlight cells. It can also burn out the ammeter unless the appropriate high-ampere scale on the meter is used.) The internal resistance is then given by V_B/I_{sc}. A less risky procedure is to connect a variable resistance across the battery and measure the voltage v_L. Continue varying the resistance until the voltage is half of V_B. At this point the variable resistance is equal to R_i. If this is still too risky—as it puts a too low of a resistance across the battery—consider the procedure in the following example.

EXAMPLE 1.3   Determine the R_i of an alkaline battery (size C) by loading the cell with a 1 Ω resistor.

---

[13]Overheating results when the voltage for an electric motor decreases, thereby increasing the current in the motor so as to preserve the power ($p = vi$) of the motor. The increased current leads to increased I^2R losses in the windings of the motor, which in turn leads to an increase in generated heat that must be dissipated to the surroundings.

Consider Fig. 1.11*a*. It is known that V_B for an alkaline cell is 1.5 V. Measuring the voltage across the 1 Ω resistor, we obtain 1.3 V, which must leave a voltage drop of 0.2 V across R_i. As the current in the circuit is given by $i = 1.3$ V/1 $\Omega = 1.3$ A, we obtain for the internal resistance $R_i = 0.2$ V/1.3 A $= 0.15$ Ω.   ■

1.4.5   Voltage and Current Sources

Voltage sources in general provide voltages that can vary with time such as sinusoids and square waves, or that can be constant with time such as the voltage of a battery. In either case, the principles that were covered for batteries in the previous section apply equally well to voltage sources in general. That is, each type of voltage source has an ideal source in series with an internal resistance as shown in Fig. 1.12*a*. Note the new circuit symbol for an independent voltage source, which includes a battery as a special case by simply specifying that $v_s = 12$ V for a 12 V battery, for example.

A second type of source, known as a *current source*, whose symbol is shown in Fig. 1.12*b*, produces a constant current output independent of voltage, as shown in Fig. 1.12*c*. A vertical *v-i* graph implies that the internal resistance of a current source is *infinite* (in contrast to a voltage source for which it is *zero*), i.e., if we somehow could turn a dial and reduce the amplitude i_s to zero, we would be left with an open circuit.

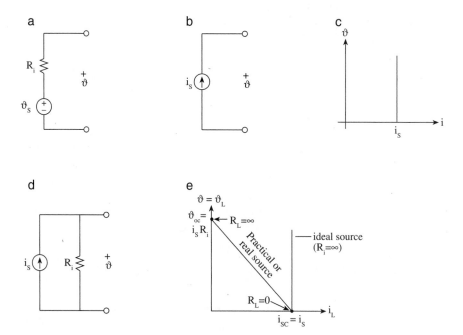

FIGURE 1.12   (a) A practical voltage source. (b) An ideal current source. (c) The *v-i* characteristics of an ideal current source. (d) A practical current source. (e) Load voltage v_L and load current i_L variation as the load resistor R_L changes, for the case when R_L is connected to the practical source of Fig. 1.12*d*.

This, of course, is again an ideal source, nonexistent in the real world, as it appears to supply infinite power. For example, connecting a load resistor R_L of infinite resistance (that is, an open circuit) to a current source would produce power $p = i_s^2 R_L$, which is infinite, as by definition the ideal current source will maintain i_s current through the open circuit. Therefore, a practical current source always appears with an internal resistance which parallels the ideal current source, as shown in Fig. 1.12d. Now, leaving the output terminals open-circuited, as shown in Fig. 1.12d, i_s simply circulates through R_i. Practical examples of current sources are certain transistors which can maintain a constant current, similar to that shown in Fig. 1.12c, for a wide range of load resistances. However, when R_L begins to exceed a certain value, the current drops sharply.

To obtain the output characteristics, we connect a load resistance R_L to the real current source of Fig. 1.12d. Now the source current i_s will divide between resistors R_i and R_L. If we vary R_L and plot the v_L-i_L graph, we obtain Fig. 1.12e. This graph, like the respective one for a voltage source, Fig. 1.11b, shows that as we decrease R_L, the load voltage v_L decreases and drops to zero for $R_L = 0$, at which point the current through the load resistor, which is now a short circuit, becomes $i_L = i_{sc} = i_s$. On the other hand, when R_L is infinite, i.e., an open circuit, the load voltage is $v_L = v_{oc} = i_s R_i$.

1.4.6 Source Equivalence and Transformation

From the standpoint of the load resistor, it is immaterial if a current or a voltage source is delivering power to R_L. If, for example, 10 W is being delivered to a load resistance by a source that is enclosed in a black box, there is no way of knowing if the hidden source is a voltage or a current source. An equivalence between current and voltage sources must therefore exist, which we now define by stating that if two separate sources produce the same values of v and i in R_L, then for electrical purposes the two sources are equivalent. The equivalence must hold for any load resistance, including $R_L = 0$ and $R_L = \infty$; in other words, if two sources produce the same short-circuit current, I_{sc}, when $R_L = 0$, and the same open-circuit voltage, V_{oc}, when $R_L = \infty$, then the sources are equivalent.

With the above statement of equivalence, we now have a convenient and quick way to transform between sources. For example, if we begin with the practical voltage source of Fig. 1.13a, we readily see that $I_{sc} = V/R$, and from (1.25), $V_{oc} = V$. Therefore, the equivalent practical current source, shown in Fig. 1.13a, has a current source of strength $I = V/R$ in parallel with a resistance R. Similarly, if we start out with a current source and would like to find the equivalent voltage source, Fig. 1.13b shows that the current source of strength I in parallel with R gives $I_{sc} = I$ when short-circuited and $V_{oc} = IR$ when open-circuited. Therefore, the equivalent voltage source is easily obtained and is shown in Fig. 1.13b.

Summarizing, we observe that under open-circuit conditions, V_{oc} always gives the voltage element (emf) of an equivalent voltage source, whereas under short-circuit conditions, I_{sc} always gives the current element of an equivalent current source. Furthermore, we easily deduce that the source resistance is always given by $R = V_{oc}/I_{sc}$. If we examine Fig. 1.13, we note that the source resistance is R and is the same for all four equivalents. That is, looking back into the terminals of the voltage source we see only resistance R, because the voltage source element, which is in series with R, is equivalent to a short (see Fig. 1.9c). Similarly, looking into the terminals of the current source, we see R, because the current source element itself, which is in parallel with R, is equivalent to an open circuit.

Open-circuit and short-circuit conditions, therefore, provide us with a powerful tool to represent complicated sources by the simple, equivalent sources of Fig. 1.13. For example, an audio amplifier is a source that provides amplified sound and therefore can be represented at the output terminals of the amplifier by one of the equivalent sources. To be able to view a complicated piece of equipment such as an amplifier simply as a voltage source in series with a resistance aids in the understanding and analysis of complex electronics. In the case of the audio amplifier, the equivalent source resistance is the output resistance of the amplifier, which for maximum power output to the speakers needs to be matched[14] to the impedance of the speakers that will be powered by the audio amplifier.

1.5   SERIES AND PARALLEL CIRCUITS

Although we have already presented such circuits when discussing Kirchhoff's laws (see Fig. 1.3), we will now consider them in detail. The series circuit in Fig. 1.3a was drawn in the shape of a loop. However, from now on we will use rectangular shapes for circuits as they appear neater and are easier to trace in complicated circuits. Figure 1.14a shows a voltage source and three resistors in a series circuit which we will now show to be equivalent to the one-resistor circuit of Fig. 1.14b by observing that the current is the same in every component of the circuit. Using Kirchhoff's voltage law,

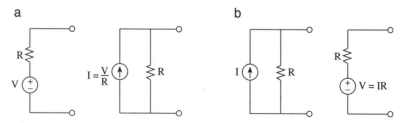

FIGURE 1.13   (a) A voltage source and its current source equivalent. (b) A current source and its voltage source equivalent.

---

[14]Matching will be considered in detail in Section 1.6.5, Maximum Power Transfer and Matching.

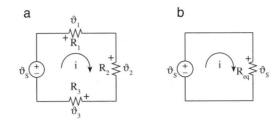

FIGURE 1.14   (a) A series circuit and (b) its equivalent.

we obtain

$$
\begin{aligned}
v_s &= v_1 + v_2 + v_3 & (1.26)\\
&= R_1 i + R_2 i + R_3 i\\
&= i(R_1 + R_2 + R_3)\\
&= i(R_{eq})
\end{aligned}
$$

for the circuit of Fig. 1.14a, and similarly we obtain $v_s = iR_{eq}$ for the circuit in Fig. 1.14b. Hence, comparing, we conclude that the equivalent resistance of N resistors in series is

$$
R_{eq} = R_1 + R_2 + \cdots + R_N \qquad (1.27)
$$

To the source, a series of resistors or a single equivalent resistor is the same, i.e., the v-i relationship is the same.

In review, we should note that small-case letters are used to denote quantities that could be time-varying ($v_s = V \sin t$), whereas capital letters denote constant quantities, such as those for a 12 V battery ($V_B = 12$ V). However, constant quantities can also be expressed by using small-case letters; e.g., the 12 V battery can be equally well referred to as $v_s = 12$ V. Convention is to use the battery symbol when the voltage is produced by chemical action, but when a constant voltage is provided by a power supply or by a signal generator, which can produce time-varying voltages and constant ones, the voltage source symbol of Fig. 1.14 is appropriate. Another review point is the polarity convention, which has the current arrow pointing at the plus when the voltage is that of a sink (voltage drop), and at the minus when it is a source (voltage rise). The following example demonstrates these points.

EXAMPLE 1.4   Three sources are connected to a series of resistors as shown in Fig. 1.15a. Simplify the circuit and find the power delivered by the sources. Using Kirchhoff's voltage law to sum the voltages' drops and rises around the loop, we have, starting with the 70 V source,

$$
\begin{aligned}
-70 + 10i + 20 + 15i - 40 + 5i &= 0\\
-90 + 30i &= 0\\
i &= 3 \text{ A}
\end{aligned}
$$

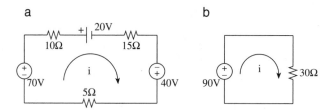

FIGURE 1.15   (a) A series circuit and (b) its equivalent.

The loop current is therefore 3 A and the equivalent circuit is the two-element circuit of Fig. 1.15*b*. The power consumed by the resistors is $i^2 R = 3^2 \cdot 30 = 270$ W, which is also the power produced by the equivalent source (90 V $\cdot$ 3 A = 270 W). The individual sources produce $70 \cdot 3 = 210$ W, $-20 \cdot 3 = -60$ W, and $40 \cdot 3 = 120$ W. Clearly, the 20 V battery acts as a load—it is being charged by the remaining two sources at the rate of 60 W.   ∎

A second way of connecting elements was shown in Fig. 1.3*b*, when discussing Kirchhoff's current law. In such a parallel arrangement we see that the voltage is the same across both resistors, but the currents through the elements will be different. Let us consider the slightly more complicated, but still two-node, circuit shown in Fig. 1.16*a*. Summing currents at the top node, we have

$$i = i_1 + i_2 + i_3 \tag{1.28}$$

i.e., the sum of the three currents $i_1 = v_s/R_1$, $i_2 = v_s/R_2$, and $i_3 = v_s/R_3$ equals the source current i. Substituting for the resistor currents in (1.28), we obtain

$$i = v_s \left(\frac{1}{R_1} + \frac{1}{R_2} + \frac{1}{R_3} \right) \tag{1.29}$$

The terms in the parentheses can be identified as the equivalent resistance of the parallel resistors. Using Ohm's law, we can define

$$i = v_s \frac{1}{R_{eq}} \tag{1.30}$$

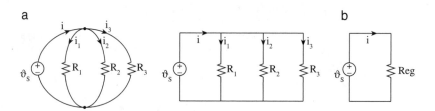

FIGURE 1.16   (a) Two ways of drawing a two-node circuit of a voltage source and three resistors in parallel. (b) The equivalent circuit.

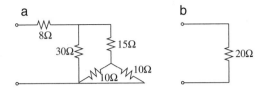

FIGURE 1.17   (a) A network of resistors and (b) its reduced equivalent.

Hence, the equivalent resistance is

$$\frac{1}{R_{eq}} = \frac{1}{R_1} + \frac{1}{R_2} + \frac{1}{R_3} \tag{1.31}$$

and the equivalent circuit is shown in Fig. 1.16*b*. Equqation (1.31) is an often-used expression that is readily extended to N resistors in parallel. Particularly useful is the two-resistors-in-parallel expression

$$R_{eq} = R_1 \parallel R_2 = \frac{R_1 R_2}{R_1 + R_2} \tag{1.32}$$

which is needed often and should be memorized. For example, a 1 kΩ and a 10 kΩ resistor in parallel are equal to an equivalent resistor of 0.91 kΩ; i.e., two resistors in parallel have a resistance that is less than the smallest resistance.

The analysis for resistors in parallel is somewhat easier if we make use of conductance G, which is defined as $G = 1/R$. Ohm's law can then be written as $i = Gv$, and (1.28) becomes

$$i = v_s(G_1 + G_2 + G_3) \tag{1.33}$$

Hence, conductances in parallel add, or

$$G_{eq} = G_1 + G_2 + G_3 \tag{1.34}$$

and (1.34) is equal to (1.31), as $G_{eq} = 1/R_{eq}$

EXAMPLE 1.5   Simplify the network of resistors shown in Fig. 1.17*a*, using the rules of series and parallel equivalents.

First, combine the 10 Ω resistors which are in parallel to give 5 Ω. Now combine this 5 Ω resistor with the 15 Ω resistor, which are in series, to give 20 Ω. Now combine the equivalent 20 Ω resistor which is in parallel with the 30 Ω resistor of the network to give 12 Ω. Finish by combining the equivalent 12 Ω resistor with the remaining 8 Ω resistor, which are in series, to give 20 Ω, which is the simplified network shown in Fig. 1.17*b*.   ■

1.5.1   Voltage and Current Division

Practical circuits such as volume controls in receivers use voltage divider circuits, such as that shown in Fig. 1.18*a* (also known as a *potentiometer*), where the tap is movable

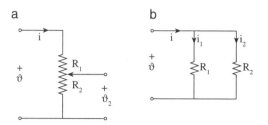

FIGURE 1.18   (a) A voltage divider and (b) a current divider circuit.

for a continuous reduction of the voltage v. The voltage source v sees a resistance which is $R_1 + R_2$, and v_2 is the voltage drop across the R_2 portion of the potentiometer. The current i flowing due to v is $i = v/(R_1 + R_2)$; therefore the output voltage v_2 is equal to iR_2, or

$$v_2 = v\frac{R_2}{R_1 + R_2} \tag{1.35}$$

which is the voltage-divider equation.

Equally useful, but more subtle, is current division. Fig. 1.18b shows a current i that divides into two components i_1 and i_2. To find the component currents we first determine i as

$$i = \frac{v}{R_1 \parallel R_2} = v\frac{R_1 + R_2}{R_1 R_2} \tag{1.36}$$

The current through R_1 and R_2 is simply given by v/R_1 and v/R_2, respectively, which by use of (1.36) is

$$i_1 = i\frac{R_2}{R_1 + R_2} \tag{1.37}$$

and

$$i_2 = i\frac{R_1}{R_1 + R_2} \tag{1.38}$$

The above two equations establish the rules of current division. Although not as straight forward as voltage division, current division at the junction of two resistors follows the rule that the larger current flows through the smaller resistor. In the limit, for example, when R_1 is a short, all the current flows through R_1 and none through R_2, in agreement with (1.38), which states that $i_2 = 0$ for this case. When analyzing circuits, we need to have at our fingertips the rules of voltage and current division, which makes these rules worth memorizing.

1.6   NETWORK SIMPLIFICATION

We have already simplified some circuits by applying the rules of series and parallel circuits. When considering more complicated circuits, usually referred to as networks, there are other, more sophisticated analysis tools at our disposal which we need to study and be able to apply. In electronics, we frequently encounter *one-port* and *two-port* devices. A one-port is a two-terminal circuit such as those shown in Fig. 1.13 or in Fig. 1.20. Two-ports are of great interest, because complicated electronic equipment can frequently be viewed as two-ports. For example, connections to an audio amplifier are made at the input and the output terminals. At the input, we connect a pickup device, whose weak signal is incapable of driving a speaker and therefore needs to be amplified. At the output, the amplifier acts as a powerful source and can easily drive a speaker that is connected to the output terminals. Hence, from the viewpoint of the user, a two-port depiction of an amplifier is all that is needed. Furthermore, without having yet studied amplifiers, we can already deduce a fundamental circuit for an amplifier: at the output port, the amplifier must look like a practical source. We now have a simple, but highly useful, three-component model of an amplifier delivering power to a load: a voltage source in series with a source resistance is connected to a load such as a speaker which is represented by R_L. This simple circuit looks like that of Fig. 1.11a (with the battery replaced by a voltage source) and is a valid representation of an amplifier at the output terminals. We will consider now several theorems, including Thevenin's, which will formalize the replacing of all or part of a network with simpler equivalent circuits.

1.6.1 Equivalence

We have already referred to equivalence in the subsection on Source Equivalence and Transformation. To repeat: two one-port circuits are equivalent if they have the same v-i characteristics at their terminals.

1.6.2 Superposition

Circuit theory is a linear analysis; i.e., the voltage–current relationships for R, L, and C are linear relationships, as R, L, and C are considered to be constants over a large range of voltage and currents. Linearity gives rise to the principle of superposition, which states that in a circuit with more than one source present, the voltage or current anywhere in the circuit can be obtained by first finding the response due to one source acting alone, then the second source acting alone, and so forth. The response due to all sources present in the circuit is then the sum of the individual responses. This is a powerful theorem which is useful since a circuit with only one source present can be much easier to solve. Now, how do we shut off all sources in the circuit but one?[15] Recall what happens to an ideal voltage source when the amplitude is cranked to zero? One is left with a short circuit (see Fig. 1.9). Similarly, when a current source is cranked down to zero, one is left with an open circuit (see Fig. 1.12). Therefore, in a circuit with multiple sources, all sources except one are replaced with their respective short or open circuits. One can then proceed to solve for the desired circuit response with only one source present. The following example illustrates the technique.

---

[15] In circuit jargon, this is also referred to as *killing a source*.

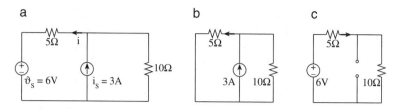

FIGURE 1.19   (a) A circuit with a voltage and current source present. (b) The voltage source is removed. (c) The current source is removed.

EXAMPLE 1.6   Use superposition to find the current i in Fig. 1.19a. For linear responses only (voltage and current, but not power), the circuit of Fig. 1.19a is a superposition of the two circuits of Figs. 1.19b and c. Hence the current is the superposition of two currents: the first, due to the current source acting alone, is flowing to the left, and the second, due to the voltage source acting alone, is flowing to the right. Hence

$$i = i|_{v_s=0} + i|_{i_s=0}$$
$$= 3\frac{10}{5+10} - \frac{6}{5+10}$$
$$= 0.5 - 0.4 = 0.1 \text{ A}$$

and a current of 0.1 A is flowing to the left in the circuit of Fig. 1.19a. Superposition, by breaking up a problem into a set of simpler ones, often leads to a quick solution and provides us with insight as to which sources contributes more.

It is tempting to calculate i^2R power dissipated in, for example, the 5 Ω resistor by adding powers. We would obtain $(0.5)^2 \cdot 5 + (0.4)^2 \cdot 5 = 2.05$ W, when in fact the actual power dissipated in the 5 Ω resistor is only $(0.1)^2 \cdot 5 = 0.5$ W. This demonstrates succinctly that superposition applies to linear responses only, and power is a nonlinear response, not subject to superposition.   ■

1.6.3   Thevenin's Theorem

This is one of the most powerful and useful theorems in circuit theory. It can greatly simplify analysis of many linear circuits and provide us with insight into the behavior of circuits. It allows replacing a complex one-port, that may contain many sources and complicated circuitry, by a practical source, i.e., a voltage source in series with a resistance. Let us consider Fig. 1.20a, which shows a general network, with two terminals for access. If the network is an amplifier, the terminals could be the output port to which a load, such as a speaker, represented by R_L, is connected.

Thevenin's theorem states that, looking into the network to the left of the dashed, vertical line, the one-port can be replaced by a series combination of an ideal voltage source V_{th} and a resistance R_{th} (as shown in Fig. 1.20b), where V_{th} is the open-circuit voltage of the one-port and R_{th} is the ratio of the open-circuit voltage to the short-circuit current of the one-port. The open-circuit voltage is obtained by disconnecting R_L

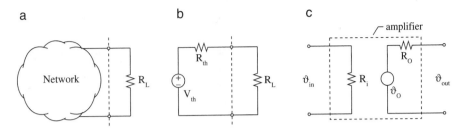

FIGURE 1.20 (a) A one-port network of arbitrary complexity, connected to a load resistor. (b) A Thevenin's equivalent circuit. (c) A Thevenin's equivalent circuit for an amplifier with gain v_{out}/v_{in} and input and output resistance shown.

and measuring or calculating the voltage, whereas the short-circuit current is obtained by shorting R_L. When it is impractical to short the output, R_{th} can also be obtained by killing all sources of the network (replacing voltage sources by shorts and current sources by open circuits) and calculating the resistance of the resulting network. This, of course, also applies to the equivalent network in Fig. 1.20b; shorting out the voltage source and looking into the network, we see R_{th}.

Insofar as the load R_L is concerned, the two networks (a) and (b) are equivalent; that is, the voltages and currents in R_L produced by the two networks are the same. This is a surprising result and implies that any two terminals (a one-port) can be viewed as a practical source (Fig. 1.13), an observation that we already made in Section 1.4 when discussing source equivalence.[16] Even though our previous development of practical sources, especially that of viewing a one-port as a practical source, was sketchy, Thevenin's theorem puts it now on a firm basis. For example, a resistor R (which is a one-port device), when viewed as a practical source, will have $R_{th} = R$ and $V_{th} = 0$ according to Thevenin's theorem.

The intent of the material covered thus far is to give us a basis for our study of electronics. One of the basic building blocks in electronics is an amplifier. Even with a limited knowledge of this subject, we can use the development in the last sections to construct an elementary circuit for an amplifier, which we have sketched in Fig. 1.20c. We will view an amplifier as a two-port device—the input port is not considered as a source for obvious reasons, and hence it will be represented by a resistance; the output port, on the other hand, is expected to deliver power to a device such as a speaker, and hence it must act as a practical source. Figure 1.20c, therefore, shows an equivalent circuit for an amplifier in its most elementary form, which Thevenin's theorem enabled us to formulate. We will use this circuit repeatedly as our study of electronics progresses.

---

[16]This far in our development, when applying Thevenin's theorem, we are primarily restricted to DC circuits, which are a combination of resistors, voltage and current sources. Capacitors and inductors are treated as open circuits and short circuits, respectively, in DC analysis. In the next chapters we will show that Thevenin's theorem is equally applicable to AC circuits, where the concept of impedance can treat capacitors, inductors, and resistors as easily as resistors in DC analysis.

1.6.4   Norton's Theorem

Norton's theorem is a dual of Thevenin's theorem. It states that the equivalent circuit for a one-port can also be a practical current source (shown in Fig. 1.13). The resistance for the Norton circuit is the same as R_{th} for Thevenin's circuit. The Norton current is given by I_{sc}, obtained by short-circuiting R_L and measuring the current. As we already have covered transformations between current and voltage sources, the relationship between Norton's and Thevenin's circuits should be clear.

EXAMPLE 1.7   The circuit shown in Fig. 1.21a is supplying power to R_L. Find the Thevenin's and Norton's equivalent for the circuit to the left of R_L.

As both sources are DC sources, we have a DC circuit that can be simplified by replacing the inductor and capacitor by a short and open circuit, as shown in Fig. 1.21b. To find Thevenin's equivalent for the circuit of Fig. 1.21b, we must find the open-circuit voltage V_{oc}, which will be the Thevenin voltage, i.e., $V_{th} = V_{oc}$. Using superposition, we first find V'_{oc} due to the 9 V battery, followed by V''_{oc} due to the 4 A current source:

$$
\begin{aligned}
V_{th} &= V_{oc} = V'_{oc} + V''_{oc} \\
&= 9\,\text{V}\frac{6}{3+6} + 4\,\text{A}\frac{3\cdot6}{3+6} \\
&= 6\,\text{V} + 8\,\text{V} = 14\,\text{V}
\end{aligned}
$$

To find R_{th}, we short the battery and open-circuit the current source and find the resistance at the terminals of Fig. 1.21b, which is a parallel combination of the 3 and 6 Ω resistors, that is, $R_{th} = 3 \parallel 6 = 2\,\Omega$. The equivalent circuit is now given in Fig. 1.21c, and as far as R_L is concerned there is no difference between the original and the equivalent circuit.

Had we started with Norton's equivalent, we would have obtained the short-circuit current by shorting the terminals in Fig. 1.21b, which would give us the Norton current I_n. Thus, using superposition again, we obtain

$$I_n = I_{sc} = 4\,\text{A} + 9\,\text{V}/3 = 7\,\text{A}$$

Hence, the Norton equivalent is a 7 A current source in parallel with a 2 Ω resistance, as shown in Fig. 1.21d. We can now double check: $R_{th} = V_{th}/I_{sc} = 14/7 = 2\,\Omega$, which checks; also, I_n can be obtained from Thevenin's circuit by $I_n = V_{th}/R_{th} = 14/2 = 7$ A, which also checks.   ∎

1.6.5   Maximum Power Transfer and Matching

We have spent considerable effort on sources up to this point, because much of electrical equipment, with the help of Thevenin's theorem, can be viewed as a practical source at the appropriate terminals—as, for example, the amplifier of Fig. 1.20c. It is now natural to ask, *how much power can a source deliver to a load that is connected to the source?* To answer this question, let us first replace an arbitrary source by its Thevenin's equivalent, as shown in Fig. 1.22a, connect a variable load to the source, vary the load, and

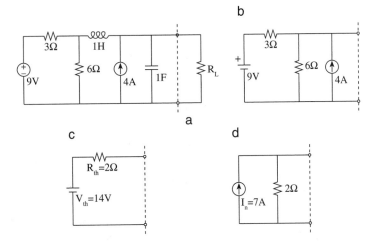

FIGURE 1.21   (a) R_L is connected to a network whose Thevenin's equivalent is desired. (b) The simplified network when both sources are DC sources. (c) Thevenin's equivalent circuit. (d) Norton's equivalent circuit.

see when maximum power is dissipated in the load. The circuit is practically identical to that in Fig. 1.11a, except that now, instead of voltage variations at the load, we are interested in power variations. The power delivered to the load is given by

$$P = i^2 R_L = \left(\frac{V}{R + R_L}\right)^2 R_L \tag{1.39}$$

and is sketched in Fig. 1.22b. To find the value of R_L that absorbs maximum power from the source, i.e., the condition of maximum power transfer, differentiate (1.39) with respect to R_L,

$$\frac{dP}{dR_L} = V^2 \frac{(R + R_L)^2 - 2R_L(R + R_L)}{(R + R_L)^4}$$

and equate the above derivative to zero, to obtain

$$R_L = R \tag{1.40}$$

This is an interesting result, otherwise known as the *maximum power transfer theorem*, which states that maximum power is transferred from source to load when the load resistance R_L is equal to the source's internal resistance R. When $R_L = R$, the load resistance is said to be matched. Under matched conditions, the maximum power delivered to the load is

$$P = i^2 R_L = \frac{V^2}{4R_L} \tag{1.41}$$

a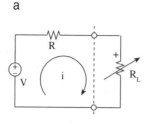

b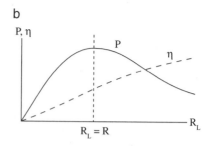

FIGURE 1.22   (a) A variable load, denoted by the arrow across R_L, is connected to a source. (b) Plot of power dissipated in a load, versus load resistance, and plot of efficiency η versus R_L.

Similarly, power $P' = i^2 R$, which is dissipated in the internal resistance, is equal to P. The power P_s generated by the voltage source V is therefore $P_s = iV = P + P' = V^2/(2R_L)$. Under matched conditions, therefore, half the power generated by the source is delivered to the load and half is lost or dissipated by the source. The efficiency of maximum power transfer is, consequently, only 50%.

We can define the efficiency η of power delivery in general by

$$\eta = \frac{P_{\text{load}}}{P_{\text{source}}} = \frac{i^2 R_L}{iV} = \frac{R_L}{R + R_L} \tag{1.42}$$

which gives the efficiency as 50%, for matched conditions, as expected; 100% efficiency is obtained for internal resistance $R = 0$ (a very powerful source), or for $R_L \to \infty$ (no power absorbed by load).

When is maximum power transfer important and when is maximum efficiency important? The answer to these questions depends on the amount of power involved and the ease with which it can be generated. For example, power utilities, which generate many megawatts of electric power, do not operate under maximum power transfer, as then half of the power would be dissipated at the power plant, a very uneconomical, inefficient, and undesirable condition. It would require very large generators merely to dissipate the heat developed. Furthermore, the terminal voltage would drop to half, which by itself would be intolerable. Power systems, therefore, tend to be operated under maximum efficiency conditions, with a goal of keeping the terminal voltage as constant as possible as the load varies.

As little use as power engineers have for maximum power transfer, communication and electronics engineers live by it. In the communications industry, signals are often faint, barely above the noise level, with powers sometimes on the order of microwatts or less. Unlike in the power industry (which can control the value of R), control over the source of a received signal such as in a radio, television, or radar transmission usually does not exist. An electronics engineer has to maximize the signal from a circuit and efficiency may not be of importance. The highest signal-to-noise ratio which leads to the best reception in a communications system is usually obtained under maximum power transfer conditions. Hence, matching a speaker to an amplifier, or matching an antenna to a receiver, gives the best performance.

To recap, the maximum power transfer theorem states that maximum power is delivered to a load by a two-terminal linear network when that load is adjusted so that the terminal voltage is half its open-circuit value. The value of the load resistance R_L will then be equal to the resistance looking back into the network, i.e., the Thevenin's resistance R_{Th}. This theorem was developed for a practical voltage source, but in view of Norton's theorem it holds as well for a practical current source.

EXAMPLE 1.8   Using Thevenin's theorem, find the power dissipated in the 10 Ω resistor in Fig. 1.19a.   Also, find what value of resistance will give maximum power dissipation.

Replacing the circuit in Fig. 1.19a to the left of the 10 Ω resistor by its Thevenin's equivalent circuit gives us a practical voltage source with $V_{th} = 21$ V and $R_{th} = 5$ Ω. The power delivered to the 10 Ω resistor is then equal to $i^2 R = (21/15)^2 10 = 19.6$ W. For maximum power transfer, the load resistor should be changed to 5 Ω. This would give $i^2 R = (21/10)^2 5 = 22.05$ W.

It is interesting to note that the power delivered to the 10 Ω resistor is not much less than the maximum, even though the resistance is far from being a matched resistance. The efficiency, though, has increased from 50% for the matched case to 66.7% for the 10 Ω load. This is fortunate for it means that to get almost the maximum power, loads need not be exactly matched. They need be only approximately equal to this value. Even a substantial mismatch, which can significantly increase the efficiency, can still produce almost maximum power.   ■

1.7   MESH OR LOOP EQUATIONS

When circuits become more complicated, the previous methods of solution—superposition and Thevenin's—might not be adequate. Two powerful techniques—*mesh* analysis and *node* analysis—which are based on Kirchhoff's laws, can be used to solve circuits of any complexity. These two methods lead to a set of linear simultaneous equations with branch currents or node voltages as the unknowns. Rarely will we try to solve more than three or four simultaneous equations by hand or with the help of a calculator. Fortunately, a general-purpose computer program, known as SPICE (*S*imulation *P*rogram with *I*ntegrated *C*ircuit *E*mphasis) is readily available to help with real complicated networks. For our purposes, we will confine ourselves to circuits with two or three unknowns which we can readily solve.

Let us first define some terms. A *node* is a junction of three or more wires. A *branch* is any type of connection between two nodes. Without going into esoteric aspects of circuit topology, we can simply state at this time that the number of unknowns in a circuit is given by $b - n + 1$, where b is the number of branches and n is the number of nodes in a circuit. Figure 1.23 shows a circuit with three branches and two nodes. Hence, the number of unknowns is 2. In the mesh method, the unknowns are the mesh

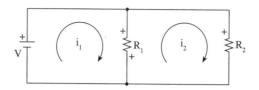

FIGURE 1.23   A two-window circuit with the mesh currents shown. The current through R_1 is to be found using mesh analysis.

or loop currents i_1 and i_2, which are assumed to flow only around the perimeter of the loop. We will use Kirchhoff's voltage law (1.10) to write two *loop equations* for the two unknown loop currents i_1 and i_2. Thus

$$\text{Loop1}: \quad V = R_1 i_1 - R_1 i_2 \tag{1.43}$$
$$\text{Loop2}: \quad 0 = -R_1 i_1 + (R_1 + R_2) i_2$$

where the voltage rises (sources) are on the left side and the voltage drops on the right side of each equation. Solving for the unknowns, we obtain

$$i_1 = \frac{V(R_1 + R_2)}{R_1 R_2} \quad \text{and} \quad i_2 = \frac{V}{R_2} \tag{1.44}$$

Should it turn out that one or both of the unknown currents in (1.44) have a negative sign, it merely means that the actual direction of the mesh currents in Fig. 1.23 is opposite to that assumed. The actual current through the voltage source is therefore i_1 in the direction shown, and the actual current through R_2 is i_2 also in the direction shown, but the current through R_1 is a combination of the two loop currents

$$i R_1 = i_1 - i_2 = V \left(\frac{1}{R_1} + \frac{1}{R_2} \right) - \frac{V}{R_2} = \frac{V}{R_1} \tag{1.45}$$

The current through R_1 thus flows in the direction of the loop current i_1. (The nature of this particular circuit is such that i_1 is always larger than i_2. Why? Another check for this particular circuit comes from the fact that the voltage across R_1 and R_2 is always V, giving V/R_1 and V/R_2 as the currents through R_1 and R_2, respectively.)

Mesh analysis is a powerful as well as a general method for solving for the unknown currents and voltages in any circuit. Once the loop currents are found, the problem is solved, as then any current in the circuit can be determined from the loop currents. One simplification should now be pointed out: instead of using the $b - n + 1$ formula we can simply count the number of windows in a circuit to determine the number of unknowns. The circuit in Fig. 1.23 has two windows, and thus two unknowns. Clearly each window is associated with a loop or mesh current.

Summarizing, we can give a series of steps that will simplify mesh analysis of a circuit with simple sources:

(1) Replace all current sources by voltage sources.

(2) Count the windows in the circuit and place a clockwise loop current in each window. The number of unknown currents is equal to the number of windows.

(3) Apply Kirchhoff's voltage law to each loop or mesh and write the loop equations. Place all source voltages in a loop on the left side of the equation and all voltage drops on the right side. To help avoid mistakes, put voltage drop polarity marks on each resistor (positive where the loop current enters the resistor).

(4) You should now have a set of equations neatly arranged and ready to be solved for the mesh currents $i_1, i_2, i_3, \ldots$. The solution will usually be carried out using determinants and Cramer's rule (detailed below), which is the standard method when solving simultaneous, linear equations. Even though the direction of the mesh currents is arbitrary, using only a clockwise direction, will result in a matrix which is symmetric, with positive diagonal terms and negative off-diagonal terms, which is nice when checking for errors. Furthermore, a diagonal term in the resistance matrix is the sum of all resistances in the respective mesh, and an off-diagonal term is the common resistance of two adjacent loop currents.

Node analysis is an alternative method which uses Kirchhoff's current law to sum currents at each node, which leads to a set of equations in which the voltages at each node are the unknowns. Since mesh or nodal analysis can be used to find unknowns in a circuit, we will not develop node analysis any further.

The following example demonstrates mesh analysis in detail.

EXAMPLE 1.9   Find the current in and the voltage across R_2 in the circuit shown in Fig. 1.24. The circuit has five branches and three nodes, implying that three independent mesh equations are needed to determine all branch currents and voltages of the circuit. Or we can conclude the same by simply noting that the circuit has three windows.

As the loop currents and the resulting polarities on each resistor are already indicated, we can proceed to writing the loop equations. Beginning with the first loop and followed

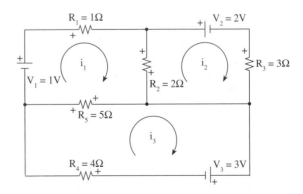

FIGURE 1.24   A three-window circuit with the loop currents sketched for each window.

by the second and third,

$$V_1 = (R_1 + R_2 + R_5)i_1 - R_2i_2 - R_5i_3$$
$$-V_2 = -R_2i_1 + (R_2 + R_3)i_2 - 0i_3$$
$$-V_3 = -R_5i_1 - 0i_2 + (R_4 + R_5)i_3$$

Rewriting in matrix form, we can readily spot errors as the resistance matrix must be symmetric with positive diagonal and negative off-diagonal terms:

$$
\begin{bmatrix} V_1 \\ -V_2 \\ -V_3 \end{bmatrix} = \begin{bmatrix} (R_1 + R_2 + R_3) & -R_2 & -R_5 \\ -R_2 & (R_2 + R_3) & 0 \\ -R_5 & 0 & (R_4 + R_5) \end{bmatrix} \begin{bmatrix} i_1 \\ i_2 \\ i_3 \end{bmatrix}
$$

Notice how nicely that checks. Furthermore, the first diagonal term represents the sum of all resistances in loop 1, the second diagonal in loop 2, and the third diagonal in loop 3, another helpful check. Substituting the values of the resistances and voltage sources into the matrix, we obtain

$$
\begin{bmatrix} 1 \\ -2 \\ -3 \end{bmatrix} = \begin{bmatrix} 8 & -2 & -5 \\ -2 & 5 & 0 \\ -5 & 0 & 9 \end{bmatrix} \begin{bmatrix} i_1 \\ i_2 \\ i_3 \end{bmatrix}
$$

The three simultaneous equations for the unknown currents can be solved using the determinant method. The solution for i_1 is obtained by first substituting the source column for the first column in the resistance matrix and dividing the resulting determinant by the determinant of the resistance matrix—this procedure is usually known as *Cramer's rule*.

$$
i_1 = \frac{\begin{vmatrix} 1 & -2 & -5 \\ -2 & 5 & 0 \\ -3 & 0 & 9 \end{vmatrix}}{\begin{vmatrix} 8 & -2 & -5 \\ -2 & 5 & 0 \\ -5 & 0 & 9 \end{vmatrix}} = \frac{-66}{199} = -0.33 \text{ A}
$$

where the determinants were evaluated by expanding in terms of their minors. Thus, the mesh current i_1 equals 0.33 A and is flowing in a direction opposite to that indicated in Fig. 1.24.

Mesh current i_2 is obtained similarly by first substituting the second column with the source column and evaluating the resultant ratio of determinants. After performing these operation we obtain $i_2 = -0.53$ A. Again, the second mesh current is opposite in direction to that assumed in Fig. 1.24.

The current through resistor R_2 can now be obtained as

$$i_{R_2} = i_1 - i_2 = (-0.33) - (-0.53) = 0.20 \text{ A}$$

and is flowing from top to bottom through resistor R_2. The voltage across R_2 is given by

$$V_{R_2} = i_{R_2} R_2 = 0.20 \cdot 2 = 0.40 \text{ V}$$

and has a polarity that makes the top of resistor R_2 positive. ■

1.8 TRANSIENTS AND TIME CONSTANTS IN RC AND RL CIRCUITS

A circuit that consists of resistors and capacitors is referred to as an RC circuit . However, most of the time when we speak of an RC circuit we mean a simple circuit with a single resistor and a single capacitor.

Except for a brief introduction to sinusoidal time variation when studying the characteristics of capacitors and inductors, we have only considered DC voltages and currents. This is not too surprising as the circuits thus far considered were driven by voltage sources such as batteries that produced constant outputs. But what happens during a brief time interval when a battery is switched into a circuit and before the circuit settles down to its steady state? During this time interval the circuit is said to be in the *transient* state. It is of great physical significance to be able to characterize a circuit during this time interval as it will show us, for example, how a capacitor charges when a RC circuit is connected to a battery or how a current builds up in an inductor when a RL circuit is connected to a battery, or how circuits respond to a battery that is repeatedly switched on and off, simulating a square wave.

1.8.1 RC Circuits

The circuit in Fig. 1.25a can charge a capacitor when the switch is in position 1, and discharge the capacitor in position 2. Resistor R must be included in this circuit as it is either part of the battery or part of a capacitor which is not ideal but is lossy, or is simply an external resistor that is added to the circuit to control the charging rate. With the switch thrown to position 1, at time $t = 0$, the voltage equation around the loop can be written as

$$V = Ri + \frac{q}{C} = Ri + \frac{1}{C} \int_0^t i(\sigma) \, d\sigma \tag{1.46}$$

where it was assumed that the capacitor was initially uncharged, that is, $\frac{1}{C} \int_{-\infty}^0 i \, d\sigma = 0$. As the battery voltage V is a constant, we can differentiate the above equation and obtain a simpler equation and its solution as

$$\frac{di}{dt} + \frac{i}{RC} = 0 \quad \text{where} \quad i = Ae^{-t/RC} \tag{1.47}$$

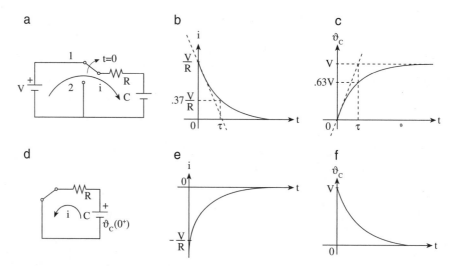

FIGURE 1.25   (a) A circuit that charges the capacitor to battery voltage V when the switch is in position 1 and discharges the capacitor when in position 2. (b) Charging current and (c) charging voltage. (d) Discharge circuit with (e) discharge current and (f) discharge voltage.

where A is an unknown constant, which must be determined from *initial conditions* of the circuit. We have learned that a capacitor has inertia for voltage, which means that if the capacitor voltage before throwing the switch was zero, the capacitor voltage immediately after the switch is thrown must remain zero, i.e.,

$$v_C(t = 0^-) = v_C(t = 0^+) = 0 \qquad (1.48)$$

where the 0^- and 0^+ imply a time just before and just after the switch is thrown. Since no voltage exists across the capacitor after the switch is thrown to position 1, we conclude that the current at that instant is given by V/R. Therefore, using (1.47), we can say that initial current is

$$i(t = 0) = Ae^{-0} \equiv \frac{V}{R} \qquad (1.49)$$

With constant A determined, we can now express the current for any time $t > 0$ as

$$i(t) = \frac{V}{R} e^{-t/RC} \qquad (1.50)$$

The current decreases as the capacitor charges which increases the capacitor voltage from initially zero volts to

$$v_C = \frac{q}{C} = \frac{1}{C} \int_0^t i \, dt = \frac{1}{C} \int_0^t \frac{V}{R} e^{-t/RC} \, dt = V \left(1 - e^{-t/RC} \right) \qquad (1.51)$$

When the capacitor voltage v_C reaches the battery voltage V, the current ceases and the capacitor is said to be fully charged. This will occur as $t \to \infty$ (or after a time $t \gg RC$ has elapsed, where RC is the *time constant* of the circuit). The current and capacitor voltage are sketched in Figs. 1.25b and c, respectively.[17] The voltage across the resistor, $v_R = Ri$, follows the shape of the current curve.

Discharge of the capacitor will be effected when the switch is thrown to position 2; the battery is then disconnected and a short placed across the RC combination as shown in Fig. 1.25d. The equation describing this state is given by (1.47), with constant A determined again by (1.48), except that now initial conditions are $v_C(0^-) = v_C(0^+) = V$. We are assuming that the capacitor was fully charged before the switch was thrown to position 2, i.e., $v_C = \frac{1}{C} \int_{-\infty}^{0} i\, dt = V$. The charged capacitor is now the source for the current, which initially is given by $i = -V/R$ and now flows in the opposite direction. The discharge current for $t > 0$ is therefore given by $i = -(V/R)e^{-t/RC}$ and is sketched in Fig. 1.25e. The capacitor voltage during discharge is

$$
\begin{aligned}
v_C(t) &= \frac{1}{C} \int_{-\infty}^{t} i\, dt = V + \frac{1}{C} \int_{0}^{t} i\, dt \\
&= V - \frac{1}{C} \int_{0}^{t} \frac{V}{R} e^{-t/RC}\, dt = V e^{-t/RC}
\end{aligned}
\tag{1.52}
$$

and is sketched in Fig. 1.25f. During discharge, the charged capacitor acts as the source for current i as well as the resultant $i_2 R$ losses in the resistor. The difference between a battery and a charged capacitor as a source is that a battery can sustain a constant current whereas a charged capacitor cannot. As the capacitor discharges, the current decreases exponentially because less charge is available subsequently. On the other hand, a battery, due to its chemistry, has a specified voltage which produces a current dependent on the load and, because it has a reservoir of chemical energy, can maintain that current. Only when the stored chemical energy is nearly depleted does the current begin to decrease (and soon thereafter we pronounce the battery to be dead).

If we throw the switch before the capacitor is fully charged, then the initial voltage when discharge begins will be the voltage $v_C(0^-)$ that existed on the capacitor before the switch was thrown to position 2. For this situation, (1.52) would become

$$
v_C(t) = v_C(0^-) + \frac{1}{C} \int_{0}^{t} i\, dt = v_C(0^-)e^{-t/RC}
\tag{1.53}
$$

which reduces to (1.52) for the case when the capacitor is fully charged ($v_C(0^-) = V$) at the time the switch is thrown to state 2. One can observe that (1.53) is therefore a more general expression than (1.52). It is hoped that no confusion arises because there are two sets of 0^- and 0^+, one for the first transient when the switch is thrown to position 1, and one for the second transient when the switch is thrown to position 2.

---

[17]Alternatively, in place of (1.46), we could have written $V = Ri + v_C = RC\, dv_C/dt + v_C$, which when solved would give (1.51).

1.8.2   Time Constant

A *time constant* τ is associated with an exponential process such as $e^{-t/\tau}$. It is defined as the time that it takes for the process to decrease to $1/e$ or 37% of its initial value ($1/e = 1/2.71 = 0.37$). Hence, when the process is 63% complete a time $t = \tau$ has elapsed. Time constants provide us with a convenient measure of the speed with which transients in circuits occur. By the time 1τ, 2τ, 3τ, 4τ, and 5τ have elapsed, 37%, 13%, 5%, 1.8%, and 0.67% of the transient remain to be completed. We can then state that for most practical purposes a transient will be completed at the end of five time constants, as only two-thirds of 1% of the original transient then remains. Consequently a knowledge of the time constant allows us to estimate rapidly the length of time that a transient will require for completion.

Referring to (1.50), we see that in a capacitor-charging circuit, the current will have decayed to $1/e$ or to 37% of its initial value in a time of $t = RC$. Hence, the time constant τ for an RC circuit is RC. We could have also examined voltage and come to the same conclusion. For example, using (1.51), which gives the capacitor voltage, we conclude that the time that it takes to charge an initially uncharged capacitor to 63% ($1 - 1/e = 1 - 1/2.71 = 0.63$) of the battery voltage V is the time constant τ.

We can now make an important observation: the time constant is a characteristic of a circuit. Therefore, in an RC circuit, the time constant $\tau = RC$ is the same for charge or discharge—which can be easily seen by looking at the charging voltage (1.51) and comparing it to the discharge voltage (1.52).

There is another aspect of time constants that should be understood. We observe that a transient would be complete in the time of one time constant τ if the current in Fig. 1.25b were to decrease at the same slope that it began. We can differentiate (1.50), evaluate it at $t = 0$, and obtain $di/dt = -i(0)/\tau$. This gives us the slope of a straight line, which if it starts at $i(0) = V/R$, intersects the t-axis at $\tau = RC$. This curve is shown as a dashed line in Fig. 1.25b. In summary, we can state the following: The time constant is the time in which the current in the circuit would reach its final value if it continued to change at the rate at which it initially started to change.

1.8.3   RL Circuits

Figure 1.26a shows an RL circuit to which a battery is connected at $t = 0$. Using Kirchhoff's voltage law, we obtain for $t > 0$

$$V = v_L + v_R = L\frac{di}{dt} + Ri \tag{1.54}$$

Rearranging the above differential equation in the form of $(d/dt)i + (R/L)i = V/L$, we can obtain the particular and general solution for current i by inspection as

$$i = Ae^{-t/(L/R)} + \frac{V}{R} \tag{1.55}$$

The unknown constant A can be determined from the initial condition of the circuit, which is assumed to be that no current flowed in R or L before the switch was closed,

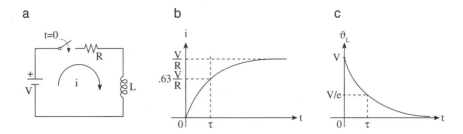

a   b   c

FIGURE 1.26   (a) A battery connected to an RL circuit and the (b) resultant current and (c) inductor voltage.

that is,

$$i|_{t=0} = 0 = A + \frac{V}{R} \tag{1.56}$$

With A determined, we can now give the complete current for time $t > 0$ as

$$i(t) = \frac{V}{R}(1 - e^{-t/\tau}) \tag{1.57}$$

where the time constant for an RL circuit is equal to $\tau = L/R$. The current response is plotted in Fig. 1.26b. The voltage response across the inductor is also given in Fig. 1.26c. It shows that all of the battery voltage is developed across the inductance at $t = 0$ and that for $t > 0$, v_L decays to zero as the transient dies out and the inductor begins to act as a short circuit.

We can now observe that when closing the switch in an inductive circuit, the current changes first rapidly and then more gradually. If the current did not taper off, but continued to change at the initial rate, it would complete the entire change in a time equal to the time constant of the circuit.

EXAMPLE 1.10   Figure 1.27a shows an inductive circuit. Before the switch is closed it is assumed that the circuit was in this state for a long time and all transients have died down, so the current through the battery and the inductor is $i(0^-) = 10/(20 + 30) =: 0.20$ A. At time $t = 0$ the switch is closed. As the inductor has inertia for current, we can state that $i_L(0^-) = i_L(0) = i_L(0^+) = 0.20$ A. It is desired to find the current i_L through the inductor for $t > 0$. In addition to the battery, the inductor becomes a source during the duration of the transient. It has energy stored in the form of $\frac{1}{2}Li^2(0)$ at the time the switch is closed. As the current i_L for $t > 0$ is desired, let us replace the circuit to the left of the inductor by Thevenin's equivalent. To do so we first remove L and find the open-circuit voltage which is the Thevenin's voltage, i.e.,

$$V_{oc} = V_{th} = 10\,\text{V}\frac{40}{20 + 40} = 6.7\,\text{V}$$

Thevenin's resistance is obtained after replacing the battery by a short as

$$R_{th} = 30 + \frac{20 \cdot 40}{20 + 40} = 43.3\,\Omega$$

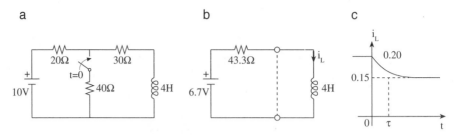

FIGURE 1.27   (a) A transient is created in an inductive circuit by closing the switch at time $t = 0$. (b) Thevenin's equivalent. (c) Inductor current after closing the switch.

The equivalent circuit valid for $t > 0$ is shown in Fig. 1.27b. The time constant is $\tau = L/R = 4/43.3 = 0.09$ s. After the transient dies down, i_L settles down to $i_L(\infty) = 6.7/43.3 = 0.15$ A. Therefore, at the time the switch is closed the inductor current is 0.20 A, decaying exponentially after that and settling down to its final value of 0.15 A. Putting this in equation form, we obtain for $t > 0$

$$i_L(t) = i_L(\infty) + \left(i_L(0^+) - i_L(\infty)\right) e^{-t/\tau}$$
$$= 0.15 + (0.20 - 0.15)e^{-t/0.09}$$
$$= 0.15 + 0.05e^{-t/0.09}$$

which is the desired answer and is plotted in Fig. 1.27c.   ■

1.9   SUMMARY

- In this chapter we developed the basics of circuit theory which are necessary for the study of electronics. We defined the circuit elements by their relationship to current and voltage, which restated is

$\overset{R}{-\!\!\!\wedge\!\!\wedge\!\!\wedge\!\!-}$	$v = Ri$	$i = Gv$
$\overset{C}{-\!\!\vdash\!\!-}$	$v = \dfrac{1}{C} \displaystyle\int_{-\infty}^{t} i\, dt$	$i = C \dfrac{dv}{dt}$
$\overset{L}{-\!\!\ell\!\ell\!\ell\!\!-}$	$v = L \dfrac{di}{dt}$	$i = \frac{1}{L} \int_{-\infty}^{t} v\, dt$

- We then classified R as an energy-converting device (electrical power i^2R to heat), and L and C as energy storage devices ($w_L = \frac{1}{2}Li^2$ and $w_C = \frac{1}{2}Cv^2$).

- Kirchhoff's laws were introduced, which enabled us to analyze circuits, and solve for currents and voltages anywhere in the circuit.

- Thevenin's equivalent circuit, when combined with the maximum power transfer condition, allowed us to view any two-terminal circuit as a practical source. This has considerable implications when studying amplifiers, allowing us to view an amplifier at the output terminals as a practical source.

- For maximum power transfer to a load, the amplifier and load should be matched, i.e., the output resistance of the amplifier should be as close as possible to that of the load.

The study of electronics will be based on these ideas, which will be developed in greater detail in the following chapters.

Problems

1. Check the dimensional correctness of

$$W = \int F \, dl \quad \text{and} \quad V = -\int E \, dl$$

2. A battery of 5 V is connected to two parallel copper plates which are separated by 1 mm. Find the force F in newtons that exists on an electron placed between the plates. Mass of an electron is $9.11 \cdot 10^{-31}$ kg.

3. How long would it take for an electron that is placed in the center between two parallel plates and then set free to arrive at one of the plates? The plates are separated by 10 cm and a battery of 12 V is connected to them.
   Ans: $6.9 \cdot 10^{-8}$ s.

4. List the three alternative forms of Ohm's law.

5. List the three alternative forms of the power expression.

6. A wire of resistance 4 Ω carries a current of 1.5 A. What is the voltage (potential drop) across the wire?
   Ans: 6 V.

7. A nichrome wire has a radius of 0.65 mm. Nichrome has a resistivity of 10^{-6} Ω-m. What length of wire is needed to obtain a resistance of 4 Ω?

8. A main reason that copper is used in household and commercial wiring is its low resistivity, which gives copper wires a low resistance. Determine the resistance per unit length of gauge No. 14 copper wire.
   Ans: $R/\ell = \rho/A = 8.17 \cdot 10^{-3}$ Ω/m. *Note* (**gauge number**, diameter in mm, area in mm²): (**4**, 5.189, 21.18), (**6**, 4.116, 13.30), (**8**, 3.264, 8.366), (**10**, 2.588, 5.261), (**12**, 2.053, 3.309), (**14**, 1.628, 2.081), (**16**, 1.291, 1.309), (**18**, 1.024, 0.8231), (**20**, 0.8118, 0.5176), (**22**, 0.6439, 0.3256).

9. A generator is delivering 300 A at a potential difference of 220 V. What power is it delivering?
   Ans: 66 kilowatts (kW).

10. A 10 Ω resistor carries a current of 5 A. Find the power dissipated in this resistor.

11. For Problem 10 (above), find the power dissipated in the resistor by using the expression $P = V^2/R$.

12. A resistor of resistance 5 Ω carries a current of 4 A for 10 s. Calculate the heat that is produced by the resistor.
    Ans: 800 J.

13. One kilowatt is equivalent to 1.341 horsepower (hp) or 0.948 British thermal units (Btu) per second or 239 calories per second. Find the electrical equivalents of 1 hp, 1 Btu/s, and 1 cal/s.

14. One joule, one newton-meter (N-m), or 1 watt-second (W-s) is equivalent to 0.738 foot-pounds (ft.-lb.). Find the equivalent of 1 kilowatt-hour (kWh) in ft.-lb.
    Ans: $2.66 \cdot 10^6$ ft-lb.

15. An electric water heater, designed to operate at 110 V, has a resistance of 15 Ω. How long will it take to raise the temperature of a 250 g cup of water from 10°C to 100°C? The specific heat of water is 1 cal/g/°C. Also, 1 W-s = 0.239 cal. Neglect the specific heat of the cup itself.

16. An electric heater is added to a room. If the heater has an internal resistance of 10 Ω and if energy costs 8 cents per kilowatt-hour, how much will it cost to continuously run the heater per 30-day month? Assume the voltage is 120 V.
    Ans: $82.94.

17. Using Kirchhoff's voltage law and Fig. 1.3, find the voltage across R_1 if the battery voltage is 12 V and $V_{R_2} = 9$ V.

18. Using Kirchhoff's current law and Fig. 1.3, find the current through R_1 if the battery current is 1 A and $I_{R_2} = 0.5$ A.
    Ans: 0.5 A.

19. (a) A DC battery of 120 V is connected across a resistor of 10 Ω. How much energy is dissipated in the resistor in 5 s?
    (b) An AC source which produces a peak voltage of 169.7 V is connected across a resistor of 10 Ω. How much energy is dissipated in the resistor in 5 s?

20. If in Problem 19 the answers to (a) and (b) are the same, what conclusions can one draw?

21. In the circuit of Fig. 1.4a, the voltage v is given by $v(t) = V_p \cos 10t$.

    (a) Find the instantaneous and average power in resistor R.
    (b) From the expression for instantaneous power, what can you say about the direction of power flow in the circuit of Fig. 1.4a?
    Ans: $p(t) = V_p^2 \cos^2 10t / R$, $P_{\text{ave}} = V_p^2 / R$.

22. (a) Find the separation between the plates of a mica-filled, parallel-plate capacitor if it is to have a capacitance of 0.05 μF and a plate area of 100 cm^2.
    (b) Can this capacitor operate at a voltage of 100 V_{DC}?
    (c) What is the maximum voltage at which this capacitor can be used?

23. A rectangular, 20 mA current pulse of 3 ms duration is applied to a 5 μF capacitor ($i = 0, t < 0$; $i = 20$ mA, $0 \le t \le 3$ ms; $i = 0, t > 3$ ms). Find the resultant capacitor voltage v. Assume the capacitor is uncharged for $t < 0$.
    Ans: $v = 0, t < 0$; $v = 4 \cdot 10^3 t, 0 \le t \le 3$ ms; $v = 12$ V, $t > 3$ ms.

24. Find the maximum energy stored in the capacitor C in Fig. 1.5a. Assume the applied voltage is $200 \sin 2\pi t$, and $C = 5$ μF.

25. A rectangular, 2 V voltage pulse of 3 ms duration is applied to a 2 mH inductor ($v = 0, t < 0$; $v = 2$ V, $0 \le t \le 3$ ms; $v = 0, t > 3$ ms). Find the resultant inductor current i. Assume initial current is zero for $t < 0$.

Ans: $i = 0, t < 0; i = 10^3 t, 0 \le t \le 3$ ms; $i = 3$ A, $t > 3$ ms.

26. A standard D cell flashlight battery is connected to a load of 3 Ω. After 6 hr of intermittent use the load voltage drops from an initial 1.5 V to a final useful voltage of 0.9 V.

(a) Calculate the internal resistance of the battery at the load voltage of 0.9 V.
(b) Calculate the average value of voltage V during the useful life of the battery.
(c) Calculate the average value of current I during the useful life of the battery.

27. For the above problem:

(a) Calculate the average power supplied by the D cell.
(b) Calculate the energy supplied in watt-hours.
(c) If the purchase price of the battery is $1.20, calculate the cost of the battery in cents/kilowatt-hour (¢/kW-h). Compare this cost with that of electric utilities, which typically sell electric energy for 8 ¢ /kW-h.
   Ans: (a) 0.48 W, (b) 2.88 W-h, (c) 41,667 ¢ /kW-h (energy supplied by the battery is 5208 times as expensive as that supplied by a typical utility).

28. A source is represented by a black box with two wires. If the open-circuit voltage at the two wires is 6 V and the short-circuit current flowing in the wires when they are connected to each other is 2 A, represent the black box by a practical voltage source; that is, find v_s and R_i in Fig. 1.12*a*.

29. A source is represented by a black box with two wires. If the open-circuit voltage at the two wires is 6 V and the short-circuit current flowing in the wires when they are connected to each other is 2 A, represent the black box by a practical current source; that is, find i_s and R_i in Fig. 1.12*d*.
   Ans: 2 A, 3 Ω.

30. Three resistors (1 Ω, 2 Ω, 3 Ω) are connected in series across an ideal voltage source of 12 V. Find the voltage drop across each resistor.

31. Three resistors (1 Ω, 2 Ω, 3 Ω) are connected in parallel to an ideal current source of 11 A. Find the current in each resistor.
   Ans: 6 A, 3 A, 2 A.

32.

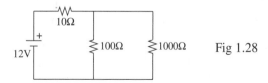

(a) Calculate the battery current in Fig. 1.28.
(b) Calculate the current through each resistor.
(c) Calculate the voltage drop across each resistor.

33.

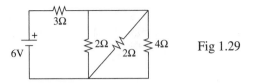

Fig 1.29

(a) Calculate the current in each resistor in Fig. 1.29.
(b) Calculate the voltage drop across each resistor.
(c) Calculate the power delivered by the battery.
Ans: (a) 1.58 A, 0.63 A, 0.32 A. (b) 4.7 V, 1.3 V. (c) 9.47 W.

34.

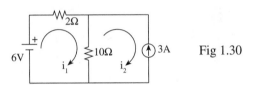

Fig 1.30

Use Kirchhoff's laws to write two mesh equations for currents i_1 and i_2 in F 1.30 (see Fig. 1.23). Solve for these currents and then:

(a) Calculate the current through and the voltage across the 10 Ω resistor.
(b) Calculate the current through and the voltage across the 2 Ω resistor.

35. Repeat Problem 34, that is, determine the unknown currents and voltages in (a) and (b), except this time use only superposition to solve for the unknowns.
Ans: (a) 1 A, 10 V. (b) 2 A, 4 V.

36. Repeat Problem 34, that is, determine the unknown currents and voltages in (a) and (b), except this time use only Thevenin's and Norton's theorems to solve for the unknowns. *Hint*: replace the circuit to the right of the 2 Ω resistor by Thevenin's equivalent, or replace the circuit to the left of the 10 Ω resistor by Norton's equivalent.

37.

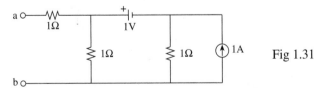

Fig 1.31

Find Thevenin's equivalent at the terminals *a-b* for the circuit shown in Fig. 1.31.
Ans: $R_{th} = 1.5\ \Omega$, $V_{th} = 1$ V.

38. If a resistor is connected to terminals *a-b* of the circuit shown in Fig. 1.31, what value should it have for maximum power transfer to it? What is the maximum power?

39.

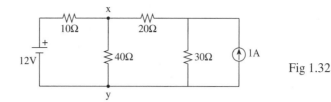

Fig 1.32

Find the Thevenin's equivalent of the circuit shown in Fig. 1.32 as viewed from the terminals x-y.

Ans: 12.414 V, 6.896 Ω?

40.

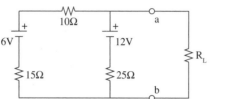

Fig 1.33

Resistor R_L is connected to the circuit shown at terminals a-b (Fig. 1.33). (a) Find the value of R_L for maximum power delivery to R_L. (b) What is the maximum power? (c) What is the power delivered to R_L when $R_L = 10\ \Omega$?

41. Discuss the meaning of matching with respect to power transfer.

42. In the circuit of Fig. 1.23 use two loop currents to solve for current $i\,R_1$ but assume current i_2 has a counterclockwise orientation.

43. Solve for current i_2 in the circuit shown in Fig. 1.24.
    Ans: −0.532 A.

44. Solve for current i_3 in the circuit shown in Fig. 1.24.

45. Use the mesh method to solve for current i_1 in the circuit shown in Fig. 1.24, except assume current i_3 is counterclockwise.

46. Use loop equations to solve for the current in resistor R_2 of the circuit shown in Fig. 1.24.
    Ans: $i_{R_5} = i_1 - i_3 = 0.19$ A.

47. When writing loop equations, is there an advantage to assuming the same orientation for all loop currents?

48. Referring to Fig. 1.25d, a 2 μF capacitor is charged to 12 V and then connected across a resistor of 100 Ω.

   (a) Determine the initial charge on the capacitor.
   (b) Determine the initial current through the 100 Ω resistor.
   (c) Determine the time constant.
       Ans: (a) $24 \cdot 10^{-6}$ C. (b) 0.12 A. (c) 200 μs.

49. Calculate the charge on the capacitor and the current through the resistor at time $t = 1$ ms for the RC circuit of Problem 48.

50. Three capacitors are connected in series across a 100 V battery. If the capacitances are 1 μF, 0.1 μF, and 0.01 μF, respectively, calculate the potential difference across each capacitor.

Hint: First show that the equivalent capacitance of three capacitors in series is $1/C_{eq} = 1/C_1 + 1/C_2 + 1/C_3$. Use Kirchhoff's law to state that battery voltage

$$
\begin{aligned}
V &= V_1 + V_2 + V_3 \\
&= 1/C_1 \int i\,dt + 1/C_2 \int i\,dt + 1/C_3 \int i\,dt \\
&= (1/C_1 + 1/C_2 + 1/C_3) \int i\,dt \\
&= (1/C_1 + 1/C_2 + 1/C_3)Q.
\end{aligned}
$$

We can now solve for the charge Q that is deposited on the equivalent capacitor C_{eq}, which is the same charge that also exists on each capacitor. Hence, the voltage on each capacitor is $V_1 = Q/C_1$, $V_2 = Q/C_2$, and $V_3 = Q/C_3$. No confusion should result because the same charge Q exists on each capacitor: the plates of the capacitors in series have opposite charges which cancel each other, leaving only the $+Q$ and $-Q$ on the outer plates.
Ans: 0.9 V, 9 V, 90 V.

51. An initially uncharged 2 μF capacitor is connected in series with a 10 kΩ resistor and a 12 V battery.

   (a) Find the charge and voltage on the capacitor after a very long time.
   (b) Find the charge and voltage on the capacitor after one time constant.

52. A 2 μF capacitor is charged to 12 V and then connected across a resistor of 100 Ω.

   (a) Determine the initial energy stored in the capacitor.
   (b) Determine the energy stored in the capacitor after two time constants have elapsed.
   Ans: (a) 144 μJ. (b) 2.64 μJ.

53. A 1 mH inductor and a 1 kΩ resistor are connected to a 12 V battery for a long time. The circuit is similar to that in Fig. 1.26. The battery is suddenly removed and a 1 kΩ resistor is substituted.

   (a) Find the initial inductor current i_0 at the time of substitution.
   (b) Find the current in the circuit after two time constants have elapsed.
   (c) Find the total heat produced in the resistors when the current in the inductor decreases from its initial value i_0 to 0.

54. Calculate the time constant of a circuit of inductance 10 mH and resistance 100 Ω.
   Ans: 100 μs.

55. Assume the switch in Fig. 1.27a has been closed for a long time so the current through the inductor has settled down to $i_L = 0.15$ A. Suppose the battery is suddenly disconnected. (a) Find and sketch the voltage across the 30 Ω resistor. (b) What is the time constant?

56. In the circuit of Fig. 1.27a the switch is closed at time $t = 0$. Find the current in the switch for $t > 0$.
   Ans: 0.11–0.018 exp$(-t/0.09)$ A.

CHAPTER 2

AC Circuits

2.1 INTRODUCTION

The previous chapter was concerned primarily with DC circuits. Steady voltages and currents are the simplest and widely occurring in circuits. Therefore, a study of electronic circuits usually begins with an analysis of DC circuits. In this chapter we will continue the study of circuits with the steady-state analysis of AC circuits.

The simplest and widely occurring time-varying voltages and currents are sinusoidal. They alternate direction periodically and are referred to by the universal name of alternating currents. Even though there is a multitude of other alternating signals,[1] such as square wave, sawtooth, triangular, etc., AC is usually reserved for sinusoidally alternating signals. There is a special characteristic of sinusoids that no other periodic waveshape possesses: *if a linear circuit is driven by a sinusoidal source, then all responses anywhere in the circuit are also sinusoidal.* This observation applies *only after all initial transients have died down and a steady state is reached.* It is referred to as *steady-state AC analysis* and is of considerable practical importance. For example, in a power utility grid, all voltages and currents vary precisely at 60 Hz (50 Hz in Mexico and Europe). This variation is so precise that it is used to operate our wall clocks throughout the grid. Nonsinusoidal waveshapes that drive circuits produce responses that vary substantially—to the point that the response has no resemblance to the source variation. It is only the sinusoid that has that special characteristic and which is used to develop the *phasor method* for AC circuit analysis. Hence, a logical next step is the study of AC circuits.

---

[1]When convenient, we will use the term *signal* interchangeably with *voltage* and *current*. Obviously in the power industry signal is rarely used since a precise distinction between voltage and current needs to be made at all times. On the other hand, in the communication industry, where currents are usually in the micro- or milliamp range, but voltages can vary anywhere from microvolts at the antenna input to tens of volts at the output of an amplifier, the term *signal* is commonly used and usually denotes a signal voltage.

Practical electronic circuits must process a variety of signals, of which the steady and sinusoidal are just two. Does this imply that we are going to have an endless procession of chapters, one for square-wave, one for exponential, and so on? Fortunately, for most practical situations a study of DC and AC circuits will suffice. The reason is that many environments such as those in power and in communication operate with sinusoidal signals and thus can be modeled as AC circuits. The power utilities, for example, put out precise 60 Hz sinusoidal voltages. Narrowband signals used in AM and FM broadcasting can be treated as quasi-sinusoidal signals. Even digital communication is usually a periodic interruption of a sinusoidal carrier. DC analysis, on the other hand, can be used to analyze circuits that are connected to sources that produce a square-wave type of signal, by considering the straight-line portions of the square-wave signal as DC. Then, by pasting together the DC type responses, an accurate representation of the square-wave response can be obtained. The remaining multitude of other signals, if not directly amenable to DC or AC analysis, can nevertheless be analyzed by the Fourier method, which is a technique that represents an arbitrary periodic signal by sinusoidal terms of different frequencies. If the resultant series of sinusoidal terms converges quickly, we can again treat the circuit as an AC circuit and obtain the circuit response to the arbitrary signal as a sum of responses to the sinusoidal Fourier terms. Thus, the sinusoidal signal seems to have a special place among all signals.

It is hoped that we have established in this brief introduction that a study of AC and DC circuits provides us with fundamental tools in circuit analysis that can be used even when a circuit is driven by voltages or currents other than DC or AC.

2.2   SINUSOIDAL DRIVING FUNCTIONS

If the voltage or current source is sinusoidal, all voltages and currents anywhere in the linear circuit will also be sinusoidal. Therefore, if it is desired to know a voltage somewhere in the circuit, all that remains to be solved for are the amplitude and phase angle of the unknown voltage. For example, say we drive a circuit with

$$v_s = V_s \cos \omega t \tag{2.1}$$

where v_s is the source voltage, V_s is the amplitude, and ω is the *angular frequency* of the sinusoid.[2] A voltage anywhere in the circuit will look like

$$v = V \cos(\omega t + \theta) \tag{2.2}$$

where the amplitude V and phase θ must be determined, but where the sinusoid varies at an angular frequency ω that is already known. For example, Fig. 2.1a shows an

---

[2] As was pointed out in Section 1.4, small-case letters are instantaneous values reserved for time-varying voltages and currents, i.e., $v \equiv v(t)$; upper-case letters are for constant values; sub-p in amplitude V_p means peak value; and ω is the angular frequency in radians per second of the sinusoid. There are 2π radians in one complete cycle and it takes time T, called a *period*, to complete a cycle, i.e., $\omega T = 2\pi$. Since there are $1/T$ cycles per second, called frequency f, we have $f = 1/T = \omega/2\pi$ cycles per second or hertz (Hz).

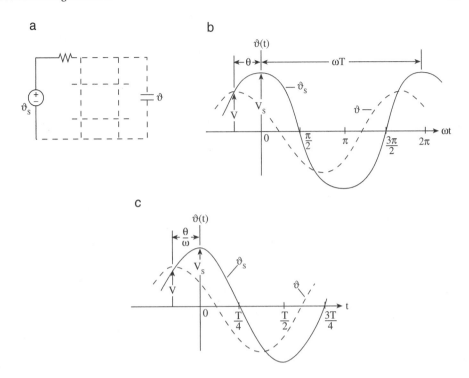

FIGURE 2.1 (a) A network with only a few elements shown. (b, c) Voltage v (for which V and ω need to be determined) and source voltage v_s, plotted versus (b) ωt and (c) t.

arbitrary network, a source voltage v_s, and a voltage v at some place in the circuit. Figure 2.1b shows what these voltages look like. Voltage amplitude V and angle θ need to be determined, but V_s and ω are known.

To clarify sinusoidal variations when ω is given, we show two plots in Fig. 2.1, one versus ωt and one versus t. With respect to v_s, voltage v is shifted to the left by θ radians (v occurs earlier in time than v_s) and we say that v leads v_s by θ radians or by θ/ω seconds. Conversely, we can say v_s lags v. Two sinusoids which are leading or lagging which respect to each other are said to be *out of phase*, and when θ is zero they are said to be *in phase*. Phase-lead and -lag networks are of considerable interest in electronics.

Phasor analysis, the subject of the next section, will teach us to determine V and θ in an easy and elegant manner without resorting to differential equation solutions which normally are required when solving circuit equations in the time domain.

2.2.1 Phasor Analysis

A circuit containing resistors R, capacitors C and inductors L is customarily referred to as an RLC circuit. If we consider a simple series RLC circuit, as shown in Fig. 2.2,

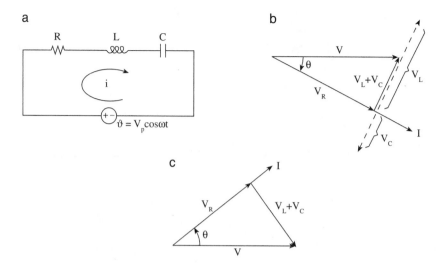

FIGURE 2.2   (a) Current $i(t)$ is to be determined when voltage $v(t)$ drives the series RLC circuit. (b) Phase relationship between the voltages $V = V_R + V_L + V_C$ and the current I when the RLC circuit is inductive. (c) Phase relationship when the circuit is capacitive.

which is connected to a time-varying voltage source $v(t)$ and want to know what the current $i(t)$ in the loop is, we start by applying Kirchhoff's voltage law around the loop and obtain the following equation:

$$v(t) = Ri(t) + L\frac{di(t)}{dt} + \frac{1}{C}\int i(t)\,dt \qquad (2.3)$$

Solving this equation is usually a formidable task. However, for sinusoidally varying voltage sources such as $v(t) = V_p \cos \omega t$, the problem becomes much simpler as we can then apply phasor analysis. The basis of phasor analysis is as follows:

(1) First, we recognize that we are dealing with a linear problem.

(2) The above equation is basically a linear differential equation (LDE) with constant coefficients.

(3) Natural solutions to LDEs with constant coefficients are exponentials (because differentiating an exponential yields the same exponential, whereas differentiating a cosine yields a sine). Hence, let us try to represent the source by an exponential. We can do this by noting that $e^{\pm jx} = \cos x \pm j \sin x$, which is a complex number[3] expression and is known as *Euler's* or as *DeMoivre's identity*. The use

---

[3]We can represent a point uniquely either by specifying the rectangular coordinates a, b or the polar coordinates r, θ. Similarly, for complex numbers, we can state that $a + jb = re^{j\theta}$, where $r = \sqrt{(a^2 + b^2)}$

of the phasor method will thus have a small penalty: all calculations will be in terms of complex numbers.

(4) Let the real source be given by $V_p \cos \omega t = \text{Re } V_p e^{j\omega t}$, where Re stands for *"the real part of."* Now, if we only could drop the Re operator we would then be representing the actual source by an exponential source $V_p e^{j\omega t}$. In fact, we can do just that. Because the system is linear, we can omit the Re operator, obtain the solution to the problem with the exponential source, and convert it to the solution for the real source by simply taking the real part of the exponential source solution.

(5) The solution for the current with an exponential source has the form $Ie^{j\omega t}$, with I as the only unknown. Hence, we have reduced the problem to finding a complex number I, which will be referred to from now on as phasor I.

(6) The problem is essentially solved once we have found phasor I. To convert phasor I to real-time current $i(t)$, we simply multiply I by $e^{j\omega t}$ and take the real part, that is,

$$i(t) = \text{Re } Ie^{j\omega t} \tag{2.4}$$

For example, if the solution is phasor $I = I_p e^{j\theta}$, then

$$i(t) = \text{Re } I_p e^{j\theta} e^{j\omega t} = I_p \cos(\omega t + \theta) \tag{2.5}$$

where I_p is the *current amplitude* (a real number) and θ is the phase angle of the current with respect to the source voltage. Thus, if we can find I_p and θ the problem is solved.

Now that we know what the phasor method[4] is, let us use it to find I_p and θ for the circuit in Fig. 2.2a. Substituting $V_p e^{j\omega t}$ for $v(t)$ and $Ie^{j\omega t}$ for $i(t)$ in (2.3) and performing the indicated differentiation and integration, we obtain

$$V_p = RI + j\omega LI + \frac{I}{j\omega C} \tag{2.6}$$

after canceling $e^{j\omega t}$ from both sides. Factoring out the unknown current I gives

$$V_p = \overbrace{\left[R + j\left(\omega L - \frac{1}{\omega C} \right) \right]}^{Z} I \tag{2.7}$$

$$= ZI$$

---

and $\theta = \tan^{-1} b/a$. Hence $1/(a + jb) = (a - jb)/(a^2 + b^2) = e^{-j\theta}/(a^2 + b^2)^{1/2}$, where a, b, r are real numbers, $j = \sqrt{-1}$, $e^{j\pi/2} = j$, and $1/j = -j$.

[4]In general, phasors are complex quantities and as such should be distinguished from real quantities. Most books do this either by bolding, starring or underlining phasors. We feel that it is sufficiently obvious when phasors are involved so as not to require special notation.

where we now identify the quantity in the square brackets as the *impedance Z*. The impedance Z is a complex quantity with a real part which is the *resistance R* and an imaginary part which is called the *reactance*. When the inductive term ωL dominates, the reactance is positive, whereas when the capacitive term $1/\omega C$ dominates, the reactance is negative.

The beauty of the phasor method is that it converts an integro-differential equation (2.3) (in the time domain) into a simple algebraic equation (2.7) (in the frequency domain) which is almost trivial to solve for I. Thus, here is the solution for the phasor current I:

$$I = \frac{V_p}{R + j(\omega L - 1/\omega C)} \tag{2.8}$$

or just simply $I = V_p/Z$. When working with complex quantities it is best to have complex numbers occur only in the numerator. Hence, we multiply the numerator and denominator of (2.8) by the complex conjugate of the denominator to obtain

$$I = \frac{V_p[R - j(\omega L - 1/\omega C)]}{R^2 + (\omega L - 1/\omega C)^2} \tag{2.9}$$

This can be converted to polar form

$$\begin{aligned} I &= \frac{V_p}{\left[R^2 + (\omega L - 1/\omega C)^2\right]^{1/2}} e^{-j\,\arctan(\omega L - 1/\omega C)/R} \\ &= I_p e^{-j\theta} \end{aligned} \tag{2.10}$$

which is the solution to the unknown current in phasor form. Now, to obtain the real-time solution we multiply by $e^{j\omega t}$ and take the real part, i.e., $i(t) = \operatorname{Re} I e^{j\omega t}$, or

$$\begin{aligned} i(t) &= \frac{V_p}{\left[R^2 + (\omega L - 1/\omega C)^2\right]^{1/2}} \cos(\omega t - \arctan(\omega L - 1/\omega C)/R) \\ &= I_p \cos(\omega t - \theta) \end{aligned} \tag{2.11}$$

The solution to this problem is now completed. The amplitude of the current is thus

$$I_p = V_p / \sqrt{R^2 + (\omega L - 1/\omega C)^2}$$

and the phase angle is $\theta = \arctan(\omega L - 1/\omega C)/R$. Note that the current amplitude and phase angle are real quantities. Since the phase angle of the source voltage is taken as zero (we began this problem with the source phasor voltage given as $V = V_p e^{j0} = V_p$) and the unknown phasor current was found to be $I = I_p e^{-j\theta}$, we see now that the current lags the voltage by θ degrees. The series RLC circuit is said to be inductive.

The lead and lag between two phasors[5] is best shown by a vector diagram like the one in Fig. 2.2b. The source voltage is represented by a phasor $V = V_p$, which is horizontal

---

[5]A phasor is a stationary vector, but is derived from a rotating vector. A phasor which is multiplied by $e^{j\omega t}$ rotates counterclockwise with time t, because the $e^{j\omega t}$ term increases the angle ωt of the phasor with time t. The rotation is frozen at $t = 0$, which drops the rotation as well as $e^{j\omega t}$ from the picture.

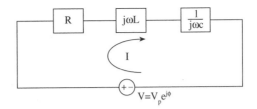

FIGURE 2.3   The phasor equivalent circuit of the circuit in Fig. 2.2a.

in Fig. 2.2b, and the current phasor $I = I_p e^{-j\theta}$ is shown at a negative angle of θ. If, in (2.7), the capacitance C or the frequency ω is decreased such that the capacitive term becomes larger than the inductive term, the phase angle θ becomes positive, as the current now leads the voltage. The series RLC circuit is now referred to as *capacitive* and the phase diagram for this situation is shown in Fig. 2.2c.

EXAMPLE 2.1   If the voltage source in Fig. 2.2a is changed to $v(t) = V_p \cos(\omega t + \phi)$, find the current $i(t)$ in the circuit.

First, let us convert the circuit to a phasor circuit by substituting impedances for the R, L, C elements and by changing the source voltage $v(t)$ to a phasor voltage, as shown in Fig. 2.3. The source phasor voltage is obtained as

$$
\begin{aligned}
v(t) &= V_p \cos(\omega t + \phi) = \operatorname{Re} V_p e^{j(\omega t + \phi)} \\
&= \operatorname{Re} V_p e^{j\phi} e^{j\omega t} = \operatorname{Re} V e^{j\omega t}
\end{aligned}
$$

where the phasor voltage V is identified as $V = V_p e^{j\phi}$. Solving for the phasor current I in the circuit of Fig. 2.3, we obtain

$$
I = \frac{V_p e^{j\phi}}{R + j(\omega L - 1/\omega C)}
$$

which is identical to (2.8) except for the additional $e^{j\phi}$ phase term. Hence the solution to the real- time current is also given by (2.11), except for the additional $e^{j\phi}$ phase term; i.e., repeating the steps that lead from (2.8) to (2.11) gives

$$
i(t) = I_p \cos(\omega t - \theta + \phi)
$$

Thus, except for the constant phase shift ϕ, this problem is virtually identical to that of Fig. 2.2a. In this problem we started out with the phasor voltage $V = V_p e^{j\phi}$ that drives the circuit and solved for the phasor current I, which turned out to be $I = I_p e^{-j\theta} e^{j\phi}$. The phase diagrams of Figs. 2.2b and c therefore also apply to this problem simply by rotating both voltage and current phasors clockwise by ϕ degrees.   ∎

2.2.2   Impedance and Phasor Relationships for $R, L,$ and C

We have seen that the relationships $v = Ri, v = L \, di/dt$, and $v = 1/C \int i \, dt$ for $R,$ $L,$ and C in the time domain changed to the phasor relationships $V = RI, V = j\omega L I,$

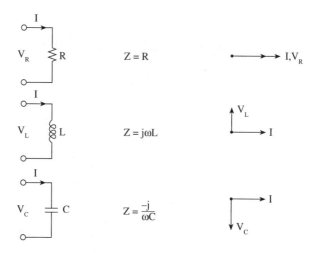

FIGURE 2.4 Impedance Z and current and voltage phasors for R, L, and C.

and $V = I/j\omega C$ in the frequency domain. Fig. 2.4 gives the impedances and shows the phase relationships between voltage and current for the three circuit elements. With the current assumed as the horizontal reference phasor, we see that the voltage is in phase with I for a resistor R, that the voltage leads I by 90° for an inductor, and that the voltage lags I by 90° for a capacitor.

Impedance in AC analysis takes the place of resistance in DC analysis. Hence for AC circuits, Ohm's law becomes $V = IZ$, except for Z being a complex number. As a matter of fact, we can generalize and state that AC circuit analysis is DC analysis with complex numbers.[6] This is a powerful statement and means that all the circuit laws that were derived in the previous chapter apply equally to AC circuits. The equivalence theorem, source transformation, Thevenin's theorem, and loop equations all apply. The next example will demonstrate that.

EXAMPLE 2.2 Find voltage $v_2(t)$ when the circuit shown in Fig. 2.5a is driven by a voltage source with $v = \cos 2t$. The time-domain circuit is first converted to a phasor circuit by changing all circuit elements to impedances as shown in Fig. 2.5b. For example, the 1 H inductor becomes an element of impedance $j\omega L = j2$, and the capacitor tranforms to $1/j\omega C = 1/j = -j$, where ω is given as 2 rad/s. The phasor source voltage is obtained from $v = 1\cos 2t = \text{Re}\,1e^{j2t}$ and is $V = 1e^{j0} = 1$. The output voltage is given by $v_2 = 2i_2$. Hence we solve for i_2 first. Basically the circuit in Fig. 2.5b is like a DC circuit and we proceed to solve for the unknowns on that basis.

---

[6]Strictly speaking this is true only for AC steady state, meaning that all transients have died down. For example, when a sinusoidal source is switched into a circuit, it could create a brief transient in addition to the forced response. AC steady state refers to the time when all transients have died down and only the response to the sinusoidal driving function remains.

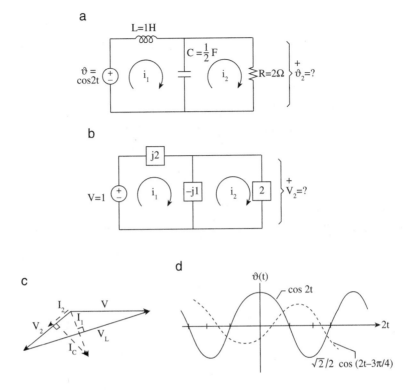

FIGURE 2.5 (a) A two-mesh circuit in the time domain, and (b) the equivalent phasor circuit in the frequency domain. (c) A phasor diagram for the circuit, showing it to be a phase-lag network. (d) A real-time sketch of input and output voltages.

Writing the mesh equations for I_1 and I_2, we obtain

$$
\begin{aligned}
1 &= j1I_1 + j1I_2 \\
0 &= j1I_1 + (2 - j1)I_2
\end{aligned}
$$

These two simultaneous equations in two unknowns are easily solved for I_2, thus

$$
I_2 = \frac{-j}{2 + j2} = \frac{\sqrt{2}}{4}e^{-j3\pi/4}
$$

The phasor diagram in Fig. 2.5c shows that I_2 lags the source voltage V by 135°, and has a magnitude of $\sqrt{2}/4$. Voltage V_2 is simply obtained by multiplying I_2 by the resistance of the 2 Ω resistor, i.e., $V_2 = \sqrt{2}/2e^{-j3\pi/4}$. The real-time voltage can now be obtained by multiplying by $e^{j\omega t}$ and taking the real part, that is,

$$
v_2(t) = \text{Re}\, V_2 e^{j2t} = \text{Re}\, \frac{\sqrt{2}}{2} e^{j(2t-3\pi/4)}
$$

$$
= \frac{\sqrt{2}}{2} \cos(2t - 3\pi/4)
$$

A sketch of the real-time voltage and current is shown in Fig. 2.5d. Hence, a 1 V sinusoidal source voltage produces a 0.7 V output voltage that lags the input voltage by 135°.

The phasor currents in Fig. 2.5c were sketched without performing any additional calculations. After finding I_2 and after sketching the three phasor voltages (which close upon themselves—see (1.10)), we observe that phasors I_1 and V_L must be orthogonal (inductor current lags inductor voltage by 90°). Furthermore, capacitor voltage V_2 lags capacitor current I_C by 90°; hence we can sketch I_C at right angles to V_2. The fact that phasor diagram currents at a node must close on themselves (current summation at the top node: $I_1 = I_2 + I_C$, see (1.11)) allows us to complete the sketch for the currents. Figure 2.5c shows that voltage and current in the 2 Ω resistor are in phase and that magnitude V_2 is twice as large as magnitude I_2. ∎

2.2.3 Admittance

In addition to impedance we often use *admittance Y*, which is defined as the *reciprocal of impedance*, i.e., $Y = 1/Z$. The advantage is that circuit elements connected in parallel have an admittance which is the sum of the admittances of the individual circuit elements. Thus, if we take a resistor, an inductor, and a capacitor and instead of connecting them in series as shown in Fig. 2.3, we connect them in parallel, we obtain

$$
\begin{aligned}
Y &= 1/R + 1/j\omega L + j\omega C \\
&= 1/R + j(\omega C - 1/\omega L) \\
&= G + jB
\end{aligned}
\tag{2.12}
$$

for the admittance, where we refer to G as *conductance* and to B as *susceptance*. A capacitive circuit for which $\omega C > 1/\omega L$ has a positive susceptance, but when $1/\omega L$ is larger than ωC the susceptance is negative and the circuit is said to be inductive.

Expressing impedance $Z = R + jX$ in terms of admittance $Y = G + jB$, we obtain

$$
Z = \frac{1}{Y} = \frac{1}{G + jB} = \frac{G - jB}{G^2 + B^2}
\tag{2.13}
$$

showing that negative susceptance corresponds to positive reactance, as expected.

2.3 HIGH-PASS AND LOW-PASS FILTERS

Since the impedance of inductors and capacitors depends on frequency, these elements are basic components in networks that are frequency sensitive and frequency selective.

We will now consider some often-used but simple two-element filters: the RC and RL filters. Even though RL filters can perform the same filtering action as RC filters, in practice the RC filter is preferred as inductors can be heavy, bulky, and costly.

2.3.1   RC Filters

An RC filter with input and output voltages is shown in Fig. 2.6a. According to Kirchhoff's laws the sum of voltages around the loop must be equal to zero, or $V_i = IR + V_o$. But $V_o = IZ_C = I/j\omega C$, which gives for the voltage gain[7] of the filter

$$\frac{V_o}{V_i} = \frac{I/j\omega C}{V_i} = \frac{V_i/(R + 1/j\omega C)j\omega C}{V_i} \tag{2.14}$$

$$= \frac{1}{1 + j\omega RC} = \frac{1 - j\omega RC}{1 + (\omega RC)^2}$$

$$= \frac{1}{\sqrt{1 + (\omega RC)^2}} e^{-j \arctan \omega RC} = \frac{1}{\sqrt{1 + (\omega/\omega_0)^2}} e^{-j\theta}$$

where $\omega_0 = 1/RC$ and is usually referred to as the *corner, cutoff,* or *half-power* frequency.[8] The magnitude $|V_o/V_i| = 1/\sqrt{1 + (\omega/\omega_0)^2}$ and the phase $\theta = \tan^{-1} \omega RC = \tan^{-1} \omega/\omega_0$ are plotted in Figs. 2.6b and c. From the magnitude plot we conclude that this is a low-pass filter and from the phase plot we conclude that it is a phase-lag network. At the corner frequency (also known as the 3 decibel (dB)[9] frequency) the output voltage is down by $\sqrt{2}$ and the output phase lags the input phase by $45°$.

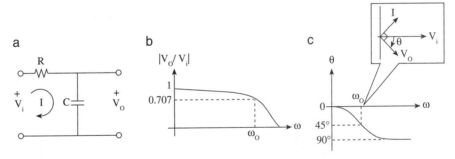

FIGURE 2.6   (a) An RC low-pass filter. (b) The magnitude plot of an RC low-pass filter on log–log paper. (c) The phase plot shows it to be a phase-lag network.

---

[7] We will use the term voltage gain for the ratio V_0/V_i, even though for the passive filter considered here it would be more appropriate to say voltage loss. However, in electronics, V_0/V_i is usually referred as gain and sometimes as a transfer function.

[8] At $\omega = \omega_0 = 1/RC$ the voltage amplitude is $1/\sqrt{2}$ of its maximum magnitude. Since power is proportional to voltage squared, the power at ω_0 is $1/2$ of its maximum power. Hence, the name half-power frequency. Strictly speaking, the half-power frequency is f_0 which is related to ω_0 by $f_0 = \omega_0/2\pi$.

[9] Power gain in terms of dB or decibel is defined as $10 \log_{10} |P_o/P_i|$, where P_o is power out and P_i is power in, or in terms of voltage as $20 \log_{10} |V_0/V_i|$.

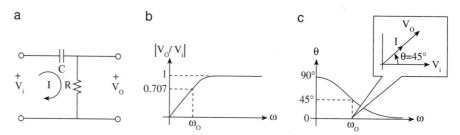

FIGURE 2.7 (a) An RC high-pass filter. (b) The magnitude plot of an RC high-pass filter on log–log paper. (c) The phase plot shows it to be a phase-lead network.

If we examine the equations and figures in more detail we find that at low frequencies the current is limited to small values by the high impedance of the capacitor. Hence, the voltage drop across R is small and most of the V_i voltage appears across the V_o terminals. But at high frequencies, most of the V_i is dropped across R because the impedance $Z_C = 1/j\omega C$ of C becomes small, practically short-circuiting the output. Hence V_o drops off sharply at high frequencies. The transition frequency between the band of frequencies when $V_o \approx V_i$ and when V_o becomes negligible can be considered to be the half-power frequency f_0. Therefore, f_0 is useful in identifying the boundary between these two regions. We call such a filter a *low-pass filter*. Thus, whenever the voltage is taken across a capacitor, the inertia property of capacitors reduces fast variations of voltage while leaving the DC component of V_i unaffected. Such a filter is thus ideal in DC power supplies when voltage smoothing is needed after rectifying an AC voltage.

2.3.2   High-Pass RC Filter

If we interchange R and C as shown in Fig. 2.7a, we obtain a filter that passes high frequencies while attenuating the low ones. Summing the voltages around the loop we obtain $V_i = I/j\omega C + V_o$. But $V_o = IR$, which gives for voltage gain

$$
\begin{aligned}
\frac{V_o}{V_i} &= \frac{1}{1 + \frac{1}{j\omega RC}} = \frac{1 - 1/j\omega RC}{1 + 1/(\omega RC)^2} \\[2mm]
&= \frac{1}{\sqrt{1 + 1/(\omega RC)^2}} e^{j\,\arctan 1/\omega RC} \\[2mm]
&= \frac{1}{1 + (\omega_0/\omega)^2} e^{j\theta}
\end{aligned}
\tag{2.15}
$$

where $\omega_0 = 1/RC$ is the half-power frequency and $\theta = \tan^{-1} \omega_0/\omega$. The magnitude and phase are plotted in Figs. 2.7b and c. For angular frequencies much larger than ω_0, the magnitude $|V_o/V_i| = 1$ and phase $\theta = 0°$, whereas for frequencies much less than ω_0, we have $|V_o/V_i| \approx \omega/\omega_0 \ll 1$ and $\theta \approx 90°$. Frequencies below $f_0 = \omega_0/2\pi$

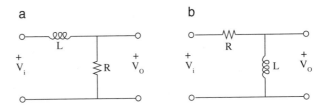

FIGURE 2.8   (a) A low-pass RL filter. (b) A high-pass RL filter.

are attenuated and frequencies above f_0 pass through the filter. A filter of this type is referred to as a *high-pass and phase-lead network*.

Such a filter is typically used as a coupler between amplifier stages: it allows the AC signal to pass from amplifier to amplifier while blocking the passage of any DC voltage. Thus the AC signal is amplified and any undesirable effects of a DC voltage—such as changing the amplifier bias or driving the amplifier into saturation—are avoided.

2.3.3   RL Filters

A low-pass RL filter is shown in Fig. 2.8a. Summing voltages around the loop, we obtain $V_i = j\omega L I + V_o$, where the output voltage is $V_o = RI$. The voltage gain is then given by

$$\frac{V_o}{V_i} = \frac{1}{1 + j\omega L/R} = \frac{1}{\sqrt{1 + (\omega/\omega_0)^2}} e^{-j\theta} \tag{2.16}$$

The last expression of (2.16) was obtained by comparing the middle term of (2.16) to the corresponding one in (2.14). Thus ω_0 in (2.16) is $\omega_0 = R/L$ and $\theta = \tan^{-1} \omega L/R = \tan^{-1} \omega/\omega_0$. We see now that this filter is a *low-pass, phase-lag network*, similar in characteristics to the RC filter of Fig. 2.6. Hence, there is no need to draw amplitude and phase plots; we can simply use those in Figs. 2.6b and c.

2.3.4   High-Pass RL Filter

Interchanging R and L, we obtain the high-pass filter shown in Fig. 2.8b. Summing voltages around the loop, we obtain $V_i = RI + V_o$, where the output voltage is $V_o = j\omega L I$. The voltage gain is thus

$$\frac{V_o}{V_i} = \frac{1}{1 + \frac{1}{j\omega L/R}} = \frac{1}{\sqrt{1 + (\omega_0/\omega)^2}} e^{j\theta} \tag{2.17}$$

Again, comparing this expression to (2.15), we conclude that $\omega_0 = R/L$ and $\theta = \tan^{-1} \omega_0/\omega$. This filter is therefore a high-pass and a phase-lead network, similar to that of Fig. 2.7a. The magnitude and phase plots are shown in Figs. 2.7b and c.

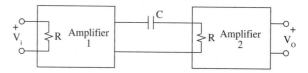

FIGURE 2.9   Two amplifier stages are coupled by a high-pass RC filter. The input resistance of the second amplifier serves as the R of the filter.

EXAMPLE 2.3   A primary purpose of an amplifier is to amplify an AC signal. To obtain sufficient gain, an amplifier consists of several amplifier stages in cascade. The amplifier stages, however, cannot just simply be connected because the output of a stage is usually characterized by a large DC voltage and a smaller, superimposed AC voltage. Connecting the output of a stage directly to the input of the next stage, the large DC level could easily saturate that stage and make it inoperative. What is needed is a high-pass filter between the stages that would stop a transfer of DC but would allow AC to proceed to the next stage for further amplification. A high-pass RC filter is ideally suited and is shown in Fig. 2.9. Design a filter that would pass AC signals above 20 Hz and block DC. The input resistance of the amplifiers is 0.1 MΩ. The cutoff frequency of the RC filter is given by $\omega_0 = 1/RC$. Solving for the capacitance, we obtain $C = 1/2\pi f_o R = 1/6.28 \cdot 20 \cdot 10^5 = 0.08 \ \mu$F. Using Fig. 2.7$b$, we see that at 20 Hz the voltage gain will be down by 3 db but that higher frequencies will be passed by the filter to the following amplifier without loss. Such a filter is also known as a coupling circuit. One can readily see that by making C larger, even lower frequencies will be passed. However, there is a limit: increasing the values of C generally increases the physical size and the cost of capacitors.   ∎

2.4   RESONANCE AND BAND-PASS FILTERS

In the previous section we combined a resistive element with one energy-storing element (RL and RC) and obtained filters with low- and high-pass action. If we add both energy-storing elements (RLC), we can obtain band-pass or band-rejection action. Band-pass filters can be used as tuning circuits where they allow one station or one channel out of many to be selected. For example, the VHF television channels 2 to 6 are located in the 54 to 88 MHz band, with each channel occupying a 6 MHz bandwidth. To receive a channel, a band-pass filter is used which allows frequencies of that channel to pass while discriminating against all other frequencies. The simplest band-pass filters are resonant circuits, which will be studied next.

2.4.1   Series Resonance

The circuit in Fig. 2.2a can be considered as a series resonant circuit. *Resonance* is defined as the condition when *current* and *voltage* are *in phase*. From (2.8) we see that

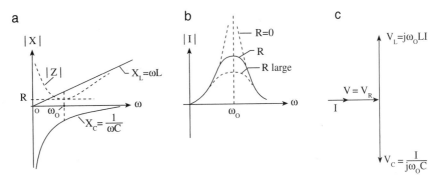

FIGURE 2.10   (a) A plot of inductive and capacitive reactance, X_L and X_C, showing that resonance occurs when they are equal. (b) Current I, in the series circuit, peaks at resonance. (c) Phasor diagram at resonance, showing that inductor and capacitor voltages are equal and opposite.

this happens when the imaginary part of the denominator becomes zero, i.e., $j(\omega L - 1/\omega C) = 0$, which gives for the resonant frequency

$$\omega_0 = \frac{1}{\sqrt{LC}} \tag{2.18}$$

The same result can be obtained by setting the phase angle θ in (2.10) equal to zero (when θ equals zero, current and voltage are in phase). At resonance, which occurs at ω_0, the inductive reactance in the series circuit is equal and opposite to the capacitive reactance; this is shown in Fig. 2.10a.

Since the reactances now cancel, what remains is the resistance. At resonance, the impedance of the series circuit is therefore a minimum and is equal to R, that is,

$$Z = R + j(\omega_0 L - 1/\omega_0 C) = R \tag{2.19}$$

As a rule, the resistance in series resonant circuits is quite small; often there are no other resistors in the circuit and R is then simply the winding resistance of the inductor coil.

Interesting things happen at resonance. For example, the current I in the series circuit becomes a maximum at resonance and from (2.8) is equal to $I = V_p/R$. A plot of the current at and near resonance is shown in Fig. 2.10b. The voltages across the inductance L and capacitance C are equal and opposite (which means that $V_L + V_C = 0$) and each of these voltages, V_L or V_C, can be much larger than the source voltage V_p. This is shown in the phasor diagram of Fig. 2.10c, which is similar to Fig. 2.2b or c, except for the phase angle θ, which at resonance is equal to zero. If, for example, we consider the inductor voltage, which at resonance can be stated as $V_L = j\omega_0 LI = j\omega_0 LV_p/R$, we can readily see that $V_L \gg V_p$ when $\omega_0 L/R \gg 1$, which is usually the case for practical situations.[10] Hence, we can view the series resonance circuit as a voltage amplifier: V_L peaks at ω_0 and drops off sharply either side of ω_0. A similar observation holds for V_C.

---

[10]The factor $\omega_0 L/R$ will soon be identified with the quality factor Q, which for practical circuits is normally larger than 5. Had we considered the voltage across the capacitor $V_C = I/j\omega_0 C = V_p/j\omega_0 RC$, we would find that $V_C \gg V_p$ when $1/\omega_0 RC \gg 1$. The Q-factor for a series resonant circuit is then

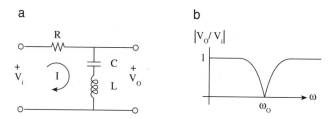

FIGURE 2.11 (a) A band-elimination or notch filter. (b) The voltage gain response as a function of frequency.

EXAMPLE 2.4 It is desired to eliminate an interfering signal of frequency 90 MHz. Design a filter to "trap" this frequency. To accomplish this, we can either insert a parallel resonant circuit tuned to 90 MHz in series or a series resonant circuit tuned to 90 MHz as a shunt. Choosing the latter, we have a notch filter as shown in Fig. 2.11a. The voltage gain of this filter is easily calculated as

$$\frac{V_o}{V_i} = \frac{Ij(\omega L - 1/\omega C)}{V_i} = \frac{j(\omega L - 1/\omega C)}{R + j(\omega L - 1/\omega C)}$$

The magnitude of this gain is given by

$$\left|\frac{V_o}{V_i}\right| = \frac{(\omega L - 1/\omega C)}{\sqrt{R^2 + (\omega L - 1/\omega C)^2}}$$

which is plotted in Fig. 2.11b. Therefore, at the resonance frequency $\omega_0 = 1/\sqrt{LC}$, the impedance of the series LC goes to zero, essentially short-circuiting any signal near the resonance frequency. If a 10 pF capacitor is used, the required inductance for this notch filter will be $L = 1/(2\pi f_0)^2 C = 3 \cdot 10^{-7}$ H $= 0.3$ μH. These are both small values for inductance and capacitance, indicating that it becomes more difficult to build resonant circuits with lumped circuit elements for much higher frequencies. The value of R is related to the Q of this circuit ($Q_o = \omega_0 L/R$): the smaller the R, the sharper the notch. ∎

2.4.2 Parallel Resonance

If the three components are arranged in parallel as shown in Fig. 2.12a, we have a parallel resonant circuit, sometimes also referred to as a *tank* circuit or a *tuned* circuit. The parallel circuit is almost exclusively used in communications equipment as a tuning

---

$Q_o = \omega_0 L/R = 1/\omega_0 RC$ (note that substituting the resonance frequency $\omega_0 = 1/\sqrt{LC}$ in $Q = \omega_0 L/R$ will give $Q_o = 1/\omega_0 RC$). In a series resonance circuit, the voltage across the capacitor or across the inductor can be significantly larger then the source voltage if $Q \gg 1$.

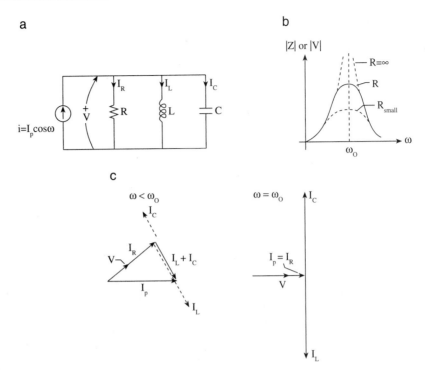

FIGURE 2.12   (a) A parallel resonant circuit, driven by a constant current source.   (b) A sketch of impedance near resonance (a sketch of voltage would be similar). (c) Phasor diagram near and at resonance.

circuit to select a band of desired frequencies. Unlike the series circuit, the parallel resonant circuit has minimum current, maximum impedance, and maximum voltage at ω_0, which makes it very desirable in practical circuits. When combined with an amplifier, a tuned amplifier gives frequency-dependent gain, so that selected frequencies are amplified. A parallel resonant circuit that is driven by a current source is readily analyzed using Kirchoff's current law (KCL). If in Fig. 2.12a the sinusoidal current has a peak value I_p, the corresponding phasor current is also I_p and a summation of phasor currents at the top node gives $I_p = I_R + I_L + I_C$. A voltage V is produced across the tank circuit by the source current such that $I_p = V/Z = VY$, where Z and Y are the impedance and admittance of the tank circuit, respectively. Since admittances in parallel add, we can state that

$$Y = G + j(\omega C - 1/\omega L) \tag{2.20}$$

where $G = 1/R$. The magnitude of admittance and impedance is

$$|Y| = \sqrt{(1/R)^2 + (\omega C - 1/\omega L)^2} \tag{2.21}$$

and

$$|Z| = \frac{R}{\sqrt{1 + (\omega RC - R/\omega L)^2}} \tag{2.22}$$

For resonance, the impedance or admittance must be real. Therefore, at the resonance frequency ω_0, the imaginary part of Y in (2.20) must vanish, that is, $\omega_0 C - 1/\omega_0 L = 0$, which gives for the resonance frequency

$$\omega_0 = \frac{1}{\sqrt{LC}} \tag{2.23}$$

which is identical to the resonance frequency for a series resonant circuit as given by (2.18). The admittance at resonance is therefore $Y = 1/R$. At parallel resonance, large currents oscillate in the energy-storing elements of L and C, which can be substantially larger than the source current I_p. We can see this by looking at the inductor current $I_L = V/j\omega_0 L = I_p R/j\omega_0 L$. Thus $I_L \gg I_p$ if $R/\omega_0 L \gg 1$, which is normally the case for practical circuits. The same reasoning shows that $I_C \gg I_p$ if $\omega_0 RC \gg 1$. We will show in the next section that the Q-factor for a parallel resonant circuit is $Q_o = R/\omega_0 L = \omega_0 RC$.

Unlike a series resonant circuit which is characterized by a small series resistance R (for an ideal resonant circuit with lossless inductors and capacitors it would be zero), in a parallel resonant circuit the shunting resistance is usually very large (an ideal parallel resonant circuit would have an infinite R). Since at resonance the capacitive and inductive susceptances cancel, the impedance consists entirely of the large shunting resistance R. For frequencies other than ω_0, the impedance (2.22) decreases as either L or C provides a decreasing impedance path. This means that at resonance, the voltage V peaks, giving a large signal voltage across the LC combination, which is a desirable condition when one frequency is to be emphasized over all others. By adjusting the values of the inductance or the capacitance in (2.23), the circuit can be tuned to different frequencies—hence the name tuned circuit. A sketch of impedance (or V) as a function of frequencies near resonance is shown in Fig. 2.12b.

In continuously tuned radios, a variable air or mica capacitor is employed. In step-tuned radios, which have a digital display of the received frequencies, the variable capacitor is a voltage-controlled diode, a *varactor*, which has the property of changing capacitance when the applied voltage is varied. Varying the voltage in steps varies the capacitance and hence the frequency in proportion. The variation is usually in 10 kHz increments in the AM band and 100 kHz increments in the FM band. Varactors have large tuning ranges, from a few picofarads to nanofarads.

Further insight into the behavior of a parallel resonant circuit is obtained by plotting phasor diagrams near and at resonance. Figure 2.12c shows the current and voltage phasors at a frequency less than ω_0, when the tank circuit is inductive because more current flows through the inductor than the capacitor. An inductive circuit is characterized by a current that lags voltage (I_p lags V). The dashed arrows show that I_L is at $-90°$ to V and is much larger than the dashed vector for I_C which is at $+90°$ to V. At resonance

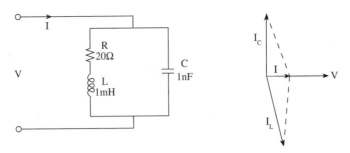

FIGURE 2.13 A practical tuned circuit and its phase diagram at resonance. R denotes the irreducible winding resistance of the inductor.

$\omega = \omega_0$, current is in phase with voltage and the inductive and capacitive currents, I_L and I_C, cancel each other but can be much larger than I_p.

EXAMPLE 2.5 A practical parallel resonant circuit is shown in Fig. 2.13. It is referred to as practical because even though the losses in a capacitor can be reduced to practically zero, $I^2 R$ losses in an inductor are always present as they are associated with the intrinsic winding resistance of the coil. Such a tuned circuit is found in radio and TV tuners where a variable air-capacitor is used to select different frequencies. Find the resonant frequency, the Q, and the bandwidth of the circuit shown.

The admittance $Y = 1/Z = I/V$ for the circuit can be written as

$$
\begin{aligned}
Y &= j\omega C + \frac{1}{R + j\omega L} = j\omega C + \frac{R - j\omega L}{R^2 + \omega^2 L^2} \\
&= \frac{R}{R^2 + \omega^2 L^2} + j\omega\left(C - \frac{L}{R^2 + \omega^2 L^2}\right)
\end{aligned}
$$

Resonance occurs when I and V are in phase or when the imaginary part of the above expression is equal to zero. Therefore, the resonance condition is

$$
C = LR^2 + \omega_0^2 L^2
$$

Solving for ω_0, we obtain

$$
\omega_0 = \frac{1}{\sqrt{LC}}\sqrt{1 - \frac{R^2 C}{L}}
$$

For high-quality coils, the condition that $L \gg R^2 C$ is more often than not met, and we have the standard resonance frequency $\omega_0 = 1/\sqrt{LC}$. Substituting the values given in the figure, we obtain for the resonant frequency,

$$
\omega_0 = \frac{1}{\sqrt{10^{-3} 10^{-9}}}\sqrt{1 - \frac{20^2 10^{-9}}{10^{-3}}} \cong 10^6 \text{ rad/s or } \omega_0/2\pi = 159 \text{ kHz.}
$$

Notice that the second square root contributes negligibly, which also allows us immediately to express Q simply as the Q for a series resonant circuit, i.e., $Q_o = \omega_0 L/R = 10^6 \cdot 10^{-3}/20 = 50$. The bandwidth, using (2.34), is therefore $B = \omega_0/Q_o = 10^6/50 = 2.104$ rad/s or 3.18 kHz.   ∎

2.4.3   Q-Factor and Bandwidth

We have alluded to the fact that the voltages across L and C in a series resonant circuit can be much higher than the source voltage and that the currents in L and C in a parallel resonant circuit can be much higher than the source current. In that sense the series resonant circuit can be considered as a voltage amplifier and the parallel resonant circuit as a current amplifier.[11]   A measure of the amplification is the *quality factor* $Q_o = \omega_0 L/R = 1/\omega_0 RC$ for the series circuit and $Q_o = R/\omega_0 L = \omega_0 RC$ for a parallel circuit (note that the Q for a series and a parallel circuit are inverses, i.e., $O_s = 1/Q_p$). The other use for Q will be as a measure of "sharpness" at resonance, or equivalently the *bandwidth* at resonance—as already indicated in Figs. 2.10*b* and 2.12*b*.

Let us now derive Q from fundamentals. Q is defined as

$$Q = 2\pi \frac{\text{maximum energy stored}}{\text{energy dissipated per period}} \tag{2.24}$$

Let us use the parallel circuit Fig. 2.12*a* at resonance ($\omega = \omega_0$) when $I_L + I_C = 0$, leaving the entire source current to flow through R, i.e., $I_p = I_R$. The average power associated with the circuit is $P = \frac{1}{2}I_p^2 R$. Hence the dissipated energy over one period T is

$$W = PT = \frac{1}{2}I_p^2 R \frac{2\pi}{\omega_0} \tag{2.25}$$

where $T = 1/f = 2\pi/\omega_0$. To calculate the stored energy is slightly more difficult. The energy stored in the inductor and capacitor is $\frac{1}{2}Li^2$ and $\frac{1}{2}Cv^2$, respectively. The instantaneous voltage across the resonant tank circuit is $v = RI_p \cos \omega_0 t$, which allows us to write for the energy stored in the capacitor

$$w_C = \frac{1}{2}Cv^2 = \frac{1}{2}CR^2 I_p^2 \cos^2 \omega_0 t \tag{2.26}$$

and for the energy stored in the inductor

$$w_L = \frac{1}{2}Li^2 = \frac{1}{2}L \left(\frac{1}{L}\int_0^t v\,dt\right)^2 = \frac{1}{2}CR^2 I_p^2 \sin^2 \omega_0 t \tag{2.27}$$

---

[11]Of course, neither is a power amplifier as they are passive circuits. Many practical amplifiers are multistage devices, with power amplification taking place at the last stage. Before power amplification can take place, the signal that drives a power amplifier should be on the order of volts. As the input to an amplifier can be a signal of insignificant power that measures in microvolts (μV), it is evident that the first stages of an amplifier are voltage amplifiers that must magnify the input signal from a level of microvolts to volts. Hence a practical amplifier consists of several stages of voltage amplification, followed by one or two stages of power amplification.

The total energy stored at any instant is therefore[12]

$$W_s = w_C + w_L = \frac{1}{2}CR^2 I_p^2 (\cos^2 \omega_0 t + \sin^2 \omega_0 t) = \frac{1}{2}CR^2 I_p^2 \qquad (2.28)$$

The implication of the sin and cos terms in the above equation is that at resonance the energy in the LC circuit oscillates between L and C, building to a maximum in L with zero in C; then as the energy in L decays, it builds in C until it is a maximum in C with zero in L. At any given time, the energy stored in L and C remains constant at $\frac{1}{2}CR^2 I_p^2$. The Q is then

$$Q_o = 2\pi \frac{W}{W_s} = 2\pi \frac{\frac{1}{2}CR^2 I_p^2}{\frac{1}{2}I_p^2 R(2\pi/\omega_0)} = \omega_0 RC \qquad (2.29)$$

which by substituting $\omega_0 = 1/\sqrt{LC}$ can also be written as $Q_o = R/\omega_0 L$. A similar procedure can be used to calculate the Q of a series resonant circuit. Qs for practical radio circuits are typically 10–100 but can be as high as several hundreds for low-loss coils.

Now let us show how Q is used to express the bandwidth of a resonant circuit. We have already shown that current, voltage, impedance, etc., peak or dip at resonance. The width of the peaking curves at the -3 dB points (or half-power points) is defined as the bandwidth. We can either use the parallel resonant circuit of Fig. 2.12a, for which the admittance is

$$\begin{aligned} Y &= G + j(\omega C - 1/\omega L) \qquad (2.30) \\ &= G\left[1 + jQ_o\left(\frac{\omega}{\omega_0} - \frac{\omega_0}{\omega}\right)\right] \end{aligned}$$

or the series resonant circuit of Fig. 2.2a, for which the impedance is

$$\begin{aligned} Z &= R + j(\omega L - 1/\omega C) \qquad (2.31) \\ &= R\left[1 + jQ_o\left(\frac{\omega}{\omega_0} - \frac{\omega_0}{\omega}\right)\right] \end{aligned}$$

We see that the expressions are analogous—hence results obtained for one apply to the other. The second expression in (2.30) was obtained by multiplying the numerator and denominator of the imaginary term by ω_0, identifying the resultant terms with $Q_o = R/\omega_0 L = \omega_0 RC$, and factoring Q_o out. Similarly for (2.31).

Let us choose the series resonant circuit of Fig. 2.2a. At resonance the impedance Z is a minimum and equal to $Z_o = R$ and the current I is a maximum and equal to

---

[12]Note that we are using small-case letters to denote time-dependent values, whereas capital letters are reserved for constant values such as DC, phasors, effective values, etc.

$I_o = V_p/R$. If we normalize the current $I = V_p/Z$ with respect to the current at resonance we have a dimensionless quantity[13]

$$\frac{I}{I_o} = \frac{1}{1 + jQ_o\left(\frac{\omega}{\omega_0} - \frac{\omega_0}{\omega}\right)} \tag{2.32}$$

which when plotted versus frequency ω shows the influence Q has on bandwidth at resonance. The -3 dB points or the half-power points are obtained when the normalized current falls to $1/\sqrt{2}$ of its maximum value. This occurs in (2.32) when the imaginary term is equal to ± 1, i.e., $Q_o\left(\frac{\omega}{\omega_0} - \frac{\omega_0}{\omega}\right) = \pm 1$. Then we have $I/I_o = 1/(1 \pm j1)$, which has an absolute value of $1/\sqrt{2}$. The two frequencies, ω_1 and ω_2, obtained by solving

$$Q_o\left(\frac{\omega_1}{\omega_0} - \frac{\omega_0}{\omega_1}\right) = -1 \quad \text{and} \quad Q_o\left(\frac{\omega_2}{\omega_0} - \frac{\omega_0}{\omega_2}\right) = 1 \tag{2.33}$$

are

$$\omega_1 = \omega_0\left[\sqrt{1 + \left(\frac{1}{2Q_o}\right)^2} - \frac{1}{2Q_o}\right]$$

and

$$\omega_2 = \omega_0\left[\sqrt{1 + \left(\frac{1}{2Q_o}\right)^2} + \frac{1}{2Q_o}\right]$$

The difference between ω_2 and ω_1 defines the bandwidth B

$$B = \omega_2 - \omega_1 = \frac{\omega_0}{Q_o} \tag{2.34}$$

which decreases with increasing Q_o. Figure 2.14 shows a plot of (2.32) for several values of Q_o. Note that the higher Q_o circuits have a narrower bandwidth; the sharper response curve gives the resonant circuit a better frequency selectivity to pass only signals of frequencies within a narrow band and attenuate signals at frequencies outside this band. To have a small frequency width between half-power points is not always desired. At times we need to broaden the frequency response so as to allow a wider band of frequencies to pass. To accomplish this, we need to lower the Q by increasing the resistance in a series resonant circuit and decreasing it in a parallel resonant circuit. The band-pass required depends on the information content of the signal. In general, signals

---

[13]The same expression holds for the normalized voltage V/V_o across the parallel resonant circuit of Fig. 2.12a. At resonance the admittance Y is a minimum and equal to $Y_o = G = 1/R$ and the voltage V is a maximum and equal to $V_o = I_p/Y_o$. Hence V/V_o and Z/Z_o are given by (2.32). As the parallel resonant circuit for $\omega < \omega_0$ is inductive (voltage leads current by $90°$), the phase of V/V_o tends to $+90°$, whereas for $\omega > \omega_0$ it tends to $-90°$. Of course at $\omega = \omega_0$, the circuit is resistive and the phase angle is zero.

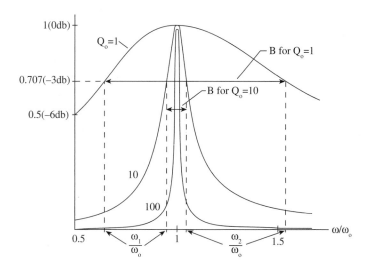

FIGURE 2.14   Frequency response of a resonant circuit. For a series resonant circuit this curve applies to I/I_o and to Y/Y_o. For a parallel circuit it is for V/V_o and Z/Z_o.

with more information require more bandwidth. For example, signals in a telephone conversation require 3 kHz of bandwidth, signals in an AM broadcast require 10 kHz of bandwidth, signals in an FM broadcast require 200 kHz of bandwidth, signals in a TV broadcast require 6 MHz of bandwidth, and signals for an 80-column computer screen display require 12 MHz of bandwidth. Example 2.5 calculated the resonant frequency and bandwidth for a typical tuned RF circuit.

The I/I_o phase of the series resonant circuit in Fig. 2.2a is $-90°$ for $\omega \ll \omega_0$ when the circuit is capacitive (see Fig. 2.10a), is $0°$ for $\omega = \omega_0$ when the circuit is resistive, and is $+90°$ for $\omega \gg \omega_0$ when the circuit is inductive. To visualize the phase relationship, recall that the source for the series circuit is a voltage source at zero phase, i.e., a horizontal vector as shown in Fig. 2.2b or c. The I/I_o is measured with respect to that vector.

EXAMPLE 2.6   (a) For the circuit shown in Fig. 2.13, find the impedance at resonance. (b) If the current I flowing into the circuit is 1 mA, find the voltage V across the tank circuit and the capacitor current at resonance.   ∎

(a) The admittance Y_o at resonance is the real part of the expression for Y in Example 2.5. Therefore,

$$Z_o = \frac{1}{Y_o} = \frac{R^2 + \omega_0^2 L^2}{R} = R(1 + Q_o^2)$$
$$= 20(1 + 50^2) = 50.02 \text{ k}\Omega$$

where Q_o is defined and calculated in Example 2.5. Note that the resonance impedance is much larger than the ohmic resistance of the coil.

(b) The voltage at resonance is $V_o = IZ_o = 1$ mA $\cdot$ 50.02 kΩ = 50.02 V. The capacitor current

$$I_C = \frac{V_o}{Z_C} = V_o \omega_0 C = 50.02 \cdot 10^6 \cdot 10^{-9} = 50.02 \text{ mA}$$

Hence, the capacitor current is 50 times larger than the current flowing into the circuit.

2.5   POWER IN AC AND RF CIRCUITS

Strictly speaking, a shorter title, "Power in AC Circuits," is sufficient, as sinusoidally excited circuits at 60 Hz or at 100 MHz *are* AC circuits. However, common usage is that circuits at 60 Hz are referred to as *AC* circuits, circuits in the 100 kHz–1 GHz range are *RF (radio frequency)* circuits, and circuits above 1 GHz are usually referred to as *microwave* circuits. Needless to say the following analysis holds for all these ranges.

2.5.1   Average Power

If a sinusoidal voltage $v(t) = V_p \cos \omega t$ applied to a circuit results in a current $i(t) = I_p \cos(\omega t + \theta)$, the *instantaneous power* is

$$\begin{aligned} p(t) &= v(t)i(t) = V_p I_p \cos \omega t \cos(\omega t + \theta) & (2.35) \\ &= \frac{V_p I_p}{2}[\cos \theta + \cos(2\omega t + \theta)] \end{aligned}$$

To obtain the average power P we can either average $p(t)$ by calculating $P = (\int p \, dt)/T$, where $T = 2\pi/\omega$ is the period (see also (1.12)), or simply by inspection of (2.35) conclude that the first term is a constant (with respect to time) and the second is a pure sinusoid which averages to zero. Therefore, the *average power* is

$$P = \frac{V_p I_p}{2} \cos \theta \qquad (2.36)$$

If the circuit is purely resistive, the phase difference θ between v and i is zero and (2.36) reduces to $P = V_p I_p/2 = RI_p^2/2$, just as in (1.12). For a purely capacitive or inductive circuit, $\theta = \pm 90°$, and $P = 0$.

Since we use phasor analysis when working with AC circuits, we can construct an expression in terms of phasor voltage and current which gives $p(t)$ as[14]

$$p(t) = \frac{1}{2} \text{Re} \left[\mathbf{VI}^* + \mathbf{VI}e^{2j\omega t} \right] \qquad (2.37)$$

---

[14] As it might be confusing if we do not differentiate explicitly between phasors and other quantities, we will bold all phasors in this section.

where from (2.35) the phasor for v is $\mathbf{V} = V_p$ and that for i is $\mathbf{I} = I_p e^{j\theta}$, $\mathbf{I}^*$ denotes the complex conjugate of $\mathbf{I}$, and Re means "take the real part." By inspection, we see that (2.37) reduces to (2.35). The average power is again given by the first term of (2.37),

$$P = \frac{1}{2} \operatorname{Re} \mathbf{VI}^* \tag{2.38}$$

which is the common expression for power calculations involving phasors and gives the same results as (2.36). Using Ohm's law $\mathbf{V} = \mathbf{IZ}$, where $\mathbf{Z} = R + jX$ is the impedance, we can also express (2.38) as

$$P = \frac{1}{2} \operatorname{Re} |\mathbf{I}|^2 \mathbf{Z} = \frac{1}{2}|\mathbf{I}|^2 R = \frac{1}{2} |I_p|^2 R$$

As expected, only the real part of the impedance is involved in power consumption. Alternatively, substituting $\mathbf{I} = \mathbf{V}/\mathbf{Z}$ in (2.38) we obtain

$$P = \frac{1}{2} \operatorname{Re} \mathbf{VV}^*/\mathbf{Z}^* = \frac{1}{2} \operatorname{Re} |\mathbf{V}|^2/\mathbf{Z}^* = \frac{1}{2} \operatorname{Re} |\mathbf{V}|^2 \mathbf{Z}/|\mathbf{Z}|^2 = \frac{1}{2}|\mathbf{V}|^2 R/(R^2 + X^2)$$

which reduces to $\frac{1}{2} \operatorname{Re} |\mathbf{V}|^2/R$ if $\mathbf{Z}$ is a purely resistive element.

EXAMPLE 2.7 A voltage source is connected through a line to a load as shown in Fig. 2.15. Calculate the power delivered to the load. The current I flowing through source, line, and load is

$$I = \frac{100}{Z} = \frac{100}{21 + j13} = 3.44 - j2.13 = 4.04e^{-j31.8}$$

The load voltage is therefore

$$\begin{aligned} \mathbf{V}_L &= \mathbf{IZ}_L = (3.44 - j2.13)(20 + j10) \\ &= 90.1 - j8.2 = 90.5e^{-j5.2} \end{aligned}$$

Using (2.38), the power dissipated in the load is

$$\begin{aligned} P_L &= \frac{1}{2} \operatorname{Re} \mathbf{V}_L \mathbf{I}^* = \frac{1}{2} \operatorname{Re} 90.5e^{-j5.2} \cdot 4.04e^{+j31.8} \\ &= \frac{1}{2} 365.5 \cos 26.6° = 163.5 \text{ W} \end{aligned}$$

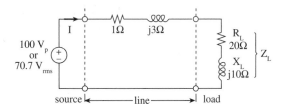

FIGURE 2.15 A source delivers power to a load, represented by impedance $Z_L = R_L + jX_L$.

The power lost in the line is

$$P_{\text{line}} = \frac{1}{2}|\mathbf{I}|^2 R_{\text{L}} = \frac{1}{2}|4.04|^2 \cdot 1 = 8.2 \text{ W}$$

and the total power delivered by the source is

$$\begin{aligned}
P_S &= \frac{1}{2}\operatorname{Re}\mathbf{V}_S\mathbf{I}^* = \frac{1}{2}\operatorname{Re} 100 \cdot 4.04 e^{+j31.8} \\
&= \frac{1}{2}404\cos 31.8° = 171.7 \text{ W}
\end{aligned}$$

Hence the number of watts produced by the source equals the number of watts dissipated by the line and the source.   ■

2.5.2   Effective or Root Mean Square (RMS) Values in Power Calculations

We have shown that a DC current of I amps flowing in R delivers I^2R of average power to the resistor. For AC, the sinusoidal current with a peak value of I_p amps delivers $I_p^2 R/2$ watts of average power. If we define an *effective value* of $I_{\text{eff}} = I_p/\sqrt{2}$ for the AC current, we can avoid having to write the $\frac{1}{2}$ factor in AC power calculations. The power is simply I^2R for AC or DC, as long as we recognize that for AC, I is the effective value.

What are the effective values for other waveshapes? Suppose we apply a square wave or a triangular wave current to R. How much average power will R absorb? In (1.12) we developed effective values for sinusoids. Following similar reasoning, we define an average power as

$$P = I_{\text{eff}}^2 R \tag{2.39}$$

where I_{eff} is the effective value for any periodic current of arbitrary waveshape. The effective current is a constant which is equal to a DC current that would deliver the same average power to a resistor R. In general, averaging the instantaneous power of a periodic current (see (1.12)), we obtain

$$P = \frac{1}{T}\int_0^T i^2 R\, dt \tag{2.40}$$

Solving for I_{eff} by equating these two expressions, we obtain

$$I_{\text{eff}} = \sqrt{\frac{1}{T}\int_0^T i^2\, dt} \tag{2.41}$$

Observe that to find the effective value, we first determine the square of the current, then calculate the average value, and finally take the square root. We are determining the *root mean square* of i, which explains why the term I_{rms} is often used for I_{eff}.

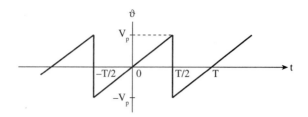

FIGURE 2.16   A triangular voltage resembling a periodic sawtooth.

We can now express (2.36), the average power dissipated in R for any periodic current or voltage, as

$$P = V_{rms}^2/R = I_{rms}^2 R \qquad (2.42)$$

Since steady DC has a constant value, the rms value of DC is the constant value. Similarly, a square wave which has a positive peak voltage V_p for half a period and negative V_p for the other half has $V_{rms} = V_p$ (squaring the square-wave voltage flips the negative peaks up, giving the same effect as a steady DC voltage of V_p volts). In Chapter 1, we also showed that a sinusoid[15] with amplitude V_p has an effective voltage $V_{rms} = V_p/\sqrt{2}$. In the following example, we will calculate the effective value for a triangular waveshape.

EXAMPLE 2.8   The periodic triangular voltage shown in Fig. 2.16 is used to sweep the electron beam in the picture tubes of televisions and oscilloscopes. Calculate the rms value of the triangular wavehape, which has a peak-to-peak value of $2V_p$. The analytical expression for v in the interval $-T/2$ to $T/2$ is $v = (2V_p/T)t$, where the term in the parentheses is the slope of the straight line. The effective voltage using (2.41) is then

$$\begin{aligned} V_{\text{eff}} &= \sqrt{\frac{1}{T}\int_{-T/2}^{T/2}\left(\frac{2V_p}{T}t\right)^2 dt} \\ &= \sqrt{\left.\frac{4V_p^2}{3T^3}t^3\right|_{-T/2}^{T/2}} = \frac{V_p}{\sqrt{3}} \end{aligned}$$

A DC voltage of $V_p/\sqrt{3}$ volts would therefore deliver the same heating power to a resistor R as the triangular voltage.   ∎

2.5.3   Power Factor

In (2.36) we showed that in the AC steady state, average power in a load is equal to $P = VI\cos\theta$, where V and I are constant rms values. As $\cos\theta$ decreases, the average

_____

[15]The voltages at electrical outlets that power utilities provide are effective voltages of 120 V_{AC}. This means that the peak voltages are $V_p = 120\sqrt{2} = 170$ V and the average values are zero.

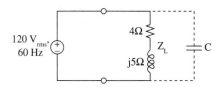

FIGURE 2.17   An inductive load $Z_L = 4 + j5$ or $Y_L = 4/41 - j5/41$ has a lagging power factor pf which can be corrected by placing a capacitor C in parallel with the load.

power decreases. If we refer to the VI as *apparent power*, we can define a *power factor* *pf* as a ratio of the average to the apparent power:

$$\text{pf} = \frac{P}{VI} = \cos\theta \tag{2.43}$$

The phase angle θ is the angle between V and I, or alternatively θ is the phase angle of the load admittance Y_L. For inductive loads ($Z_L = R + jX = |Z_L|e^{+j\theta}$ or $Y_L = 1/(R + jX) = |Y_L|e^{-j\theta}$), when I lags V, θ is negative and we speak of lagging pf's. The values for this angle range between $\theta = 0$ for a purely resistive load (pf $= 1$) and $\theta = \pm90°$ for purely reactive loads (pf $= 0$). Note also that pf can be equal to one for a circuit with reactive elements such as a series RLC or a parallel RLC circuit at resonance (of course a circuit at resonance is purely resistive).

As most loads in industrial and home use are inductive (motors), the pf at many sites can be substantially less than unity. This has undesirable implications. As the power utilities supply a constant effective voltage, a decreasing pf at a site causes the effective current to increase in order to maintain the horsepower output of electric motors at a constant level. The increased current in turn causes larger I^2R losses, which can cause the operating temperatures of the equipment to rise excessively. A power factor correction, by placing a capacitor across the inductive loads (essentially forming a parallel resonant circuit), would increase the efficiency of operation (see Fig. 2.17). Ideally, the average power should equal the apparent power. Power utilities expect large industrial users to operate at pf's exceeding 0.9 and can impose monetary penalties on noncompliers.

EXAMPLE 2.9   A load of $Z_L = 4 + j5$ is connected to a voltage source of 120 V_{rms}, 60 Hz. Since this is an inductive load it is operating at a lagging power factor of 0.6247. The pf can be changed by placing a capacitor C across the load. (a) Find the value of C to correct the pf to 0.9 lagging. (b) Find the value of C to change the pf to 0.9 leading. ∎

(a) The power factor is pf $= \cos(\tan^{-1} 5/4) = 0.6247$ lag. The average power absorbed by the load is

$$P = \frac{V_R^2}{R} = I^2R = \frac{120^2}{4^2 + 5^2} \cdot 4 = 1405 \text{ W}$$

To correct the pf by placing C across Z_L we create a new load impedance Z_L'. Since it is easier to use admittances when placing elements in parallel, we will work with

$$Y_L' = \frac{1}{Z_L'} = Y_L + j\omega C = \frac{1}{Z_L} + j120\pi C$$
$$= 4/41 - j5/41 + j120\pi C$$

If the new load admittance corrects the pf to 0.9 lagging, the new phase angle between current I and voltage V will be $\cos^{-1} 0.9 = 25.84°$. This is a negative angle as I lags V by $25.84°$. The tangent of this phase angle is therefore $\tan(-25.84°) = -0.4843$, which also must equal the ratio of imaginary and real parts of the new load admittance Y_L', that is,

$$-0.4843 = \frac{-5/41 + 120\pi C}{4/51}$$

which allows us to solve for the value of capacitance C with the result that $C = 198 \ \mu\text{F}$.

(b) For this case the pf is $\cos(\tan^{-1} 5/4) = 0.6247$ lead. Placing C across Z_L results in the same expression for Y_L' as above. However now the angle $25.84°$ is positive as I leads V; hence $\tan(25.84°) = 0.4843$, which also must be equal to the ratio of imaginary and real parts of Y_L', that is,

$$0.4843 = \frac{-5/41 + 120\pi C}{4/51}$$

Solving this for C gives us $C = 448.8 \ \mu\text{F}$.

Note: Had we corrected Z_L for a unity power factor, which would correspond to a zero angle between I and V, then $\tan 0° = 0$ and the resulting equation for C would be

$$0 = \frac{-5/41 + 120\pi C}{4/51}$$

which when solved gives $C = 324 \ \mu\text{F}$. It is interesting to observe the increasing values of C as the power factor correction for the load Z_L increases. To make a modest correction of a pf of 0.625 lag to a pf of 0.9 lag required a C of 198 μF, to correct to a pf of 0 required a C of 324 μF, and to overcorrect to a pf of 0.9 lead required a C of 448.8 μF.

Also note that a correction of pf to zero corresponds to changing the load to a purely resistive load. This implies that an inductive load to which a capacitance has been added in parallel has been transformed to a parallel resonant circuit. This approach is explored in the next example.

EXAMPLE 2.10   Find the power factor of the load $Z_L = 20 + j10$, shown in Fig. 2.15. Determine the capacitance that would have to be placed in parallel with the load so as to correct the pf to unity.

The load pf, which is $\cos\theta$, where θ is the phase angle between the load current and load voltage (or the angle of the load impedance $Z_L = R_L + jX_L$, i.e., $\theta = \tan^{-1} X_L/R_L$), is equal to

$$\text{pf} = \cos\theta = \cos\tan^{-1}\frac{X_L}{R_L} = \cos\tan^{-1}\frac{10}{20} = \cos 26.6° = 0.89$$

As calculated in Example 2.6, the load dissipates 163.5 W, and the AC current has a peak value of $I = 4.04$ A.

We can correct the pf to 1 by placing a capacitor across the load and resonating the resulting parallel circuit. Note that no additional power will be consumed by the new load as the capacitor, being purely reactive, consumes no power. With the capacitor in place, we obtain a parallel resonant circuit like that shown in Fig. 2.13. For unity power factor, the resonance condition worked out in Example 5 gives the value of the capacitor as $C = L/(R^2 + \omega_0^2 L^2)$, where R and L are now the resistance and inductance of the load. Since the frequency is not specified, we can only give the reactance of the capacitance. Therefore, multiplying both sides of the equation for C by ω, we obtain

$$X_C = \frac{R_L^2 + X_L^2}{X_L} = \frac{20^2 + 10^2}{10} = 50 \ \Omega$$

Hence, a capacitor with reactance $X_C = 1/\omega C = 50 \ \Omega$, in parallel with the load, will result in a load that has unity power factor. If the frequency is known, the capacitance can be calculated.

The current circulating in source and line now has a smaller peak value of $I = |100/(1 + j3 + (20 + j10)) \parallel (-j50)| = 3.82$ A), where the symbol $\parallel$ means "in parallel." Hence, the line loss is now smaller and is equal to $(3.82)^2(1/2) = 7.3$ W, whereas the line losses before were 8.2 W. Furthermore, after the pf correction, the load consumes more power:

$$P_L = I^2 R_{eq} = I^2 \frac{R_L^2 + X_L^2}{R_L} = 3.82^2 \frac{20^2 + 10^2}{20} = 182.5 \ \text{W}$$

If the load is a motor, it can be throttled back to the previous consumption of 163.5 W, further increasing the efficiency of the system.   ■

2.6   TRANSFORMERS AND IMPEDANCE MATCHING

Transformers are used for transmitting AC power, for changing AC voltages and currents to higher or lower values, and for insulating equipment from the power lines. They are single-frequency devices (60 Hz) with efficiencies approaching 100%. Stray magnetic fields, produced by transformer windings, are minimized by use of ferromagnetic iron cores which confine the magnetic fields to the iron medium of the transformer. If, in addition, winding losses and core losses (which increase with frequency) are kept small (which they invariably are in well-designed transformers), the result will be exceptionally high efficiencies of transformers, normally stated as $W_{out} \cong W_{in}$. Operation at low frequencies reduces core losses but requires transformers with large cross sections, which in turn increases the size and weight of transformers. In situations where weight is a constraining factor such as in aircraft, operation is usually at 400–800 Hz (see Example 2.12 on transformer design).

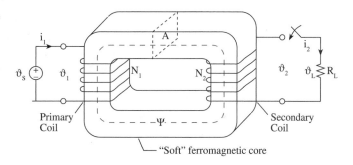

FIGURE 2.18 An iron-core transformer of cross section A in which the *primary* and *secondary* windings are tightly coupled by the *common magnetic flux ψ*. The turns ratio is N_2/N_1, where N_1 and N_2 are the number of turns in the primary and secondary coils, respectively.

Transformers are also widely used at audio frequencies in coupling, input, output, interstage, modulation, and impedance matching applications. Construction is similar to power transformers, but instead of operating at a single frequency, audio transformers must operate over a wide band of frequencies, typically 20 Hz to 20 kHz. As transformers are usually bulky, even when miniaturized, a principle in modern circuit design is to replace, whenever possible, transformer-coupled circuits by direct-coupled circuits in interstages or output stages of amplifiers.

A class of special-purpose transformers are high-frequency transformers, such as the pulse transformer. These devices must operate over a broad range of frequencies and are expected to transmit square waves or trains of pulses, while maintaining as closely as possible the original shape. An example is the television flyback transformer, which operates at 15.75 kHz and after rectification generates the high DC voltages (10 kV or more) needed by cathode ray tubes (CRTs).

2.6.1   Flux Linkages and the Ideal Transformer

A typical transformer has primary and secondary coils wound on a soft[16] iron core. If, as shown in Fig. 2.18, the primary is connected to a voltage source of $v_s = V_p \cos \omega t$ and the secondary is left open-circuited, a small current will flow in the primary which will induce a flux in the core and which in turn will lead to an induced voltage in the primary windings. The induced voltage, according to Faraday's law, is

$$v_1 = -N_1 \frac{d\psi}{dt} \tag{2.44}$$

---

[16]The magnetic properties of soft ferromagnetic iron are such that the material can be easily magnetized and demagnetized, which makes it useful for transformers, relays, tapeheads, etc., which are all devices in which the magnetic field changes rapidly and reverses at 60 Hz in transformers and faster in tapeheads. Hard ferromagnetic material, in contrast, is difficult to magnetize and demagnetize, which makes it useful in permanent magnets and recording materials, such as audio and videotape.

where N stands for the number of winding turns and ψ is the magnetic flux induced in the transformer core by the current that flows in the primary winding. This voltage is usually referred to as back-emf or counter emf (electromotive force) and is necessary to counter the applied voltage; otherwise an abnormally large current would flow in the primary winding (the winding resistance in practical transformers is much too small to limit the current).

Because time-varying flux, which is almost perfectly confined to the soft ferromagnetic iron core, links also the secondary winding, it will induce a voltage there, which again by Faraday's law is given by

$$v_2 = -N_2 \frac{d\psi}{dt} \tag{2.45}$$

We can readily see that, depending on the turns ratio N_2/N_1, we either have a voltage step-up or step-down, that is,

$$\frac{V_2}{V_1} = \frac{N_2}{N_1} \tag{2.46}$$

According to (2.46), obtained by dividing (2.45) by (2.44), the ratio of secondary to primary voltage is equal to the ratio of secondary to primary turns, and because it is a ratio, it is equally valid for instantaneous (v) or rms values (V). As the secondary is open-circuited, no power is dissipated there. Hence, any current in the primary need not deliver any power and is therefore 90° out of phase with the primary voltage. A small lagging primary current, called a magnetizing current, is necessary to induce the magnetic flux in the core (for most purposes we can even ignore the magnetizing current as losses in practical transformers are small (a few percent), which makes the zero-loss ideal transformer such a useful model). The power factor, $\cos\theta$, is zero in the primary circuit, as no power is transferred from the voltage source to the transformer. In applications where a transformer is working into a large impedance and a voltage step-up is needed, the open-circuited secondary case is a useful model.

Another practical situation is when power is transferred through the transformer, from source to load. The switch in the secondary is now closed, current flows, and the resistor R_L absorbs power which must be supplied by the source. However, now the voltage balance in the primary circuit is upset. The current I_2 in the secondary winding induces a new flux in the core which causes a new voltage to be induced in the primary, upsetting the balance between source and primary voltage, which was $v_s = v_1$. As the source voltage remains constant, the voltage imbalance in the primary causes a new current to flow in the primary winding which in turn will induce a new voltage (equal and opposite to that caused by I_2) that will bring back balance to the primary circuit. The new primary current I_1 accounts for the power $I_2^2 R_L$ delivered to the load and hence is in phase with the source voltage[17] (in the primary circuit, power factor $\cos\theta$

---

[17]Capital letters denote rms or effective values. For example, $v_s = V_p \cos\omega t$ can be represented by its rms voltage V_s. All the induced voltages are given by Faraday's law, which states that a time-varying magnetic

is now unity). In other words, the power supplied by the source $V_s I_1 = W_1$ is equal to $V_L I_2 = I_2^2 R_L = W_2$ consumed by the load. Since $V_s = V_1$ and $V_L = V_2$, we have

$$V_1 I_1 = V_2 I_2 \tag{2.47}$$

The above power expression, $W_1 = W_2$, valid for lossless or ideal transformers, is a useful approximation for practical transformers, which, when well designed, can be almost 100% efficient. The current transformation[18] in a transformer can now be readily obtained by combining (2.47) and (2.46) to give

$$\frac{I_2}{I_1} = \frac{N_1}{N_2} \tag{2.48}$$

which shows it to be inverse to that of voltage transformation: a step-up in the turns ratio gives a step-up in voltage but a step-down in current.

Summarizing, an ideal transformer is lossless with perfect coupling between the primary and secondary, a consequence of the magnetic flux being completely confined to the core of the transformer. Practical iron-core transformers are well approximated by ideal transformers.

EXAMPLE 2.11   An ideal transformer (see Fig. 2.18) is rated at 3600/120 V and 10 kVA. The secondary winding has 60 turns.

Find the turns ratio N_2/N_1, the current ratio, primary turns N_1, current I_1, and current I_2.   ∎

Since this is a step-down transformer ($V_2 < V_1$), from (2.46), $N_2/N_1 = V_2/V_1 = 120/3600 = 1/30 = 0.0333$.

From (2.48), $I_2/I_1 = N_1/N_2 = 30$; therefore, $I_2 = 30 I_1$ and current is stepped up by a factor of 30, whereas voltage is stepped down, i.e., $V_2 = V_1/30$.

Since $N_1 = 30 N_2$ and N_2 is given as 60, we obtain for $N_1 = 1800$ turns. To find I_1 we first note that $V_1 I_1 = V_2 I_2 = 10,000$ and that $V_1 = 3600$ is given. Therefore, $I_1 = V_1 I_1/V_1 = 10,000/3600 = 2.78$ A. Using (2.48), we obtain $I_2 = I_1(N_1/N_2) = 2.78(30) = 83.33$ A.

---

flux will induce a voltage in a coil that it links. Hence, a DC voltage cannot be transformed. For example, if the primary circuit contains a DC and an AC component, only the AC component is passed on to the secondary. Keeping the DC component, especially if it is large, out of the load might be a critical function of the transformer. Another safety function is isolation of secondary from primary: no direct connection exists to a common 440 V_{AC} power line, for example, after a transformer reduces the voltage to 120 V_{AC} for household use.

[18]Current transformation can be obtained another way. For a given frequency, Faraday's law states that flux induced in a core is proportional to current I and number of turns N in a winding. Closing the switch, a new current flows in the secondary, which will induce a new flux in the core, upsetting the voltage balance in the primary circuit. Therefore a new current must flow in the primary to cancel the additional flux due to I_2. Voltage balance is reestablished when $I_1 N_1 = I_2 N_2$.

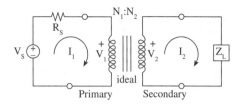

FIGURE 2.19   A transformer connects a load Z_L to a source V_s.

EXAMPLE 2.12   An iron-core transformer, shown in Fig. 2.19 (the three vertical bars are a symbol for a tightly coupled, iron-core transformer for which the ideal transformer approximation is valid), is used to deliver power from a source to a load. Because an ideal transformer is lossless, all of the power delivered to the primary circuit will also be delivered to the secondary circuit. Typically, the primary coil is connected to an energy source and the secondary coil is connected to a load. If the load is $Z_L = 500 - j400 \ \Omega$, $R_s = 10 \ \Omega$, and the windings are $N_1 = 100$ turns and $N_2 = 1000$ turns, find the average power delivered to the load Z_L when (a) $I_1 = 5e^{j\Pi/4}$ and (b) $V_s = 120e^{j0}$.

(a) Since $|I_1| = 5$ A we obtain from (2.48) that $|I_2| = |I_1| N_1/N_2 = 5(100/1000) = 0.5$ A. Power delivered to the load is therefore $P_L = |I_2|^2 R_L = (0.5)^2 500 = 125$ W.

(b) To calculate power dissipated in the load, we must find I_2 or V_2 because $P_L = |I_2|^2 R_L$ or $P_L = |V_R|^2/R_L = |V_2 R_L/Z_L|^2/R_L = |V_2/Z_L|^2 R_L = |I_2|^2 R_L$. To calculate V_2, we first find V_1, and then using the turns ratio, we obtain $V_2 = V_1 N_2/N_1$. Voltage V_1 is across the primary of the transformer and is a fraction of the source voltage V_s. In other words, V_1 is the product of V_s and the ratio of R_s and Z'_L. The impedance Z'_L is the load impedance Z_L projected into the primary and can be obtained by (2.50) as $Z'_L = Z_L(N_1/N_2)^2 = (500 - j400)(100/1000)^2 = 5 - j4 \ \Omega$. The primary voltage is then $V_1 = V_s Z'_L/(R_s + Z'_L) = 120(5 - j4)/(15 - j4)$. The power delivered to the load is therefore

$$
\begin{aligned}
P_L &= |V_2/Z_L|^2 R_L = |(V_1 N_2/N_1)/Z_L|^2 R_L \\[2mm]
&= \left| 120\frac{5 - j4}{15 - j4} 10\frac{1}{500 - j400} \right|^2 500 \\[2mm]
&= \left| \frac{12}{15 - j4} \right|^2 500 = \frac{144}{241} 500 = 298.7 \text{ W}
\end{aligned}
$$

An alternate, somewhat simpler method is to perform the calculations in the primary circuit remembering that in an ideal transformer power consumed in the secondary circuit is equal to that in the primary circuit. Hence, once the load impedance Z'_L which is projected into the primary circuit is known, the power consumed by Z_L is also equal to $P_L = |I_1|^2 R'_L = |120/(15 - j4)|^2 5 = 298.7$ W.   ■

EXAMPLE 2.13   Design a transformer to furnish 5 V and 10 A from a 120 V, 60 Hz supply. The core is to have a cross section of 1 square centimeter (cm^2). The maximum core flux density is to be 1 *tesla* (1 T = 1 weber per square meter). Find (a) the number of turns of each winding and (b) the full-load current in the primary.   ∎

(a) Derive the relationship between voltage, frequency, and flux. If the applied voltage is sinusoidal, so is the flux, i.e. $\psi(t) = \psi_p \sin \omega t$, where ψ_p is the peak value of the magnetic flux. The induced or counter emf from (2.44) is given as

$$v = -N\psi_p \omega \cos \omega t$$

and must be very nearly equal to the applied voltage in a well-designed transformer. Expressing the above voltage in terms of rms values, we obtain

$$V = 4.44 N f \psi_p = 4.44 N f A B_p \qquad (2.49)$$

where f is the frequency ($\omega = 2\pi f$), A is the cross section of the core in meters squared (m^2), B is the flux density in teslas, and V is in volts if flux ψ is in webers. Equation (2.49) is an often-used equation in transformer design.

The induced volts per turn is $V/N = 4.44 \cdot 60 \text{ Hz} \cdot 10^{-4} m^2 \cdot 1 \text{ T} = 0.0267$, which gives for the primary $N_1 = V/ \text{(volts/turn)} = 120/0.0267 = 4505$ turns, and for the secondary, $N_2 = 5/0.0267 = 187$ turns. Note that an increase in the cross-sectional area of the core would be accompanied by a corresponding decrease in winding turns.

(b) Using (2.48), we have $I_1 = N_2 I_2 / N_1 = 187.10/4505 = 0.42$ A. Hence at full load the transformer draws 0.43 A and deliveres 50 W to the load, assuming the load is resistive and transformer losses are negligible.

2.6.2   Impedance Transformation

In addition to stepping voltages and currents, a transformer can also change impedances. By using (2.46) and (2.48), the expression for impedance transformation is given by

$$\frac{Z_1}{Z_2} = \frac{V_1/I_1}{V_2/I_2} = \frac{V_1}{V_2} \cdot \frac{I_2}{I_1} = \left(\frac{N_1}{N_2}\right)^2 \qquad (2.50)$$

which shows that a small impedance Z_2 connected to the secondary will appear as a larger impedance $Z_1 = Z_2(N_1/N_2)^2$ at the primary, if $N_1/N_2 > 1$.

Examining (2.50), it appears that a transformer with its many turns of wire in the primary and secondary—normally suggesting large inductances—does not contribute any inductance of its own to the primary or the secondary circuit. A pure resistance R_2 transforms into a pure resistance R_1. This somewhat surprising result is due to the cancellation of the two fluxes in the core which are induced by the currents I_1 and I_2. Hence, a transformer acts as an inductance-free device that can change the value of a pure resistance. This property is very useful when maximum power transfer is desired between a mismatched source and load.

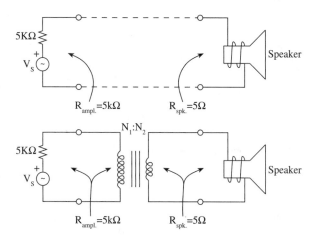

FIGURE 2.20 (a) A speaker directly connected to an amplifier would usually be badly mismatched. (b) For maximum power transfer, an iron-core (denoted by the vertical bars) transformer is used to provide matched conditions.

EXAMPLE 2.14 It is desired to transfer maximum power from an audio amplifier that has a large internal resistance to a speaker with a small internal resistance. Speakers typically have small resistances (4, 8, 16 Ω) because the wire winding, which is attached to the speaker cone and which moves the cone back and forth (thus creating acoustic pressure waves), has to be light in weight to enable the speaker to respond well to high frequencies. If the amplifier is characterized by an output impedance of 5000 Ω and the speaker by 5 Ω, find the turns ratio of an audio transformer for maximum power transfer to the speaker. Figure 2.20a shows the mismatch if the speaker were connected directly to the amplifier. Only a small fraction of the available power would be developed in the speaker. We can show this by calculating the ratio of power dissipated in the speaker to power dissipated in the amplifier: $P_{5\Omega}/P_{5000\Omega} = I^2 5/I^2 5000 = 0.001$ or 0.1%. Thus, an insignificant amount of power reaches the speaker. ∎

We know from Section 1.6 that for maximum power transfer the load resistance must be equal to the source resistance. This matching can be achieved by placing an iron-core audio transformer between amplifier and speaker as shown in Fig. 2.20b. Using (2.50), we can calculate the turns ratio needed to produce a matched condition. Thus

$$\frac{Z_1}{Z_2} = \frac{5000}{5} = 1000 = \left(\frac{N_1}{N_2}\right)^2$$

which gives that $N_1 = 33N_2$. Therefore, a transformer with 33 times as many turns on the primary as on the secondary will give a matched condition in which the source appears to be working into a load of 5000 Ω and the speaker appears to be driven by a source of 5 Ω.

2.7 SUMMARY

In this chapter we developed necessary circuit tools needed to analyze electronic circuits. As pointed out at the beginning, electronics is an interaction of the three RLC circuit elements with active components such as transistors, operational amplifiers, etc.

- Phasor analysis, by introducing complex quantities, gave us a method to analyze single-frequency circuits as easily as analyzing DC circuits. By the use of complex quantities such as impedance $Y = R + jX$ and admittance $Y = G + jB$, we were able to obtain forced responses of current and voltage anywhere in a circuit when the forcing function was a sinusoid.

- The frequency response of filters was characterized in terms of cutoff or half-power frequency ($\omega_C = 1/RC$ or $\omega_C = R/L$) and resonance frequency. Resonance—and resonant frequency $\omega_0 = 1/\sqrt{LC}$—was defined in circuits that contained inductance L and capacitance C as the condition when input current and voltage were in phase. At the resonant frequency the input impedance and admittance are purely real quantities. The quality factor Q was defined and it was observed that for practical RF circuits $Q \geq 5$.

- For a series resonant circuit $Q = \omega_0 L/R_s = 1/\omega_0 R_s C$, where R_s is the series resistance and the voltage across the inductor and the capacitor was equal but opposite and can be larger than the source voltage V_s, i.e., $V_L = V_C = QV_s$.

- Similarly for a parallel resonant circuit, $Q = \omega_0 R_p C = R_p/\omega_0 L$, where R_p is the parallel resistance and the current in the inductor and in the capacitor was equal but opposite and can be larger than the source current, i.e., $I_L = I_C = QI_s$. Hence, the series (parallel) resonant circuit acts as a voltage (current) amplifier.

- Frequency selectivity of resonant circuits was determined by the bandwidth B, which is related to Q by $B = \omega_0/Q$, where the bandwidth $B = \omega_2 - \omega_1$ and ω_2, ω_1 are the half-power frequencies, i.e., $\omega_2 = \omega_0 + \omega_0/2Q$ and $\omega_1 = \omega_0 - \omega_0/2Q$.

- Transformers were shown to be very efficient devices for changing the levels of AC currents, voltages, and impedances according to $V_2/V_1 = N_2/N_1$, $I_2/I_1 = N_1/N_2$, and $Z_2/Z_1 = (N_2/N_1)^2$, where N_1 (N_2) is the number of turns in the primary (secondary) winding.

Problems

1. Determine the angle by which i leads or lags $v = 10\cos(\omega t - 10°)$ if

(a) $i = 5\cos(\omega t - 20°)$,
(b) $i = 5\cos(\omega t - 5°)$,

(c) $i = -5\cos(\omega t - 30°)$.

   Ans: (a) i lags v by 10°. i leads v by 350° is also correct because $\cos(x - 20°) = \cos(x + 340°)$. However, it is customary to express phase differences by angles less than 180°.

2. A sinusoidal voltage, expressed by $v(t) = V_p \cos(\omega t + \theta)$, has a peak value of 50 V. At $t = 0$ it is decreasing and has a value of 40 V. Find θ.

3. Represent the following complex numbers in polar form: $2 + j3, 3 - j5$, and $-7 + j9$.

   Ans: $3.6\exp(j56.5°)$, $5.8\exp(-j26.2°)$, and $11.4(j127.9°)$.

4. State the corresponding phasors for the real-time voltages: $v(t) = 5\cos\omega t$, $v(t) = 5\sin\omega t$, $v(t) = V_p\cos(\omega t + \theta)$, and $v(t) = 120\cos(\omega t + \theta)$.

5. Given the phasor currents $j10$, $10 + j10$, and $10 - j10$, find the corresponding real-time currents. Assume the frequency f of the currents is $f = \omega/2\pi$.

   Ans: $i(t) = -10\sin\omega t$, $14.1\cos(\omega t + 45°)$, and $14.1\cos(\omega t - 45°)$.

6. Given the phasor currents $j10$, $10 + j10$, and $10 - j10$, find the instantaneous currents for $\omega = 377$ rad/s and $t = 1$ ms.

7. Representing $v_1 = 5\cos(\omega t - 20°)$ and $v_2 = 7\cos(\omega t + 30°)$ by phasors, find $v_1 + v_2$.

   Ans: $10\cos(\omega t + 9.4°)$.

8. What is the impedance of a 10 Ω resistor and 5 H inductor in series?

9. A current source of $4\cos\omega t$ is connected across a 10 Ω resistor and 20 mH inductor in series. Find the phasor voltage and the real-time voltage across the series combination if the angular frequency is 377 rad/s.

   Ans. $50.1\exp(j37.6°)$, $v(t) = 50.1\cos(377t + 37.6°)$.

10. What is the impedance of the parallel combination of a 1 kΩ resistor and a 100 mH inductor at a frequency of 1 kHz?

11. What is the impedance of a resistor R in series with a parallel combination of a capacitor C and inductor L. Assume angular frequency ω.

    Ans: $Z = R + j\omega L/(1 - \omega^2 LC)$.

12. For the circuit shown in Fig. 2.3, draw the phasor diagram for all voltages. Assume $\omega = 100$, $L = 1$ H, $C = 50$ μF, $R = 100$ Ω, and source voltage $v = 200\cos\omega t$. At this frequency, is this an inductive or capacitive circuit?

13.

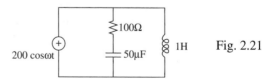

In the circuit shown in Fig. 2.21, use phasor analysis to calculate the real-time voltage $v_R(t)$ across resistor R. Assume $\omega = 1000$ rad/s.

   Ans. $v_R = 196.1\cos(1000t + 11.3°)$.

14. For the two-mesh circuit shown in Fig. 2.21, use mesh analysis and the phasor method to find the phasor and real-time current in the inductor.

15. In the circuit of Fig. 2.5a interchange the inductor and capacitor, and then solve for the phasor current in the resistor.
    Ans: $I_R = 0.5 + j0.5$.

16. Similar to impedance, resistance, and reactance, which are measured in ohms (Ω), admittance, conductance, and susceptance are measured in siemens (S). Find the admittance, conductance, and susceptance of a 100 Ω resistor in series with a 10 μF capacitor at an angular frequency of 1000 rad/s.

17. A low-pass filter is shown in Fig. 2.6a. If the corner frequency is to be 400 Hz,

    (a) Find R if $C = 0.5$ μF,
    (b) Find the voltage gain at 600 Hz.
        Ans: (a) 795.8 Ω; (b) 0.55.

18. Find the half-power frequency (in hertz) for the high-pass filter shown in Fig. 2.7a, given that $C = 1$ μF and $R = 10$ kΩ.

19. A high-pass filter is shown in Fig. 2.7a. If the cutoff frequency is to be 400 Hz,

    (a) Find C if $R = 1$ kΩ,
    (b) Is this a phase-lead or phase-lag network?
    (c) What is the power gain in decibels at 200 Hz?
        Ans: (a) 0.398 μF; (b) phase-lead; (c) −6.99 dB.

20. A series resonant circuit is used to tune in stations in the AM broadcast band. If a station, broadcasting on 870 kHz, is to be received and the fixed inductor of the resonant circuit is 20 μH, find the capacitance of the variable capacitor.

21. To eliminate an interfering frequency of 52 MHz, design a series resonant circuit to be inserted as a shunt. You have available a 10 pF capacitor. Determine the required inductance.
    Ans: 0.94 μH.

22. The applied voltage to a series resonant circuit like that shown in Fig. 2.2a is 1 V.

    (a) Determine how much larger the voltages across L, C, and R are at resonance. Use the values $R = 0.1$ Ω, $L = 0.1$ mH, and $C = 0.01$ μF, and first calculate the resonance frequency.
    (b) Should V_L and V_C be larger than 1 V, explain how this is possible.
        Ans: (a) $V_L = j1000$ V, $V_C = -j1000$ V, and $V_R = 1$ V.

23. Repeat Problem 22, except now change R to 10 Ω. What conclusion can you draw by comparing the two answers?

24. For the resonant circuit of Problem 22, calculate the power delivered by the voltage generator.
    Ans: $10 \cos^2 \omega_0 t$ W.

25. In the parallel resonant circuit of Fig. 2.13, $R = 5$ Ω, $L = 1$ mH, and $C = 100$ pF.

    (a) Find the resonant frequency, the Q of the circuit, and the bandwidth.
    (b) If it is desired to double the bandwidth, what changes in the parameters of the circuit must be made?

26. In the FM radio band, which spans the frequencies from 88 MHz to 108 MHz, stations must be separated by 0.20 MHz. To avoid overlapping, stations must broadcast with a bandwidth less than that. If the bandwidth of an FM station, which has a carrier frequency of 100 MHz, is 70 kHz, calculate the Q-factor of the receiving circuitry.
   Ans: 1429.

27. Parallel resonant circuits are used to select frequencies because the voltage and impedance peak at resonance. Calculate the impedance of the resonant circuit specified in Problem 25.

28.

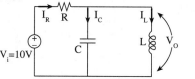

V$_i$=10V

Fig. 2.22

For the parallel resonant circuit shown in Fig. 2.22, find V_o/V_i for all frequencies.
   Ans: $j\omega L/[R(1 - \omega^2 LC) + j\omega L]$.

29. For the parallel resonant circuit shown in Fig. 2.22,

   (a) Find the resonance frequency ω_0.
   (b) Find V_o/V_i at resonance and show that the output voltage V_o (V_L or V_C) rises to equal the input voltage V_i at the resonant frequency.

30. For the parallel resonant circuit shown in Fig. 2.22,

   (a) Find I_R, I_C, and I_L at resonance.
   (b) What is the source current at resonance and what is the impedance (V/I_R) of the circuit at resonance?
   Ans: (a) $I_R = 0$ at ω_0, $I_C = j\omega_0 C V_i$, $I_L = -j\omega_0 C V_i$; (b) $I_i = 0$, $Zi = \infty$.

31. For the values $V_i = 10$ V, $R = 5$ kΩ, $C = 100$ pF, and $L = 10$ μH in Fig. 2.22,

   (a) Calculate the resonant frequency ω_0.
   (b) Calculate I_R at $\omega = 2.83 \cdot 10^7$.

32.

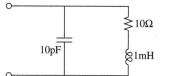

10pF

10Ω

1mH

Fig. 2.23

For the resonant circuit shown in Fig. 2.23 find the resonant frequency f, the Q, and the bandwidth of the circuit at resonance.
   Ans: 1.59 MHz, 10^3, and 1.59 kHz.

33. In Example 2.5, which utilizes what is basically a parallel resonant circuit, the calculation for Q was carried out in terms of the expression $Q = \omega L/R$, which is the Q-expression for a series resonant circuit. Starting with the Q-expression

for the parallel resonant circuit shown in Fig. 2.12a, which is $Q = R/\omega L$, show that for the resonant circuit of Fig. 2.13 it is valid to use $\omega L/R$ to calculate Q.

34. For the circuit shown in Fig. 2.23,

(a) Find the impedance at resonance.
(b) Find the voltage across the circuit at resonance when a current of 1 μA is flowing into the circuit.
(c) Find the capacitor current at resonance.
    Ans: (a) 10 MΩ, (b) 10 V, and (c) 1 mA.

35. Calculate the power dissipated in the transmission line shown in Fig. 2.15 by using the power expression given by (2.36).

36. In Example 2.5, is the instantaneous power supplied by the source ever negative? If yes, for how many degrees out of every 360° cycle?
    Ans: 64°. *Hint:* review Figs.1.4, 1.5, and 1.7.

37. In the series circuit of Fig. 2.2a, assume the rms source voltage is 120 V, 60 Hz and that $R = 20$ Ω, $L = 2$ H, and $C = 2$ μF.

(a) Calculate the average power delivered by the source.
(b) Calculate the power factor.

38.

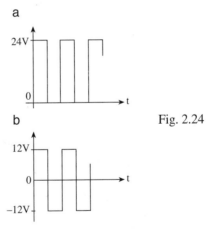

Fig. 2.24

(a) A square wave is shown in Fig. 2.24a. Find the rms voltage of this waveform.
(b) If this waveform is passed through a series capacitor, which removes its DC level so it looks like that in Fig. 2.24b, find the rms voltage of this waveform.
(c) If the Fig. 2.24b waveform is rectified, leaving only the positive voltages, find the rms voltage of the rectified waveform.
    Ans: (a) 16.97 V, (b) 12 V, and (c) 8.49 V.

39. A pure inductance of 10 Ω is placed across a 220 V rms AC generator. What resistive load can be in parallel with the inductance if the current in the circuit is limited to 40 A rms?

40. A motor is connected to a 120 V rms, 60 Hz line. The motor is rated at 2 kVA and operates with a power factor of 0.8.

    (a) Find the apparent and real power.
    (b) What is the current flowing into the motor?
    (c) Find the resistance R and inductance L of the motor.
        Ans: 2000 VA, 1600 W; 16.67 A; and 5.76 Ω, 0.11 H.

41. It is desired to run the motor in the previous problem with a power factor of unity. To achieve this, a capacitor C is connected in parallel across the motor, which results in a circuit like that shown in Fig. 2.13. Calculate the value of C to achieve this condition.

42.

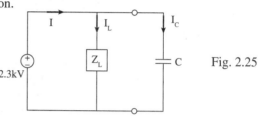

Fig. 2.25

In Fig. 2.25 the load Z_L is 27 kW at a power factor of 0.75. The voltage at the load is 2.3 kV, 60 Hz. For power factor correction a capacitor is connected in parallel with the load. What size capacitor C must be used to change the power factor to a more favorable 0.93?

Ans: 6.6 μF.

43. A motor can be represented by a resistance R and inductance L in series. If this motor has a power factor of 0.866 at a frequency of 60 Hz, what will the power factor be for a frequency of 440 Hz?

44. Design a transformer that will provide at the secondary 12 V at 5 A when the primary is connected to 120 V, 60 Hz. Determine the number of turns in the primary and secondary windings if the core has a cross section of 2 cm^2 and the core flux is not to exceed 0.5 T.

    Ans: 4505, 451.

45. In the previous problem, if the load is a resistive 2.4 Ω, how much power is delivered to it?

46. In a 60 Hz transformer, the maximum flux density is 1.5 T. What area of core is needed to produce 2 V per turn of winding?

    Ans: 50 cm^2.

47. Audio amplifiers use transformer coupling between the final amplifier stage and a speaker to match impedances and in order to avoid having the DC current of the final stage flow through the speaker coil. We wish to connect an 8 Ω speaker to an 8000 Ω amplifier. Calculate the turn ratio of the output transformer.

48. A doorbell is connected to a transformer whose primary contains 3000 turns and which is connected to 120 V$_{AC}$. If the doorbell requires 0.2 A at 10 V, find the number of turns necessary for the secondary winding and the current that flows in the primary.

    Ans: 250, 16.7 mA.

CHAPTER 3

Diode Applications

3.1 INTRODUCTION

A *diode* is our first encounter with a nonlinear element. Recall that R, L, and C are linear elements, meaning that a doubling of an applied voltage results in a doubling of current in accordance with Ohm's law. A diode, which has two terminals, or two electrodes (hence *di-ode*), acts more like an on–off switch. When the diode is "on," it acts as a short circuit and passes all current. When it is off, it acts as an open circuit and passes no current. The two terminals of the diode are different and are marked as plus and minus. If the polarity of an applied voltage matches that of the diode (referred to as forward bias), the diode turns "on" and functions as a short circuit (it mimics a switch in the on positon). When the applied voltage polarity is opposite (reverse biased), the diode is off. Another good analogy to a diode is the plumber's check valve which allows water in a pipe to flow in one direction but not in the other. To explain this fascinating behavior of a diode requires some solid-state physics, which we will leave for the next chapter. In this chapter we will explore practical applications of the diode.

A diode is also referred to as a *rectifier*. For example, placing a diode in series in a circuit that carries an alternating current will result in a current that flows in only one direction, determined by the forward bias. Hence the current is rectified. Perhaps the greatest use of diodes is in power supplies where an AC source, typically 120 V_{AC} supplied by the power utilities, is converted to a DC source.

3.2 RECTIFICATION

3.2.1 Ideal and Practical Diodes

Figure 3.1a shows the v-i characteristics of an ideal diode which are also part of the characteristics of an on–off switch. We will use these characteristics to approximate

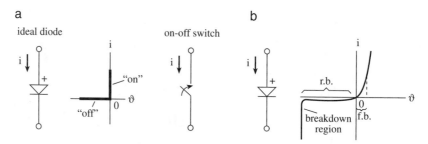

FIGURE 3.1   (a) Symbol and current flow direction of an ideal diode. The on–off states of an ideal diode mimic those of an on–off switch. (b) A practical diode and its v-i characteristics.

those of a practical diode, Fig. 3.1b, such as the popular IN4002, which is used in small power supplies.

The v-i characteristics of the IN4002 are shown in Fig. 3.1b. Note that although in general the ideal diode is a good approximation to the practical diode, significant differences exist. The differences relate to the operating range of the practical diode and provide guidance when a more accurate model than the ideal diode is needed. The most obvious differences are:

(a) Under forward bias conditions it takes a finite forward voltage of approximately 0.7 V for a practical diode to conduct. This voltage drop is maintained during conduction.

(b) The maximum forward current is limited by the heat-dissipation ability of the diode; for the IN4002, for example, maximum forward current is 1000 mA = 1 A.

(c) There is a small but finite reverse current. The reverse current is usually of no significance in a practical circuit as it is in the nanoampere range (note the different scales in Fig. 3.1b for forward and reverse current).

(d) Every diode has a maximum reverse voltage that cannot be exceeded. If it is, the diode breaks down and shorts, and a large reverse current flows that causes the diode to overheat and be permanently damaged. For the IN4002, the reverse breakdown voltage is 100 V. If there is a chance that the breakdown voltage will be exceeded in a circuit, a diode with a larger breakdown voltage must be used.

3.2.2   Half-Wave Rectifier

If a diode and load resistor are connected in series with an AC source as shown in Fig. 3.2a, the resultant voltage $v_o = i_o R_L$ and current i_o through the load resistor R_L are as pictured in Fig. 3.2b (assuming the diode is ideal, i.e., ignoring the 0.7 V drop when the diode conducts, and assuming the diode is open-circuited when it does not conduct). It is a pulsating current, but it is DC. Note that the term DC voltage can mean

a

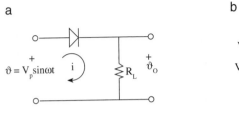

b

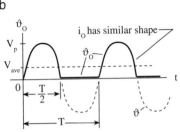

FIGURE 3.2   (a) A half-wave rectifier circuit. (b) The pulsating DC voltage v_o across load resistor R_L.

constant voltage (a battery voltage, for example) or a variable voltage with constant polarity. Hence, the circuit of Fig. 3.2a changes AC into DC. If we could smooth out the pulsating DC, we would obtain a DC voltage of

$$V_{ave} = V_{DC} = \frac{1}{T} \int_0^{T/2} V_p \sin \omega t \, dt = \frac{V_p}{\pi} \tag{3.1}$$

where the applied voltage is $v = V_p \sin \omega t$, period $T = 1/f$, and $\omega = 2\pi f$. Thus, the average or DC voltage, if the applied voltage is 120 V_{AC}, would be $120\sqrt{2}/\pi$ or 54 V. The DC current flowing through the load resistor R_L is $I_{DC} = V_p/\pi R_L$.

EXAMPLE 3.1   The input to the half-wave rectifier of Fig. 3.2a is 120 V_{AC}. If the resistance R_d of the diode during conduction is 20 Ω and the load resistance $R_L = 1000$ Ω, find the peak, DC, and rms load currents and the power dissipated in the load and the diode.

The peak load current is $I_p = V_p/(R_L + R_d) = 120\sqrt{2}/(1000 + 20) = 0.166$ A. The DC current from (3.1) is $I_{DC} = I_p/\pi = 0.053$ A. The rms current, from (2.41), is

$$I_{eff} = \sqrt{\frac{1}{T} \int_0^{T/2} i^2 \, dt} = \sqrt{\frac{1}{T} \int_0^{T/2} (I_p \sin \omega t)^2 \, dt} = I_p/2 = 0.083 \text{ A}$$

The total input power to the circuit is $P_t = P_L + P_d$. The power dissipated in the load is $P_L = I_{rms}^2 R_L = 0.083^2 \cdot 1000 = 6.92$ W, and the power dissipated in the diode is $P_d = I_{rms}^2 R_d = 0.14$ W. The total power supplied by the source is therefore 7.06 W. ∎

Note that, in the above calculations, we are neglecting the 0.7 V voltage drop across the diode during the conducting phase. Also note that the peak reverse voltage across the diode during the nonconducting phase is $120\sqrt{2} = 170$ V, which the diode must be capable of withstanding.

3.2.3   Full-Wave Rectifier

The half-wave rectifier uses only half the input waveform. An arrangement that can use all of the input waveform is the full-wave bridge rectifier, shown in Fig. 3.3a. Current

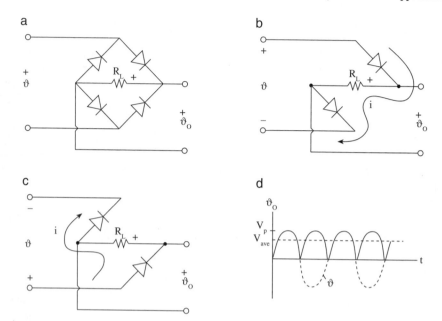

FIGURE 3.3   (a) A full-wave bridge rectifier. (b,c) Conduction path when the input polarity is as shown. (d) Output voltage v_o of the full-wave rectifier.

flows in the same direction through the load resistor, for both polarities of the input voltage. This is accomplished by having two forward-biased diodes in series with R_L at any time, as shown in Figs. 3.3b and c. In Fig. 3.3b, the input voltage has a polarity that makes the top input terminal positive; hence, only the two diodes shown are forward biased and are conducting. As the input voltage reverses, the bottom terminal becomes positive. Now the remaining two diodes conduct, while the previous two are open-circuited, as shown in Fig. 3.3c. The result is that current flows through R_L at all times, as shown in Fig. 3.3d. One should remember that in a full-wave rectifier two diodes are always in series with a resultant voltage drop of 1.4 V. When designing a low-voltage power supply, this could be a substantial voltage drop that is not available to the output voltage v_o.

The average voltage of a full-wave rectifier is given by

$$V_{ave} = V_{DC} = \frac{1}{T/2} \int_0^{T/2} V_p \sin \omega t \, dt = \frac{2V_p}{\pi} \tag{3.2}$$

If 120 V_{AC} is the input voltage, the full-wave rectifier has the potential to produce twice the voltage of a half-wave rectifier, i.e., 108 V_{DC}.

3.2.4 Rectifier Filters

The pulsating waveforms generated by rectification are not very useful, but can be smoothed to produce almost perfect DC. For that purpose we can use the inertia properties of capacitors and inductors. Recall that a capacitor smoothes the voltage across it and an inductor smoothes the current through it.

Another way of looking at rectifiers is as follows: the pulsating voltage which is produced has high-frequency components as well as DC. What is therefore needed is a low-pass filter, which would pass the DC but would limit passage of the high frequencies. The simplest low-pass filter (an RC filter, Fig. 2.6) is a capacitor in parallel with the load resistor as shown in Fig. 3.4a. The capacitor voltage, shown in Fig. 3.4b, which is also the output voltage, is now much smoother than the pulsating waveform of a simple half-wave rectifier. It is interesting to see how this comes about. Basically the capacitor stores energy which decays exponentially during the discharge phase when the capacitor provides energy to R_L. The energy in the capacitor is periodically replenished during the charging phase when the diode conducts. It is analogous to maintaining a steady flow of beer from a tap that is attached to the bottom of an open barrel of beer. As the level of beer in the barrel decreases, imagine that a bucket of beer is periodically dumped into the barrel, maintaining the steady flow at the tap.

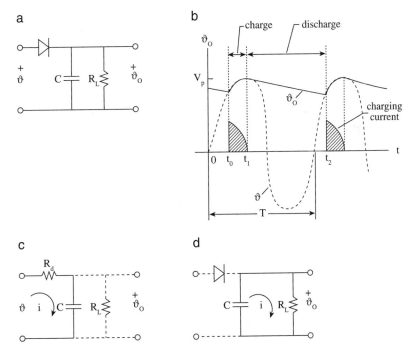

FIGURE 3.4 (a) A capacitor across the load resistance will smooth the pulsating DC. (b) The smoothed voltage. The circuit during (c) charge and (d) discharge.

Figure 3.4c shows the equivalent circuit during the charging phase. Diode current i_d flows during the time interval $t_1 - t_0$ and provides the charging current $i_C = C \, dv_o/dt$ as well as the load current $i_L = v_o/R_L$. The small resistance (fraction of an ohm to several ohms) of the diode during conduction is given as R_d. This makes for a small time constant $R_d C$, implying that the capacitor is being charged quickly and hence the capacitor voltage v_o can easily follow the sinusoidal input voltage v. Hence, v_o increases to the peak level V_p of the input voltage $v = V_p \sin \omega t$. Then, as v decreases but v_o holds to V_p, the diode becomes reverse-biased and opens.

Figure 3.4d shows the discharge phase. During the time interval t_2-t_1 the diode is open, essentially disconnecting the input voltage from the capacitor. The capacitor is left alone to supply energy to R_L, which it does while discharging with a time constant of $R_L C$. As R_L is typically much larger than R_d, the time constant $R_L C$ is now large and the exponential drop in capacitor voltage from V_p is small.

Practical filters are usually designed to have a negligible drop in capacitor voltage during the discharge phase. If this is the case, the output voltage can simply be approximated as

$$v_o = V_{DC} \cong V_p \tag{3.3}$$

and the load current as $I_L = I_{DC} = V_p/R_L$. In that sense a rectifier and a rectifier with a filter are fundamentally different. A rectifier by itself can only produce a DC level, given by (3.1) and (3.2), which is substantially less than V_p, whereas a rectifier when combined with a capacitor filter can substantially increase the DC output voltage to approximately the peak input voltage V_p.

3.2.5   Ripple Voltage Remaining after Filtering

The capacitor filter does a creditable job of producing DC. Nevertheless, after smoothing the pulsating waveform, as shown in Fig. 3.4b, a significant ripple voltage remains which we would now like to quantify. During the discharge period, the capacitor voltage decays exponentially: it begins with V_p and at the end of the discharge period, $t = t_2$, is equal to

$$v_o = V_p e^{-(t_2-t_1)/R_L C} \tag{3.4}$$

The discharge time is on the order of the period T of the input voltage; hence we can approximate $t_2-t_1 \approx T$. The ripple voltage v_r is then the difference between V_p and the voltage at the end of the discharge period, i.e.,

$$v_r = \Delta v_o = V_p - V_p e^{-T/R_L C} \cong V_p \frac{T}{R_L C} = \frac{V_p}{f R_L C} \tag{3.5}$$

where it is assumed that we have chosen the time constant to be much longer than the period of the input voltage, i.e., $T/R_L C \ll 1$, which will guarantee a small decay during discharge and hence a small ripple voltage. Furthermore, we have used the approximation $e^\Delta \cong 1 + \Delta$ when $\Delta \ll 1$ and $T = 1/f$, where T and f are the period and frequency of the input voltage.

The last term in (3.5) can be considered as a design equation for capacitor filters. It states that the ripple voltage is inversely proportional to capacitance. The other terms in (3.5) usually cannot be changed as the peak voltage V_p, the frequency f, and the load resistance R_L are fixed in practical situations. Hence it pays to use the largest capacitance possible as this will give the smoothest DC. A word of caution though: at the time when the power supply is turned on, the uncharged capacitor presents a short which can result in unacceptably high initial currents passing through the diode. Such currents could easily burn the diode out unless we place a small resistance in series with the diode which would limit the initial current to values within the diode specifications.

EXAMPLE 3.2 Find v_o and the ripple voltage v_r when the output of a full-wave rectifier, shown in Fig. 3.3d, is applied to a load resistor R_L with a capacitor filter. That is, replace the half-wave rectifier with a full-wave rectifier in Fig. 3.4a.

The output voltage v_o of a filtered rectifier is shown in Fig. 3.4b. In the case of the full-wave rectifier, the negative part of the input voltage is flipped up, so the input to the capacitor is the waveshape given in Fig. 3.3d. Hence, the DC input pulses occur at twice the rate of the half-wave rectifier, allowing the capacitor only half the time to discharge before the diode conducts and recharges the capacitor. The implication is that the DC output voltage is even better approximated by V_p than in the case of the half-wave rectifier. However, an even more important implication is that the discharge period is now approximately one-half the input period, or $t_2 - t_1 \approx T/2$. Thus, the ripple voltage in (3.5) becomes

$$v_r = V_p \frac{T}{2R_L C} = \frac{V_p}{2f R_L C} \tag{3.6}$$

and is seen to be one-half as large as that for the half-wave rectifier. Furthermore, if a 60 Hz input voltage is used, the ripple voltage has a frequency of 60 Hz for a half-wave rectifier but a frequency of 120 Hz for the full-wave rectifier, which usually is less objectionable and is easier to smooth if additional filtering is needed.[1]

In a practical situation, we might need to supply a DC voltage of 170 V to a load represented by $R_L = 1\ \text{k}\Omega$. If the ripple voltage must be less than 3 V, we need a capacitor with a capacitance of $C = 170/3 \cdot 2 \cdot 60 \cdot 1000 = 472 \cdot 10^{-6}\ \text{F} = 472\ \mu\text{F}$, assuming 120 V_{AC} at 60 Hz and that a full-wave rectifier was used. The load current flowing in the load resistor is $I_L = I_{DC} = V_p/R_L$. If other DC voltages are needed, a step-up or step-down transformer is used to supply the correct input voltage. ∎

3.2.6 Voltage Doubler

An easy way to double DC voltage is to use the circuit shown in Fig. 3.5. Two half-wave rectifiers in series charge the top capacitor to V_p when the input voltage swings

---

[1]Note that the period of full-wave rectifier output is one-half the period of sinusoidal input. Hence the

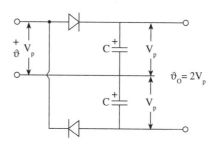

FIGURE 3.5   The DC output voltage of a voltage-doubler rectifier circuit is equal to twice the peak input voltage.

positive and charge the bottom capacitor to V_p when the input voltage swings negative. As the polarities of the charged capacitors are in phase, the voltage at v_o is double the peak input voltage. The voltage at v_o is like that of a full-wave rectifier; that is, since both halves of the input voltage are used, the ripple frequency is 120 Hz, if the input frequency is 60 Hz. The applicable formula for ripple voltage is therefore (3.6). As in the rectifiers considered above, when a load resistor is connected, current is supplied to the load resistor by the discharge of the capacitors.

3.3   CLIPPING AND CLAMPING CIRCUITS

We will now see how diodes make elementary waveshaping circuits possible. At times it is desirable to limit the range of a signal or to remove an unwanted portion of a signal.

3.3.1   Clipping

A typical function of a clipping circuit is to cut off part of an input waveform. For example, the clipper shown in Fig. 3.6a cuts off the input waveform that goes above the value V_1 and below $-V_2$. Since both diodes are reverse biased by batteries V_1 and V_2, whenever input voltage v exceeds V_1, diode one conducts and places battery voltage V_1 at the output. The difference voltage $v - V_1$ is dropped across R. A similar process occurs when the input signal swings negatively. Hence the sinusoidal waveshape is altered and becomes the clipped output voltage v_o, shown in Fig. 3.6b. If V_1 and V_2 are much smaller than V_p of the input, a waveshape is produced that looks like a square wave. The transfer characteristics of such a circuit are shown in Fig. 3.6c.

---

frequency of full-wave rectifier is twice that of the sinusoidal input. A half-wave rectifier; on the other hand, has the same period as the sinusoidal input.

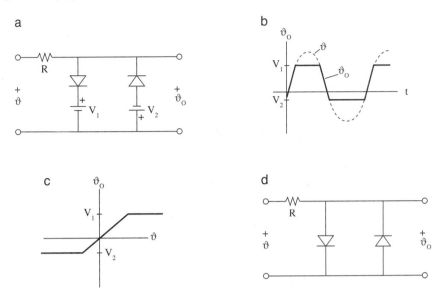

FIGURE 3.6   (a) A clipper circuit that limits the input voltage to $+V_1$ and $-V_2$. (b) The clipped output voltage. (c) The transfer characteristics v_o vs v of the clipper circuit.

3.3.2   Limiters

Clippers can be used as protection circuits against voltage overshoot. For example, it is clear that the clipper of Fig. 3.6a will not allow the voltage v_o to go past $+V_1$ and $-V_2$. Therefore, if both batteries are removed (i.e., replaced by shorts in Fig. 3.6a), the output will be zero as one of the diodes will always be conducting. The validity of that statement is based on the diodes acting as ideal diodes (see Fig. 3.1a). But we have learned that a diode is characterized by a forward voltage of about 0.7 V, which must be exceeded before the diode will conduct. Therefore, a clipping circuit, Fig. 3.6d, with both external batteries removed can serve as a protection circuit in the beginning stages of a high-gain amplifier when the input voltages are on the order of millivolts. Such high-gain amplifiers saturate easily and need the protection that such a simple circuit provides: essentially two opposite diodes in parallel, connected from input to ground, which limits the input voltage swings to ± 0.7 V.

Another use of clippers is in noise limiting. For example, in a radio receiver a selected signal could be subject to strong noise pulses (sparks, lightning, etc.) which distort and add the familiar crackling to the signal. Choosing the bias voltages of the batteries in Fig. 3.6a to be somewhat larger than the desired signal will clip the noise pulses at that level but pass the signal.

EXAMPLE 3.3   Show a battery backup system for an electronic clock or timer which would engage when the power supply fails.

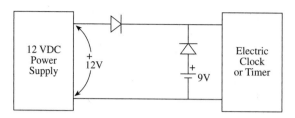

FIGURE 3.7   A battery backup system.

Let us assume the clock has a built-in power supply that delivers 12 V to the clock circuitry. Should the power fail, we would like a 9 V battery to be automatically connected to the clock circuitry. Figure 3.7 shows such a circuit. When there is no power failure and the internal power supply is delivering 12 V, the top diode conducts, connecting 12 V to the clock circuitry. The diode in series with the battery has a 3 V reverse bias and is open. During a power failure, the 12 V changes to zero. The top diode then becomes reverse biased and opens. This in turn forward biases the diode that is in series with the 9 V battery; the diode conducts, connecting the battery to the clock. Of course this assumes that 9 V is adequate to keep the clock running properly.   ∎

3.3.3   Clamping

If the DC value of a signal needs to be changed (the signal as a whole shifted up or down along the vertical axis), a capacitor can be charged to the desired value, and when connected in series with the signal source, will give the signal the desired level. Such a circuit is called a clamping circuit. Figure 3.8a shows a circuit that clamps the peak value of a signal at zero. For positive values of the input voltage $v = V_p \sin \omega t$, the diode conducts and rapidly charges the capacitor to the peak value V_p as the input voltage increases to V_p. The RC time constant of the circuit is very small (the only R is that of the forward-biased diode, typically less than 1 Ω); hence the capacitor voltage increases as rapidly as the input voltage until the capacitor is charged to V_p. Then, as the input voltage decreases sinusoidally from V_p, the diode becomes reverse biased and opens. The capacitor voltage remains at V_p because the open diode prevents the capacitor from discharging. The output voltage is now the constant capacitor voltage in series with the input voltage, i.e.,

$$v_o = -V_p + V_p \sin \omega t = V_p(\sin \omega t - 1) \qquad (3.7)$$

Whatever is connected to the output v_o usually has a sufficiently large resistance that the capacitor discharge is negligible. Should there be some discharge, the capacitor will be recharged on the next cycle of the input voltage.

Figures 3.8b and c show circuits that clamp the top of the input signal at the battery voltage $+V$ and $-V$, respectively. In case of Fig. 3.8b we have

$$v_o = -V_p + V + V_p \sin \omega t = V_p(\sin \omega t - 1) + V \qquad (3.8)$$

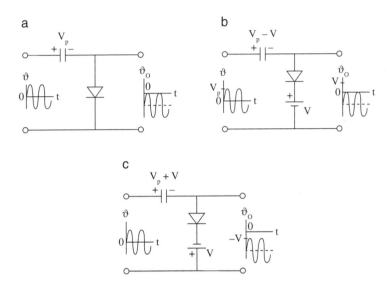

FIGURE 3.8 These circuits clamp the peak of the input voltage at (a) zero, (b) $+V$, and (c) $-V$.

and in case of Fig. 3.8c,

$$v_o = -V_p - V + V_p \sin \omega t = V_p(\sin \omega t - 1) - V \qquad (3.9)$$

Similarly, the circuits shown in Figs. 3.9a, b, and c clamp the bottom of the input signal at 0, $-V$, and $+V$ volts, respectively. The output voltage for the three circuits is $v_o = V_p(\sin \omega t + 1)$, $v_o = V_p(\sin \omega t + 1) - V$, and $v_o = V_p(\sin \omega t + 1) + V$.

3.4 ZENER DIODE VOLTAGE REGULATION

Even though the capacitive filters in a power supply smooth the rectified voltage, the output voltage can still vary with changes in line voltage which can be due to a variety of causes, for example, surges when large loads such as motors are switched on and off. If some circuits in electronic equipment must have an absolute steady voltage, we resort to *Zener diodes*. Zener diodes are a special kind of diode that can recover from breakdown caused when a reverse-bias voltage exceeds the diode breakdown voltage. Recall that in Fig. 3.1b we observed that ordinary diodes can be damaged when the breakdown voltage is exceeded. Zener diodes, on the other hand, are designed to operate in the breakdown region (provided the current does not become too excessive) and recover completely when the applied reverse voltage changes to less than the breakdown voltage. Breakdown occurs always at the same, precise values of voltage. The attraction of the Zener diode is that the voltage across it stays very nearly constant for any current within its operating range. This feature makes it a good voltage regulator or

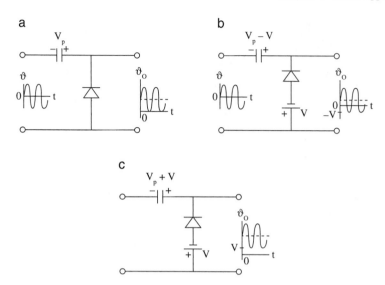

FIGURE 3.9   Reversing the diode from that in Fig. 3.8 clamps the bottom of the waveshape at 0, $-V$, and $+V$, as shown in (a), (b), and (c), respectively.

voltage reference element. Since Zener diodes are manufactured with a great variety of voltages (2 to 200 V), the need of any circuit for a constant voltage can be easily met. Typically, a TV set or a quality stereo receiver might have several sensitive subcircuits which need a steady voltage supply.

Figure 3.10a shows a Zener diode voltage regulator. It is a simple circuit consisting of a series resistor R_s across which the excess voltage is dropped and a Zener diode, chosen for a voltage V_z at which it is desired to maintain the load R_L. As the input voltage v fluctuates between the values V_{min} and V_{max} (both must be higher than V_z for regulation to take place), the load is maintained at the constant voltage V_z. Figure 3.10b shows the variation ($I_{max} - I_{min}$) of the Zener current, which is in response to the input voltage variations ($V_{max} - V_{min}$). Therefore, the current through the Zener diode varies[2] so as to keep the current through R_L constant. The varying Zener current (and the steady load current) flows through R_s, which causes a varying voltage across R_s. This in turn allows the voltage across the load to remain constant.

EXAMPLE 3.4   It is desired to hold a load resistance R_L at a constant voltage of 100 V as the input voltage varies between 120 and 110 V. If a voltage regulator of the type

---

[2]The avalanche mechanism is such that when the breakdown begins at the voltage V_z, the current begins to flow precipitously with even minor increases in voltage. Hence, for all practical purposes, the voltage remains constant at V_z. Of course, the current cannot increase beyond a certain value without overheating and damaging the diode. Each Zener diode, besides V_z, specifies a maximum Zener current. The fact that a reverse-bias breakdown current flows in a Zener diode is denoted by the special symbol in Fig. 3.10a, which is similar to an upside down diode symbol.

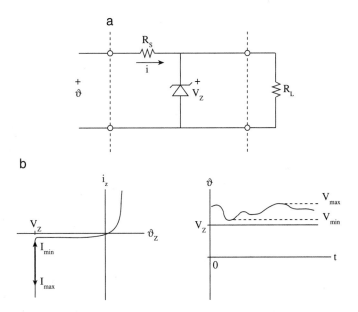

FIGURE 3.10 (a) A voltage regulator is inserted between the input voltage and the load. (b) The Zener current i_Z varies between I_{max} and I_{min} in response to the varying input voltage so as to keep the load current and load voltage constant.

shown in Fig. 3.10a is to be used, find the best value of R_s to accomplish that purpose, given that $R_L = 10$ kΩ.

First, we choose a Zener diode with $V_z = 100$ V. Second, we must find what the maximum current through the Zener diode under normal operation will be and to make sure that it does not exceed the maximum allowed Zener diode current. Then we determine R_s.

To begin with let us assume that the input voltage is fixed at $V_{min} = 110$ V; then a 10 V voltage drop across series resistance R_s would leave R_L with a voltage drop of 100 V—the desired condition. For this to happen a current of 10 mA must flow through R_L and R_S, which would determine the series resistance as $R_s = 10$ V/10 mA $= 1$ kΩ. A Zener diode would not be needed if the voltage were to remain at 110 V as no Zener current would flow even if a Zener diode were present. However, the input voltage changes as shown in Fig. 3.10b. The change from 110 V to 120 V is usually not rapid but can take place in a matter of seconds, minutes, or even hours.

As the input voltage rises to 120 V, the current through R_s will rise proportionally. To keep R_L at 100 V, the current through R_L must remain at 10 mA and any excess current should flow through the Zener diode. When the input voltage is at $V_{max} = 120$ V, 20 V is dropped across R_s and 20 mA flows through R_s (10 mA through R_L and 10 mA through the Zener diode). Hence as shown in Fig. 3.10b, the Zener current varies between $I_{z, min} = 0$ and $I_{z, max} = 10$ mA in response to input voltage variations, while the load voltage remains constant at 100 V.

The condition $I_{z, \text{min}} = 0$ can be used to define an optimum value for R_s, i.e.,

$$R_{s, \text{optimum}} = \frac{V_{\text{min}} - V_z}{I_L}$$

which for our example gives $R_{s,\text{opt}} = (110\ V - 100\ V)/10\ \text{mA} = 1\ \text{k}\Omega$.

If we know the maximum current $I_{z, \text{max}}$ that the Zener diode can tolerate, we can specify a minimum value of R_s that can be used in a Zener diode voltage regulator circuit as

$$R_{s, \text{min}} = \frac{V_{\text{max}} - V_z}{I_{z, \text{max}} + I_L}$$

If we assume that $I_{z, \text{max}} = 30$ mA, we obtain for $R_s = (120 - 100)/(30 + 10) = 0.5$ kΩ = 500 Ω. The advantage of using a smaller resistance for R_s is that if the input voltage drops below 110 V, regulator action can still take place. The disadvantage is that (i) $R_{s,\text{min}}$ dissipates more power than $R_{s,\text{opt}}$, (ii) the Zener diode current varies between $I_{z,\text{min}} = 10$ mA and $I_{z, \text{max}} = 30$ mA, whereas with $R_{s,\text{opt}}$ the Zener current varies only between 0 and 10 mA, and (iii) if the input voltage exceeds 120 V, the Zener diode current will exceed the maximum allowable current $I_{z, \text{max}}$ and most likely damage the diode. ■

There is always some danger that the maximum diode current will be exceeded, either by an unexpected upward fluctuation of the input voltage or by a sudden removal of the load, which would cause all of the input current to flow through the diode. The latter case, that of a sudden open-circuited load ($R_L = \infty$), would usually ruin the Zener diode as it is most likely that $I_{z, \text{max}}$ would be exceeded.

3.5   SILICON-CONTROLLED RECTIFIERS (SCRS)

3.5.1   Introduction

A device with widespread application in industry is the silicon-controlled rectifier. It is used for speed control of motors, for dimming of lights, for control of heating furnaces, and in general wherever control of power is needed.

Recall that in a rectifier, current starts to flow immediately when the diode becomes forward-biased. An SCR is basically a rectifier to which a gate has been added. By delaying the application of an enabling signal to the gate, the SCR can delay the onset of the rectified current. Figure 3.11 shows three examples of power control by an SCR when three different gating pulses are applied (the input is a sinusoidal voltage $v = V_p \sin \omega t$). Figure 3.11a shows voltage and current waveshapes that are basically those of a half-wave rectifier: the short gating pulses occur at the start of the sinusoid. Figure 3.11b shows the gating pulses delayed by 90°. Because the rectified current is delayed until the gating pulses trigger the SCR, only half the power would be delivered to a resistive load which is in series with the SCR. Figure 3.11c shows a pulse that is

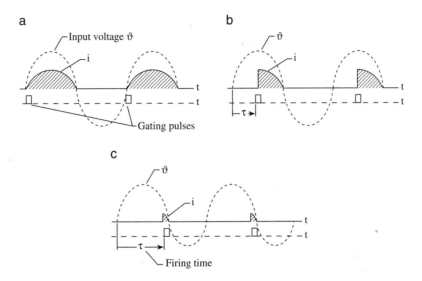

FIGURE 3.11 Short gating pulses applied to an SCR control the onset of the rectified current. (a), (b), and (c) show that progressively less power would be delivered to a load.

delayed by almost 180°. Very little current flows and very little power is delivered to a load. This is the basics of SCR control of power.

If we call the onset time of the gating pulses as firing time τ (which is related to the firing angle α by $\alpha = \omega\tau$), we can express the average (DC) value of the current as

$$I_{\text{ave}} = \frac{1}{T} \int_{\tau}^{T/2} I_p \sin \omega t \, dt = \frac{I_p}{2\pi}(1 + \cos \omega\tau) \tag{3.10}$$

To show that the average current depends on the firing angle α, we can substitute $\omega\tau = \alpha$ in the above expression. Hence, if α or τ is zero, (3.10) gives $I_{\text{ave}} = I_p/\pi$, which is the average current for a half-wave rectifier, and was already derived before as (3.1). When $\alpha = 180°$, no current and no power is delivered by the SCR.

3.5.2 SCR Characteristics

The symbol of an SCR is shown in Fig. 3.12a, where the terms anode and cathode are leftovers from the days of the vacuum tube thyristor, which performed a similar function as the solid-state SCR. Typical SCR v-i characteristics are shown in Fig. 3.12b. Note that the forward current has both an on- and an off-state. Normally the diode stays in the off-state (insignificant current flow from anode to cathode) until a current pulse (typically 5–50 mA) is injected into the gate which triggers the on-current state with significant flow of current, which can be as high as several thousand amperes for the larger SCRs. In the on-state, the voltage v_{ac} across the SCR is only a few volts. The SCR could also be triggered, without gate current injection, if v_{ac} exceeds the breakover

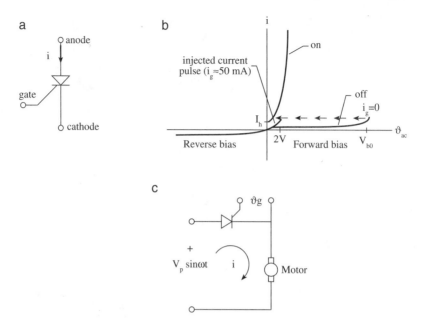

FIGURE 3.12 (a) The SCR symbol. (b) Voltage–current characteristics of an SCR. (c) A simple SCR speed control circuit for a DC motor.

voltage V_{bo} (typically in the hundreds of volts). If this happens, the SCR conducts and the voltage v_{ac} drops to one or two volts. The practical way to trigger an SCR is by gate current injection, though. Once triggered, the device will stay on irrespective of what the triggering pulse does subsequently. Gate current is required only long enough to build up full anode current, typically on the order of microseconds with resistive loads. However, there is a minimum holding current, I_h (typically 100 mA), required to keep the on-current going. Once forward current starts flowing, it can continue indefinitely until something in the external circuit reduces current flow to below the I_h value. The key feature of an SCR is that a small gate current can "fire," or trigger, the SCR so that it changes from being an "open" circuit into being a "short" circuit. The only way to change it back again is to reduce the anode current to a value less than the holding current I_h.

It is not necessary to use very short pulses in the gate circuits. Other periodic wave-shapes can also trigger the SCR and might be more practical if they are easier to gener-ate, although short pulses are the most efficient. Recall that triggering is accomplished by injecting a small amount of current into the gate at the desired time. Even a sinu-soidal waveshape, delayed correctly, can accomplish that. Frequently RC circuits of the type considered in Fig. 2.6 are used to provide a phase-delayed gate voltage (for design of such gate voltages the student is referred to SCR manuals). Remember that once conduction begins, the gate loses control and the SCR remains in the high-conduction state until the anode potential is reduced to practically zero volts. Figure 3.12c shows a simple SCR circuit to control the speed of a DC motor.

The main reason for the use of SCRs is efficiency of power control. If we were to use a variable resistance (*potentiometer*) in series with the load to reduce the voltage to the load, we would waste valuable power in the series resistance, which would be particularly serious when large amounts of power are involved. The SCR, on the other hand, wastes little power when the load receives little power. Also, if the half-wave rectifier action which the SCR provides (the SCR conducts current in one direction only) is inadequate, a full-wave rectifier (current flow in both directions) can precede the SCR. As both halves of the input sinusoid are now used, the power available to the load is doubled. Such a doubling of power is also possible without the use of a full-wave rectifier. A *Triac*, which consists of two SCRs back-to-back, is a three-terminal (gate, anode, and cathode leads, similar to Fig. 3.12a) device which permits full control of AC. It is very popular in dimmer switches and speed control of motors.

EXAMPLE 3.5 Calculate the power in watts available to a 10 W motor which is controlled by an SCR as shown in Fig. 3.12c (for simplicity, let us ignore the inductance of the motor winding). The combination is connected to a 117 V AC line. Calculate the power corresponding to the three v-i figures in Fig. 3.11.

An applied periodic gate voltage, which is progressively shifted from $0°$ to $90°$ and then to almost $180°$, results in motor current flow as shown in Figs. 3.11a, b, and c. The effective current in these figures is given by (see the equation in Example 3.1)

$$
I_{\text{rms}} = \sqrt{\frac{1}{T} \int_{\tau}^{T/2} \left(I_p \sin \omega t\right)^2 dt} = \sqrt{\frac{1}{2\pi} \int_{\alpha}^{\pi} \left(I_p \sin \omega t\right)^2 d\omega t}
$$

$$
= \frac{I_p}{2} \sqrt{1 - \frac{\alpha}{\pi} + \frac{\sin 2\alpha}{2\pi}}
$$

(remembering that the SCR is basically a half-wave rectifier which delays the flow of current by τ seconds). In the above expression we have made the transformation from time t to phase angle ωt by making the substitution $\omega t = \alpha$ and $\omega T = 2\pi$. Now, the power P delivered to the motor which is representative by a load with resistance $R = 10 \ \Omega$ is $P = I_{\text{rms}}^2 R$ or $P = V_{\text{rms}}^2 / R$. Since the voltage but not the current is specified, we can use Ohm's law, $I_p = V_p / R$, and express power as

$$
P = I_{\text{rms}}^2 R = \frac{(V_p/2)^2}{R} \left(1 - \frac{\alpha}{\pi} + \frac{\sin 2\alpha}{2\pi}\right)
$$

The power delivered to the motor, corresponding to Fig. 3.11a (when $\alpha = 0$), is $P = [(117\sqrt{2}V)/2]^2/10 \ \Omega = 684.45$ W. The power delivered to the motor, corresponding to Fig. 3.11b (when $\alpha = \pi/2$), is $P = [(117\sqrt{2}V)/2]^2/10 \ \Omega \left(1 - \frac{\alpha}{\pi}\right) = 6844.5/10 \ \Omega(1/2) = 342.23$ W. Finally, the power delivered to the motor, corresponding to Fig. 3.11c (when $\alpha \approx \pi$), is $P \approx 0$ W. Note that we are discussing starting power in this example. Once the motor starts turning, conditions will change.

In this example, we started with the calculation of the effective current I_{rms}, because the load current was explicitly shown in Fig. 3.11. We could have started as well with voltage and obtained the same results because the load voltage mimics the load current in the SCR circuit of Fig. 3.12c. This can be seen as follows: when the SCR is turned off and no current flows through it, it is an open circuit, and hence the line voltage develops across the open SCR and the voltage across the motor is zero—just as the current is. When current flows through the motor, that is, when the SCR conducts, the voltage drop across the conducting SCR is about 1 V—hence most of the line voltage is dropped across the motor. Of course the mimicking between current and voltage in a resistive load was implied when we used Ohm's law, $I_p = V_p/R$, in the above calculations.   ■

3.6   SUMMARY

- A diode was portrayed as a kind of on–off switch. Such a model has obvious limitations: an on–off switch passes current in either direction and a diode does not. In this idealized picture, a forward-biased diode conducted current (the switch was on) whereas a reverse-biased diode did not (the switch was off). A more realistic model for a diode included a forward voltage drop of 0.6–0.7 V (sometimes referred to as an offset voltage or contact potential) which stayed constant while the diode was conducting. The remarkable characteristic of a semiconductor diode is the speed with which it can switch from conducting to nonconducting. Switching times can be nanoseconds or less.

- Electronic equipment needs a constant voltage to function properly, with a battery as an ideal power supply. One of the most important uses of a diode is in rectification which converts AC to DC. All electronic equipment has a power supply which performs this function in addition to filtering the rectified voltage so a smooth and steady voltage is produced. Rectifiers can be of the half-wave or the full-wave type, with the full-wave type being more efficient. Zener voltage regulation can be used in critical parts of a circuit that need to be kept at a constant voltage even if the DC voltage of the power supply varies in response to power line voltage variations.

- Diodes are also used in waveshaping circuits, for example, when the DC level of a signal needs to be changed or in clipping circuitry when tops or bottoms of a signal need to be eliminated.

- Finally we introduce a useful modification of a diode in the form of a silicon-controlled rectifier. An SCR is a diode with a gate added, which when an appropriate signal is applied to the gate, can control the onset of the rectified current and thus can control the amount of power reaching a load. Popular use is in DC motor speed control and in light dimmers. Industrial SCRs have current ratings in the thousands of amperes and are able to control large amounts of power.

Problems

1. Calculate the DC output current from the half-wave rectifier circuit shown in Fig. 3.2a. The rms input voltage is 120 V_{AC} and $R_L = 150\ \Omega$.

2. A half-wave rectifier diode, which has an internal resistance of 20 Ω while conducting, is to supply power to a 1 kΩ load from a 110 V_{AC} (rms) source. Calculate

   (a) The peak current.
   (b) The DC load current.
   (c) The rms load current.
   (d) The total input power.

   Ans: (a) 152.5 mA, (b) 48.5 mA, (c) 76.2 mA, (d) 5.92 W.

3. Calculate the DC output current from the full-wave rectifier circuit shown in Fig. 3.3a. The rms input voltage is 120 V_{AC} and $R_L = 150\ \Omega$.
   Ans: 0.72 A.

4. Repeat Problem using four 20 Ω diodes in a full-wave bridge rectifier circuit.

5. Design a half-wave rectifier with capacitor filter to supply 40 V_{DC} to a load of $R_L = 2$ kΩ. Assume the rectifier is connected to a 120 V_{AC}, 60 Hz line by a transformer. Find the turns ratio of the transformer and the capacitance if the ripple voltage is not to exceed 1% of the DC voltage.
   Ans: $C = 833$ mF, $N_1/N_2 = 4.24$.

6. Referring to Example 3.2, redraw Fig. 3.4b for a full-wave rectifier. Sketch the output voltage v_o (identical to load voltage) and the diode charging current i_d for at least two periods of the input voltage.

7. Determine the peak voltage of the full-wave bridge rectifier, shown in Fig. 3.3a, assuming the diodes are ideal but have a forward voltage drop (offset voltage) of 0.7 V. The input voltage is 120 V_{AC}.
   Ans: 168.5 V.

8. A full-wave bridge rectifier with a capacitor filter supplies 100 V_{DC} with a ripple voltage of 2% to a load resistor $R_L = 1.5$ kΩ. If one of the diodes burns out (open circuit), calculate the new DC voltage and the ripple voltage that will be supplied to R_L.

9. The voltage-doubler circuit shown in Fig. 3.5 has a load resistance $R_L = 1$ kΩ connected to it. If the input voltage is 120 V_{AC}, 60 Hz, and if the voltage drop between charging pulses is to be less than 10% of the DC output voltage, specify the smallest value of capacitance needed for the capacitors.
   Ans: 41.7 mF.

10. Sketch the transfer characteristic and the output voltage v_o for the clipper circuit of Fig. 3.6a. Assume the input voltage is $v = 10 \sin \omega t$ and $V_1 = 3$ V and $V_2 = 0$.

11. If $V = 5$ V in the clamping circuit of Fig. 3.9b, draw the output voltage v_o if the input voltage is $v = 2 \sin \omega t$.
   Ans: The minimum point of the input voltage v will be clamped at -5 V.

12. Design a circuit that will clamp the maximum of any periodic voltage $at = 4$ V.

13. Design a simple modification to the battery backup circuit of Fig. 3.7 so that when the power supply is on, the battery is charged at a current of 5 mA.

14. The voltage regulator shown in Fig. 3.10a, in which the Zener diode has a breakdown voltage of 20 V, is used to regulate the voltage across the load resistor R_L at 20 V. The input voltage is $v = 30$ V and $R_L = 1000\ \Omega$. For R_s you have available two resistors (100 Ω and 1000 Ω). Which resistor would you choose? Justify by calculations.

    Ans: 100 Ω.

15. The Zener diode shown in Fig. 3.10a has a voltage breakdown of 100 V with a maximum rated current of 20 mA. If the supply voltage is 150 V, find the range of load resistance R_L over which the circuit is useful in maintaining R_L at 100 V if $R_s = 1.5\ k\Omega$.

16. In the SCR control circuit for a motor shown in Fig. 3.12c, we are given that $V_p = 100$ V, $R_L = 20\ \Omega$, and the conduction angle $\chi = 120°$, where $\chi = \pi - \alpha$. Calculate the average load current and the average load power.

    Ans: 1.19 A, 100.6 W.

17. Using the parameters of Problem 16, determine the power dissipated in the SCR if the on-voltage across the SCR is 1.5 V.

CHAPTER 4

Semiconductor Diodes and Transistors

4.1 INTRODUCTION

In the previous chapter we made use of diodes without explaining the underlying physics. This was acceptable as for most applications diode characteristics are well approximated by those of a simple on–off switch. However, in order to understand the extremely fast diode switching speeds (faster than a nanosecond) or the 0.6–0.7 offset voltage (also referred to as *contact potential*) we have to understand and view the diode as a *pn* junction, which in turn requires an elementary understanding of electron and hole motion in semiconducting material. Furthermore, a transistor can be modeled by two diodes back-to-back, provided we treat each diode as a *pn*-junction. An understanding of the *pn*-junction which we can simply define as a junction between *n*-type and *p*-type semiconducting material[1] is thus necessary to an understanding of diodes and transistors.

4.2 HOLE AND ELECTRON CONDUCTION IN SEMICONDUCTORS

4.2.1 Intrinsic Semiconductors

Germanium and silicon, which are Group IV atoms in the periodic table, have 4 valence electrons. If each atom could share 4 more electrons from the adjacent atoms, the outer shell would be completed (which is 8 electrons), giving the atom more stability. Figure 4.1*a* shows a two-dimensional model of the crystal lattice for silicon. The atoms in this highly ordered structure are held together by covalent bonds (shared electrons).

---

[1] In the next section we will show that *p*-type material is a semiconductor such as silicon (Group IV in the periodic table) that is doped with Group III atoms, which makes it a good conductor with positive (hence the *p*) charge carriers called holes. *n*-type semiconductor is silicon doped with Group V atoms, which makes it a good conductor with negative (*n*) charge carriers which are electrons.

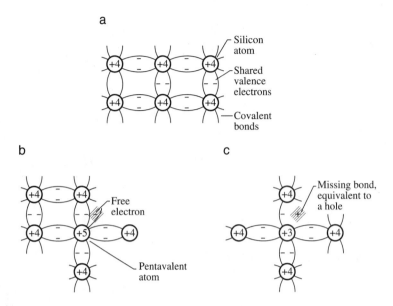

FIGURE 4.1   (a) The highly ordered, crystalline structure of the semiconductor silicon, showing silicon atoms held together by covalent bonds. (b) n-type doping creates free electrons. (c) p-type doping creates free holes.

Since there are no free electrons available one would expect intrinsic semiconductors to be poor conductors, which they are at low temperatures. As the temperature increases, the atoms in the lattice structure begin to vibrate more about their equilibrium positions, which begins to break some covalent bonds and liberate electrons.[2] At room temperature, silicon can almost be classified as a conductor,[3] albeit a poor one. For example, the intrinsic concentration of free electrons, at room temperature, in silicon is $n_i = 1.5 \cdot 10^{16}$ electrons/m^3 (with an equivalent number of holes present as well). The n_i concentration is very small when compared to the density of atoms in silicon, which is $5 \cdot 10^{28}$ atoms/m^3. As the temperature increases, more electrons are freed and silicon becomes a better conductor. This very property leads to failure of semiconductor devices if temperature rises are not controlled by heat sinks or other means (with increased temperature, current and therefore $I^2 R$ losses increase, which in turn leads to further increases of temperature).

It is interesting to observe that conduction is by electrons and by positively charged carriers, called holes, which are created when a bond is broken and an electron is freed

---

[2]Temperature T is a measure of thermal energy W. According to Boltzmann's law, the "jiggling" energy of the atoms, which are locked into place in the crystal lattice of silicon, but can vibrate about their equilibrium positions, is given by $W = kT$, where k is Boltzmann's constant and is equal to $k = 1.38 \cdot 10^{-23}$ joules/kelvin.

[3]Hence the name *semiconductor*. Conductivity of silicon is somewhere between that of a good conductor and that of a good insulator (conductivity σ of Si is $4 \cdot 10^{-4}$ S/m, whereas copper has $\sigma = 5.7 \cdot 10^7$ S/m, and a good insulator such as porcelain has $\sigma = 2 \cdot 10^{-13}$ S/m).

(commonly referred to as *production of a hole–electron pair*). This leaves behind a vacancy or a hole. Another electron from some adjacent broken bond can jump into the hole and fill the vacancy, leaving a hole somewhere else (commonly referred to as *elimination of a hole–electron pair by recombination*). The hole therefore travels and acts as a positively charged particle with an equivalent mass and velocity. We noted in the previous paragraph the electron concentration n_i. An equivalent concentration of holes exists in silicon at room temperatures, i.e., $n_i = p_i$, where p_i is the intrinsic hole concentration.

To put silicon conduction on a quantitative basis, let us restate Ohm's law (1.7), valid at any point inside a material. To obtain the point relation, we must use densities such as electric field E and current density J instead of voltage V and current I. If we use resistance $R = \rho(\ell/A)$, given by (1.6), in Ohm's law, we obtain $V = RI = \rho(\ell/A)I$. Rearranging, we obtain $V/\ell = \rho(I/A)$, which can be expressed as $E = \rho J$, where electric field is in volts per meter and current density is in amperes per meter squared. This results in the commonly used point expression

$$J = \sigma E \tag{4.1}$$

where conductivity σ is the inverse of resistivity, i.e., $\sigma = 1/\rho$. The conductivity in semiconductors can now be expressed in terms of the customary quantities, which are the carrier density n, the charge of the carriers (electronic charge $e = 1.6 \cdot 10^{-19}$ coulombs (C)), and the carrier mobility μ, which gives

$$J = e(n_i \mu_n + p_i \mu_p)E \tag{4.2}$$

The measured mobilities for silicon at room temperature are $\mu_n = 0.135$ m^2/V·s for electrons and $\mu_p = 0.048$ for holes. We can see that holes are only one-third as mobile in silicon as are electrons, which is due to the heavier equivalent mass of holes.

EXAMPLE 4.1 Determine the conductivity for intrinsic silicon (Si) at room temperature (300 K).

Since for intrinsic semiconductors, $n_i = p_i = 1.5 \cdot 10^{16}$, we have that $\sigma = en_i(\mu_n + \mu_p) = 1.6 \cdot 10^{-19} \cdot 1.5 \cdot 10^{16}(0.135 + 0.048) = 4.4 \cdot 10^{-4}$ S/m. ■

Hole travel in silicon is depicted in Fig. 4.2. Imagine a piece of silicon between two conducting plates which are charged by a battery of voltage V. An electric field thus exists between the plates and has a direction from right to left. Suppose an electron in silicon near the positive plate is freed. It will jump to the positive plate and leave a hole behind which is filled by an electron jumping in from a neighboring broken bond and so on. The result is that electrons move to the right (toward the positive plate) while the hole moves to the left (toward the negative plate). A current flows and is maintained as long as the battery has sufficient energy to maintain the potential V across the plates.

4.2.2 Extrinsic Semiconductors

The ability to vary the conductivity of semiconducting material over a large range leads directly to many useful devices, including the diode and transistor. One way to increase

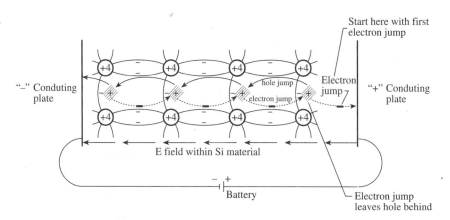

FIGURE 4.2   Conduction by holes and electrons in silicon.

the conductivity of an intrinsic semiconductor would be to heat it—neither practical nor desirable. However, there is a better way. The conductivity of a semiconductor can be substantially increased by adding some (typically 1 in 10 million) impurity atoms (called *dopents*) to the pure crystal structure. As the characteristics now depend strongly on the impurity content, we call it an extrinsic semiconductor as opposed to an intrinsic semiconductor, whose performance characteristics are dependent on a pure crystal. It may be surprising to find that even moderate doping decreases the resistivity by many orders.

4.2.3   *n*-Type Semiconductors

The atoms in silicon are arranged in a regular and highly ordered lattice. They are held in fixed positions by strong forces and only have limited movement (which increases with temperature) about their equilibrium positions. If we now replace some of the bound silicon atoms by atoms from Group V of the periodic table (phosphorus, arsenic, antimony—also known as *donor impurities*) which have a valence of 5, only four of the valence electrons will be used to complete the covalent bond with the nearby silicon atoms. The remaining electron, which is very loosely bound to the atom, becomes a free electron and is available for conduction. *n*-type doping is shown in Fig. 4.1*b*. During conduction, i.e., when current flows, the extra electron leaves the pentavalent atom, leaving behind a positively charged ion. Another electron from a nearby impurity atom will then jump in and neutralize it again (note that the semiconductor must be overall and pointwise neutral—otherwise strong electrostatic forces would exist which would destroy the semiconductor material). The continual jumping of the extra electrons, which are referred to as majority carriers, is the primary conduction mechanism in *n*-type doped semiconductors.

A small number of free holes also exist in *n*-type semiconductors and are called minority carriers. They are created when bonds are broken by thermal agitation of the lattice atoms, similar to electron–hole creation in intrinsic semiconductors. Their contribution to current flow is insignificant in comparison to that of the majority carriers.

4.2.3.1   *p*-Type Semiconductors

Replacing some Si atoms with atoms from Group III (boron, gallium, indium—also known as *acceptor impurities*), which have three valence electrons, will create a *p*-type material. Because four valence electrons are required to form and complete all adjacent electron-pair bonds, a hole is created by the missing bond. A hole can be considered a positive charge which can diffuse or drift through a crystal. This becomes important when an external voltage is applied across the semiconductor, creating an electric field inside the semiconductor which then acts on the holes, causing them to move. The resultant current in *p*-type material is thus primarily by positive charges,[4] which are also referred to as majority carriers. *p*-type doping is illustrated in Fig. 4.1*c*.

A small number of free electrons also exist in *p*-type semiconductors and are called minority carriers. They contribute insignificantly to current.

4.2.4   Conduction in Doped Semiconductors

Typical impurity concentrations are $N = 10^{22}$ donor or acceptor atoms/m^3, which is seen to be much higher than the intrinsic concentration n_i or p_i at room temperature ($n_i = p_i = 1.5 \cdot 10^{16}$ carriers/m^3). Since the impurity atoms provide free carriers, the total number of free carriers in doped semiconductors is $n = N_d + n_i \approx N_d$ in donor materials and $p = N_a + p_i \approx N_a$ in acceptor materials. An important relationship for doped semiconductors is $np = n_i^2$; that is, the product of electrons and holes in doped silicon is equal to electrons squared (or holes squared) in pure silicon. The implication of this relationship (which we are not going to derive as it involves Boltzmann statistics, Fermi levels, and so on) is that increasing the majority carriers by increasing the doping level will decrease the minority carriers proportionally. Hence for a doping level of 10^{22}, the minority concentration reduces to $(1.5 \cdot 10^{16})2/10^{22} = 2.25 \cdot 10^{10}$, which is substantially less than the intrinsic level of carriers n_i. We conclude that in doped silicon, conduction is primarily by the impurity carriers. Hence the conductivity for *n*-type semiconductors is

$$\sigma = e(n\mu_n + p\mu_p) \approx eN_d\mu_n \qquad (4.3)$$

and for *p*-type semiconductors it is

$$\sigma = e(n\mu_n + p\mu_p) \approx eN_a\mu_p \qquad (4.4)$$

---

[4]As pointed out before, hole motion takes place when a neighboring election jumps in to fill an existing hole, which in turn creates a new hole. As this process repeats, a hole moves in the direction of the electric field and toward the negative end of the semiconducting material. The main point is that there is a current which is due to motion of positive charge carriers. In that sense Benjamin Franklin is vindicated: current flow is by positive charges, which he postulated not knowing at that time that electrons are the charge carriers in metallic conductors.

EXAMPLE 4.2 (a) Find the conductance of arsenic- and indium-doped silicon if the doping level is 10^{22} atoms/m^3. (b) Find the resistance of a cube of the above material if the cube measures 1 mm on a side.

(a) Arsenic results in n-type material with a conductivity from (4.3) as $\sigma = eN_d\mu_n = (1.6 \cdot 10^{-19})(10^{22})(0.135) = 216$ S/m. Indium results in p-type material with a conductivity from (4.4) as $\sigma = eN_a\mu_p = (1.6 \cdot 10^{-19})(10^{22})(0.048) = 76.8$ S/m. We see that the doped semiconductor has a conductivity that is larger by a factor of almost a million when compared to the conductivity of intrinsic silicon ($\sigma_i = 4.4 \cdot 10^{-4}$ from Example 4.1).

(b) Ohm's law gives resistance as $R = \rho(\ell/A)$, where ρ is resistivity ($= 1/\text{conductivity}$), ℓ is the length of a material along which the current flows, and A is the cross-sectional area through which the current flows. Hence for the n-type material, $R = (1/216 \text{ S/m})$ $(0.001 \text{ m}/(0.001 \text{ m})^2) = (0.0046 \text{ }\Omega\cdot\text{m})(1000 \text{ m}^{-1}) = 4.6 \text{ }\Omega$. For p-type material we obtain $R = 12.9 \text{ }\Omega$. We see that n-type material is a better conductor than p-type, which is a direct consequence of the higher mobility of electrons than of holes. ∎

In conclusion we can state that even a small concentration of impurities (a doping level of 10^{22} atoms/m^3 is still very small in comparison to the density of atoms in silicon, which is $5 \cdot 10^{28}$ atoms/m^3) can dramatically increase the conductivity of a semiconductor. The fact that electrons move about the crystal faster than holes by a factor of $0.135/0.048 = 2.8$ makes n-type material more desirable in high-speed devices, which we will discuss further when considering n-channel FETs and npn transistors.

4.3 *PN*-JUNCTION AND THE DIODE-JUNCTION AND THE DIODE

Consider a rod of silicon in which the doping during manufacture is suddenly changed to make half the rod p-type and the other half n-type, as shown in Fig. 4.3a. We now have a pn-junction in the middle of the rod. Let us examine the charge distribution near the junction.

In Fig. 4.3a we represent the fixed or immovable lattice atoms as ions (circled minuses or pluses), each accompanied by one hole or electron (uncircled pluses or minuses) to preserve charge neutrality. The uncircled quantities (holes and electrons) are the free-charge carriers which, when in motion, constitute an electric current. Near the junction, the charge distribution as shown in Fig. 4.3a is unstable and can exist only for a very brief time during manufacture of the junction. The free charges on opposite sides of the junction will immediately combine[5] with the result that the charge distribution looks like that shown in Fig. 4.3b. If we plot the charge density q_v (C/m^3) along the rod,

---

[5]When opposite charges combine, they annihilate each other with a release of energy. It is as if a small current flowed for a brief time. The energy release can be visible light as it is in LEDs (light emitting diodes), which can emit continuous green, red, or any other colored light from a diode junction, provided a continuous diode current flows which supports a continuous combination of electrons and holes at the junction.

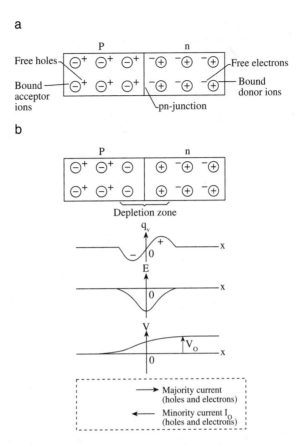

FIGURE 4.3   (a) Unstable free-charge distribution in a new junction. (b) Stable charge distribution in a *pn*-junction with graphs of charge density, electric field, and potential. The bottom graph shows the equality of drift and diffusion current in the junction of an unconnected diode.

we find that in the *p*-region near the junction we have uncovered negative acceptor ions, and across the junction, positive donor ions. These ions are part of the atomic lattice and cannot move. But what they do is prevent further free-charge motion because the holes in the *p*-region now see positive charge across the "border" and are not inclined to move into the *n*-region. Similarly, electrons in the *n*-region are not attracted to the negative charges across the junction. This is the charge distribution in every new, unconnected diode after its manufacture.

   Immediately below the graph for charge density in Fig. 4.3*b*, we plot the electric field *E*, which is obtained from the differential form of Gauss' law, that is, $dE/dx = q_v/\varepsilon$, where the derivative with respect to *x* is along the axis of the silicon rod and ε is a constant (the permittivity of silicon). Hence the charge density variation is proportional to the slope of the electric field, or conversely the *E*-field is proportional to the integration of the charge density ($E \propto \int q_v dx$). The negative *E*-field means the *E* is pointing in the negative x direction (from right to left, or from the positive ions in the n-region to the negative ions in the p-region). Hence, as concluded in the previous paragraph, the E-field in the junction opposes motion of majority carriers but supports motion of minority carriers, of which there are only few (electrons in the p-region and holes in the n-region). It appears that life is getting complicated: we now have four currents in the junction, majority current (referred to also as diffusion current) by holes and electrons, and minority current (referred to as drift current) by holes and electrons. Fortunately, in most practical situations we can ignore drift current as being negligible, although it greatly aids in the understanding of the pn-junction.

Immediately below the E-field graph, we plot also the variation of the potential or voltage field V across the junction. Using (1.3), $E = -dV/dx$, which states that the E-field is equal to the negative rate of change of V, (or conversely, $V = -\int E\,dx$). We observe that the V-field increases when moving from the p-region to the n-region. This only confirms the unwillingness of holes to move from the p-region to a region with a higher positive potential. Similarly for electrons from the n-region, they are repelled by the more negative potential of the p-region. But more importantly, we have now obtained the potential jump V_o across the junction, which is 0.7 V for silicon (0.2 V for germanium). Recall, in the previous chapter, we stated that a diode, under forward bias, has a contact potential or offset voltage of 0.7 V which must be exceeded before the diode will conduct. It should now be clear that this voltage is due to the internal electric field in the depletion zone.[6]

The region near the junction is called a depletion region or a depletion zone since it is depleted of all free carriers. In that sense it is a nonconducting region—a thin insulating sheet between the p- and n-halves of the silicon rod. We will now show that action within this zone is the key point in understanding diode and transistor action: under forward bias the diode conducts as the depletion zone will be flooded with free charge carriers, whereas under reverse bias the diode becomes an open circuit as the depletion zone will be even more devoid of carriers, i.e., the depletion zone will be enlarged. Let us now elaborate.

---

[6]A question can now arise: can we use a pn-junction as a current source? For example, if we connect a resistor or even a short (zero resistance wire) across a diode, will a current flow? The answer is *no*, because diffusion current due to majority holes equals drift current due to minority holes (these two currents are in opposite directions). Similarly, majority diffusion current due to electrons equals minority drift current due to electrons. The result in no net current flow. Again we should note that drift current is a motion (caused by the electric field of the junction) of thermally generated minority charges, whereas diffusion current is a motion of majority charges across the junction because of their large concentrations on both sides (it is similar to two gases in separate volumes diffusing when released into one volume). It is only the thermally more energetic majority carriers that will have a chance to overcome the opposing charge barrier and diffuse across the junction, balancing the thermally generated minority carriers drifting across the junction. Still not satisfied, one might raise the question as to what happens to the junction contact potential V_o when a short is placed across the pn-junction. Think of it this way: instead of the shorting wire, imagine the p-end and the n-end are bent upward until they touch. We have now created a new pn-junction whose contact potential V_o is equal and opposite to that of the original junction. Hence, there is no net potential around the loop, and therefore no current.

4.3.1   Forward Bias

If we connect a battery of voltage V across the *pn*-junction with the positive of the battery on the *p*-side of the junction (p to p) and the negative to the *n*-side (n to n), as shown in Fig. 4.4*a*, the battery will inject holes into the *p*-region and electrons into the *n*-region.[7] This is referred to as forward-biasing a *pn*-junction. The effect will be that the junction is now flooded with carriers. The electrons and holes will diffuse across the junction in response to the lowered junction potential (which is now $V_o - V$) and

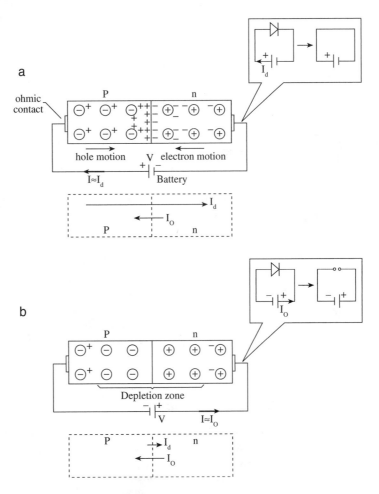

FIGURE 4.4   (a) Forward-biasing a junction will dramatically increase majority carriers and hence majority current. (b) Reverse-biasing depletes the junction completely, leaving only the minority current I_o.

---

[7] Actually the battery will cause an excess of electrons on the *n*-side and a deficiency of electrons on the *p*-side which can be considered as an excess of holes.

recombine.[8] As a result a large majority current (diffusion current) flows in the circuit composed of the *pn*-junction and battery as long as the battery has sufficient energy to maintain this current. Populating the depletion zone with an abundance of carriers changes it to a conducting region; we show this pictorially in the cutout of Fig. 4.4*a* in which a forward-biased diode (first picture) is shown as a short (second picture). In the cutout we used the diode symbol ($-\triangleright\!\!-$) to represent the *pn*-junction.

This is the fundamental mechanism by which a diode, which acts as an on–off switch, can be in the on-mode. It can switch between these two modes very fast—on the order of nanoseconds. Incidentally, the minority current is by-and-large unaffected by any outside voltage V that is placed on the *pn*-junction—the few minority carriers that are generated by thermal energy are subject to the internal junction potential (reduced or increased by V) and drift across the junction.

4.3.2 Reverse Bias

A battery connected such that the plus of the battery goes to the *n*-side and the minus goes to the *p*-side of a *pn*-junction, as shown in Fig. 4.4*b*, will cause electrons and holes to be repelled further from the junction, greatly increasing the depletion zone. A reverse bias increases the contact potential to $V_o + V$ at the junction, thus increasing the barrier height for majority carriers. The effect is to introduce an insulating region between the *p*- and *n*-sides through which no current can flow. After all, what is an insulator? It is a region devoid of free charge carriers. We refer to this in the cutout of Fig. 4.4*b* where a reverse-biased diode is shown to act as an open circuit.

The *pn*-junction would be an almost ideal diode, i.e., a voltage-controlled on–off switch, were it not for the small minority drift current which is present under forward or reverse bias. Hence, under reverse bias, the diode is not an open circuit, but the small drift current, usually referred as *reverse saturation current* I_o, which is the only current present under reverse bias, gives the diode a finite but large resistance (many megaohms). The drift current is small, typically 10^{-12}A for Si and 10^{-6}A for Ge; this fact alone has made silicon the preferred material over germanium for diodes and transistors.

4.3.3 Rectifier Equation

We are now ready to derive a quantitative relationship for current in a *pn*-junction. This relationship is better known as the *diode equation*. For reverse bias, a very small drift current, the reverse saturation current I_o, flows across the junction, as the majority diffusion current is blocked by the reverse bias. According to Boltzmann's law, the reverse saturation current is given by $I_o = K \exp(-eV_o/kT)$, where K is a constant depending on junction geometry and k is *Boltzmann's constant* ($k = 1.38 \cdot 10^{-23}$ J/K). However,

---

[8]Solid-state physics tells us that at the junction V cannot exceed V_o (hence junction voltage $V_o - V$ remains positive) even if the externally applied V is larger than V_o.

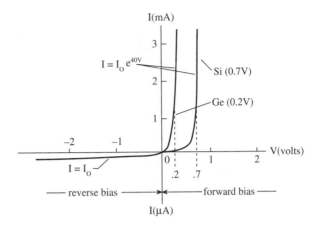

FIGURE 4.5   The *I-V* characteristics of a *pn*-junction (diode current vs applied voltage).

for a forward bias voltage V, in addition to I_o, the diffusion current is present in great strength, and according to Boltzmann's law is again given by $I_d = K \exp(-e(V_o - V)/kT)$. The diffusion and drift currents are opposite in direction, so that the total junction current is

$$I = I_d - I_o = K e^{-eV_o/kT}(e^{eV/kT} - 1) = I_o(e^{eV/kT} - 1) \qquad (4.5)$$

The total junction current I is the sum of hole and electron currents, i.e., $I = I_h + I_e$, where the expressions for I_h and I_e are the same as (4.5). For example, hole current is the difference between hole diffusion and hole drift current, that is, $I_h = I_{h,d} - I_{h,o}$, and similarly for electron current, $I_e = I_{e,d} - I_{e,o}$.

We can check the above equation at room temperature ($T = 68°F = 20°C = 293$ K), when $e/kT = 40$ V^{-1}. Equation (4.5) then simplifies to

$$I = I_o(e^{40V} - 1) \qquad (4.6)$$

For example, without an external potential applied, i.e., $V = 0$, we find that $I = 0$, as expected. For reverse bias, when $V < 0$, (4.6) reduces to $I \approx -I_o$, as the exponential term is much smaller than unity (even for small voltages such as $V = -0.1$, the exponential term is $e^{-4} = 0.02 \ll 1$). For forward bias, on the other hand, $V > 0$, the exponential term dominates, such that $I \approx I_o e^{40V}$ (even for a small voltage $V = 0.1$, $e^4 = 55 \gg 1$). Figure 4.5 shows a graph of the rectifier equation.

In Fig. 4.5 we do not show the breakdown region of a diode which occurs when the reverse bias voltage exceeds the specified maximum voltage, which for popular diodes such as 1N002, 1N004, and 1N007 is 100 V, 400 V, and 1000 V, respectively. For a graph including breakdown, refer to Fig. 3.1*b*.

A diode is thus not an ideal voltage-controlled on–off switch, but rather a device with much more current flow in one direction than in the other. A convenient way to check the state of a diode is to measure its forward- and reverse-biased resistance. Using an analog ohm-meter (which measures resistance by placing a small voltage across a device to be measured, reads the resultant current, and obtains resistance as the ratio of applied voltage and resultant current), we connect the leads of the ohm-meter to the diode and read the resistance. The leads are then reversed and the resistance read again. Typically, we will find for a small diode (1N4004) a forward resistance of several ohms and a backward resistance of many megaohms.

EXAMPLE 4.3   If the reverse saturation current I_o for a silicon diode at room temperature is 10^{-12}A, (a) find the current at biasing voltages of $V = -0.1, 0.1$, and 0.5 V. (b) Should the temperature of the diode rise by 30°C, find the new currents for the same biasing voltages.

(a) At -0.1 V, (4.6) gives $I = I_o(e^{-4} - 1) \approx -I_o = -10^{-12}$ A. At 0.1 V, (4.6) gives $I = I_o(e^4 - 1) \approx 55I_o = 55$ pA. At 0.5 V, (4.6) gives $I = I_o(e^{20} - 1) = I_o 4.85 \cdot 10^9 = 0.49$ mA.

(b) If the temperature rises by 30°C, the factor e/kT changes to $e/kT (293/(293 + 30)) = 40(293/323) = 36.3$. Hence I_o at the new temperature, using $I_o = K \exp(-eV_o /kT)$, gives $I_o = 10^{-12} \exp V_o(40 - 36.3) = 10^{-12} \exp 2.59 = 13.3 \cdot 10^{-12}$A, where $V_o = 0.7$ was used for the contact potential. Using this formula tells us that the reverse saturation current increases by a factor of 13 when the temperature increases by 30°C. We should now note that the use of the above formula for calculating changes in I_o when the temperature change is questionable. A widely accepted formula, based on extensive experimental data, is that I_o doubles for every 10° degrees increase in temperature. Therefore, a more accurate estimate of the new reverse saturation current would be $I_o = 10^{-12} \cdot 8 = 8$ pA, since for a 30°C rise the saturation current doubles three times, that is, the saturation current increases by a factor of 8.

Using this, we calculate the forward current at $V = 0.1$ V and at 50°C to be $I = I_o(e^{3.63} - 1) = 8 \cdot (37.7 - 1) = 294$ pA, which is seen to be a large increase from 55 pA at room temperature.

Similarly at $V = 0.5$ V, we obtain $I = I_o(e^{36.3 \cdot 0.5} - 1) = I_o e^{18.15} = 0.61$ mA, again an increase from 0.49 mA at room temperature.   ■

4.4   *PN*-JUNCTION AND THE TRANSISTOR

4.4.1   The Bipolar Junction Transistor (BJT)

Now that we understand how a *pn*-junction operates, mastering transistors should be straightforward. Figure 4.6 shows a *npn* and a *pnp* transistor, the majority current flow in each, and the circuit symbol for each transistor. We will refer to this transistor as a *bipolar junction transistor* (BJT)—bipolar because holes and electrons are involved in its operation (although for the most part we will ignore the contribution of the small minority current). Furthermore, the input region is referred to as the emitter, the center region as the base, and the output region as the collector. In this type of transistor, we have two junctions with the input junction always forward-biased and the output

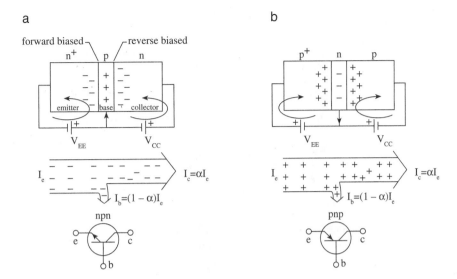

FIGURE 4.6 (a) A *npn* transistor formed by two *pn*-junctions back-to-back with biasing batteries connected (top figure). The middle figure shows the flow of majority carriers and the bottom figure shows the circuit symbol of a transistor. (b) Similar figures for a *pnp* transistor.

junction always reverse-biased. The polarities of the batteries ensure correct biasing: for forward bias in the *npn* transistor, the battery V_{EE} has its negative terminal connected to the *n*-type emitter (*n* to *n*) and its positive terminal to the *p*-type base (*p* to *p*), whereas for the reverse-biased output junction, battery V_{CC} is connected *p* to *n* and *n* to *p*.

The bottom sketch in Fig. 4.6 represents the circuit symbol of a transistor. To differentiate between a *npn* and *pnp* transistor, we draw the arrow in the direction that a forward-biased current would flow. At this time it might be appropriate to remind the student that thanks to Benjamin Franklin, the accepted tradition is that the direction of current is that of positive charge flow. The pictures for the *pnp* transistor, where holes are majority carriers, are therefore less complicated. For the *npn* transistor, on the other hand, the arrows for current and electron flow are opposite.

What basic principles are involved in transistor operation? First let us state that the emitter is expected to provide the charge carriers for transistor current; hence the emitter region is heavily doped (heavier doped regions are labeled with a + sign: n^+ in the *npn* transistor and p^+ in the *pnp* transistor). For good transistor operation, the current I_e injected into the base by the forward-biased input junction should reach the collector almost undiminished.[9] To make sure that only a few charge carriers that arrive

---

[9] If this is the case ($l_c = l_e$), there can be no current amplification, but we have the possibility for voltage and power amplification as the same current that flows through the low-resistance, forward-biased input (emitter) junction, also flows through the high-resistance output (collector) junction. This output current, therefore, represents a greater amount of power than the controlling signal. In this way the transistor is capable of amplifying.

in the base region recombine with the oppositely charged carriers in that region, the base region is lightly doped and made very thin. This is depicted by the thin downward arrow in the middle sketch of Fig. 4.6, which represents the small base current I_b due to recombination. We can now summarize differences in diodes and transistors by stating:

- *Diode*: large recombination at forward-biased junction.

- *Transistor*: small recombination at forward-biased emitter junction because center region is thin ($\approx 10^{-6}$m) and lightly doped—hence most majority carriers from the emitter do not recombine in the base region but proceed to the collector region where they are once more majority carriers.

The Grounded-Base Transistor

The majority current—electrons in the *npn* transistor and holes in the *pnp* transistor— "arrives" in the collector region as I_c. The efficiency of charge transport from emitter to collector is given by the factor α, such that

$$I_c = \alpha I_e \tag{4.7}$$

Figure 4.7*a* shows a circuit to measure the collector current of a transistor in a grounded-base configuration (like that shown in Fig. 4.6). We see that varying V_{cb} has very little effect on I_c; the collector curves are straight and uniformly spaced with little new information—essentially confirming that α is near unity ($\alpha \approx 0.99$, i.e., collector current is nearly equal to emitter current). The grounded-base transistor with emitter/collector as input/output terminals does not amplify current and is used only occasionally in special situations. Similarly, a transistor with its collector grounded has special properties that are occasionally useful.

The Grounded-Emitter Transistor

In electronic circuits, the most widely used configuration is the grounded emitter, shown in Fig. 4.7*b*. Since I_b is now the input current, current amplification is now possible (note that typically $I_b \ll I_e$). For this case, the current gain β is a more useful parameter than α. It is obtained by summing the currents into the transistor, $I_e = I_b + I_c$, and then substituting for I_e using (4.7), which gives $I_c/\alpha = I_b + I_c$ and finally $I_c = (\alpha/(1 - \alpha))I_b$. Thus, the desired relationships are

$$I_c = \beta I_b \tag{4.8}$$

and

$$\beta = \frac{\alpha}{1 - \alpha} \tag{4.9}$$

where for most transistors β has a value between 50 and 1000. With β known or read off of the collector characteristics, (4.8) is a frequently used expression in electronics.

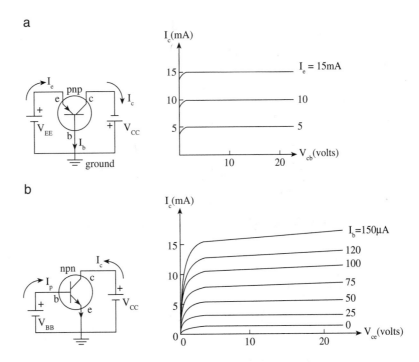

FIGURE 4.7   (a) Typical collector characteristics of a *pnp* transistor in the grounded-base configuration. (b) Typical collector characteristics of a *npn* transistor in the grounded-emitter configuration.

Figure 4.7*b* shows typical collector characteristic curves. We can see that once V_{ce} exceeds about 1 V, the collector curves become straight and almost independent of V_{ce}, that is, further increases in V_{ce} have little effect on the collector current I_c, which remains constant for a given I_b. The straight, horizontal curves suggest that the transistor acts as a constant current source. The output current I_c (in mA) can be varied by the much smaller input current I_b (in μA). Hence, not only do we have a possibility of large current amplification (I_c/I_b), but the base current I_b controls the output current I_c—i.e., when I_b increases, I_c increases and similarly when I_b decreases. Therefore, the grounded-emitter transistor acts as a current-controlled current source.

Finally, we should keep in mind that the voltage drop across the forward-biased input junction must be at least $V_{be} = 0.7$ V for a silicon transistor to be on and the biasing battery V_{BB} must be able to develop this voltage. Once the base–emitter junction exceeds 0.7 V by even a small amount, base current I_b increases rapidly as shown by the forward-biased part of Fig. 4.5 (to apply this figure to this situation, assume the vertical axis is I_b, and the horizontal axis is V_{be}). Thus, when the transistor is used as an amplifier and the input current varies over a typical range (10 to 1), the base–emitter voltage varies only little about 0.7 V. These concepts will become clearer when we consider transistor amplifiers.

There is no significant difference in the operation of *npn* and *pnp* transistors, except for the change in battery polarities and the interchange of the words "positive" and "negative," and "holes" and "electrons." In practice, however, *npn* transistors dominate as electrons respond faster to signals than the heavier holes which are the majority carriers in *pnp* transistors. For both types of transistors, we have ignored the contribution of the small minority current as the reverse saturation current I_o is only a small part of the collector current I_c.

EXAMPLE 4.4   Using the collector characteristics of the grounded-emitter transistor shown in Fig. 4.7*b*, determine the current gain β for this transistor.

Using the upper region of the graphs, we calculate the current gain as $\beta = \Delta I_c / \Delta I_b = (15.5 - 13)\text{mA}/(150 - 120)\mu\text{A} = 83.3$. If we use the lower region we obtain $\beta = (3 - 1.5)\text{mA}/(25 - 0)\mu\text{A} = 60$, showing that the transistor is not as linear as it appears from the graphs. If in some electronic circuits, transistor operation is confined to a small region of the graphs, knowing the appropriate β can be important.   ■

4.4.2   The Field Effect Transistor (FET)

A second class of transistors exists, known as *field effect transistors*. Even though they are simpler in concept, they were invented after the bipolar transistor. Unlike the BJT, which is a current amplifier (with significant current flow in the input loop), the FET is basically a voltage amplifier, i.e., the controlling parameter at the input is a voltage (with practically no current flow in the input loop). In that sense, the FET is analogous to the vacuum tubes of olden days, which also were voltage amplifiers.

Figure 4.8*a* shows a cross section of a basic FET. We start with a rod of *n*-type material and surround it by a ring of *p*-type material. The ends of the rod are referred to as drain and source, and the ring as a gate. If a battery V_{DD} is connected across the ends of the rod, current I_d will flow, because the rod acts as resistor of resistance $R = \rho\ell/A$, where ρ, ℓ, and A are the resistivity, length, and cross-sectional area of the rod. Now, unlike in the BJT, we reverse-bias the input junction by connecting a biasing battery V_{GG} as shown (*n* to *p* and *p* to *n*). The reverse-biased junction will create a nonconducting depletion region between the *n*-channel and the *p*-type ring, effectively decreasing the width of the channel, i.e., decreasing the cross-sectional area A through which current I_d can flow. As a matter of fact if the negative gate voltage V_{gs} is sufficiently large it can completely pinch off the channel, so no drain current flows; this voltage is appropriately known as the gate cutoff voltage $V_{gs(\text{off})}$ or pinch-off voltage V_p (approximately -5 V in Fig. 4.8*d*). A signal voltage applied to the reverse-biased input junction is therefore very effective in controlling the flow of majority carriers through the channel. We conclude that a FET acts as a variable source–drain resistor that controls the output current in sync with the input signal. Furthermore, as the applied input signal does not need to deliver any power to the reverse-biased input circuit (input gate resistance is many megaohms), a large voltage and power gain is possible. Figure 4.8*b* depicts a modern, small-package FET, fabricated by diffusing multiple layers of differently doped silicon. The circuit symbol for a FET is given in Fig. 4.8*c* (opposite

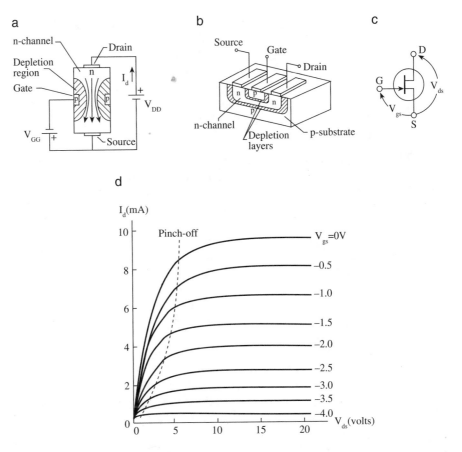

FIGURE 4.8 (a) *n*-channel junction FET. (b) Modern *n*-channel FET. (c) Circuit symbol for a FET. (d) Current–voltage characteristics (drain characteristics) of a typical *n*-channel FET.

arrow for a *p*-channel FET); note that the direction of the arrow is the direction of current flow of a forward-biased junction, which is consistent with the convention adopted previously for the BJT.

Figure 4.8*d* shows the drain characteristics of an *n*-channel FET. Different *I*-*V* curves for V_{gs} ranging from to 0 to −4 V are obtained by varying V_{ds} from 0 to 20 V for each curve. We can use the "knee" in the curves to distinguish two different regions, the ohmic region and the saturation region where the curves become flat and horizontal, signifying that the transistor acts as a voltage-controlled current source (this is the normal operating region for a transistor and is often referred to as the on-region).[10]

---

[10]Note that in the case of a *p*-channel FET, for the *pn*-junction formed by the gate and *p*-channel to be reverse-biased, the gate–source voltage V_{gs} must vary from zero to positive values.

OHMIC REGION ($V_{ds} < 4$ V). This is the region in which the FET acts as a variable resistor and obeys Ohm's law, that is, the I-V curves become less steep as the negative bias on the gate increases, indicating that the resistance of the channel increases. For example, for $V_{gs} = 0$ V, the slope of the curve gives a resistance $\Delta V/\Delta I = 4V/10$ mA $= 400$ Ω. For $V_{gs} = -3$ V, the channel is pinched off more and we obtain $\Delta V/\Delta I = 4V/1.5$ mA $= 2667$ Ω. For small drain–source voltages, the channel current is thus effectively controlled by an applied gate voltage.

SATURATION REGION OR CONSTANT CURRENT REGION (4 V $< V_{ds} < 20$ V). This region is more tricky to explain. First, we must realize that V_{ds} also reverse-biases the junction. For example, with $V_{gs} = 0$ V, that is, with the gate shorted to source, V_{DD}, as shown in Fig. 4.8a, puts the positive battery terminal on the n-channel and the negative battery terminal on the p-type collar (p to n and n to p, implying reverse bias). Next we observe that the voltage which is placed by the battery across the n-channel rod varies from V_{DD} at the drain to 0 V at the source. This being the case, the n-channel near the drain is more reverse-biased than near the source, which explains the nonuniform (more pinched off near the drain) depletion zone within the channel. Looking again at the curve for $V_{gs} = 0$ in Fig. 4.8d, we find that as V_{ds} increases from 0 V, the current I_d increases to about 7 mA as if the channel had a constant resistance; the channel width then decreases until it is almost pinched off at about 5 V. At this voltage the current is about 9 mA, and further increases of V_{ds} up to 20 V increase I_d only little. The reason for current leveling off in the saturation region is that even though voltage increases, the channel is further pinched (resistance increases), keeping the current at approximately 9 mA.

For each of the remaining curves ($V_{gs} = -1, -2, -3$, and -4 V) the current saturates at progressively lower values (about 6.5, 4, 1.5, and 0.5 mA) because each negative gate–source voltage places additional reverse bias on the pn-junction. Thus at $V_{gs} = -4$ V, the channel is already sufficiently pinched that a V_{ds} increase from 0 V to only 1 V pinches the channel off at about 0.5 mA. Further increase of V_{ds} does not increase the saturated current. We conclude that at $V_{gs} = -4$ V and for $V_{ds} > 1$ V, the transistor acts as a constant current source of magnitude $I_d = 0.5$ mA, i.e., a constant current source is turned on at $V_{ds} \approx 1$ V.

In conclusion, we observe that there are two types of totally different pinch-offs, one due to negative gate voltage V_{gs} which can pinch off the channel completely so $I_d = 0$, and the second due to drain–source voltage V_{ds} which limits the current to the value at the "knee" of the I_d-V_{ds} curves; that is, this pinch-off allows a steady current flow at the saturation level in the channel but does not allow further increases beyond that level (this pinch-off voltage is denoted by the dashed curve in Fig. 4.8d). The V_{ds} pinch-off determines the normal operation range (saturation region) for a transistor which is between the V_{ds} pinch-off voltage and 20 V (see Fig. 4.8d); in other words, we can say that at the V_{ds} pinch-off voltage or larger, the transistor is "on." We can now define a pinch-off voltage V_p as the sum of gate voltage and drain voltage, $V_p = V_{gs} - V_{ds}$.

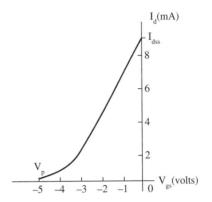

FIGURE 4.9   Transfer characteristic of an *n*-channel FET.

For the *n*-channel FET shown in Fig. 4.8*d*, $V_p \approx -5$ V (note that V_{gs} is negative but V_{ds} is positive for an *n*-channel FET). Drain current in the saturation region can be approximated by

$$I_d = I_{dss}(1 - V_{gs}/V_p)^2 \tag{4.10}$$

where I_{dss} is the saturation current when $V_{gs} = 0$ V (about 9 mA for Fig. 4.8*d*).

For the BJT, an important factor was the current gain β. A similar factor for the FET is the *transconductance* g_m, defined by

$$g_m = \Delta I_d / \Delta V_{gs} \tag{4.11}$$

which is equal to a change in drain current caused by a change in gate voltage at some specified value of V_{ds}. For example, using the central region of Fig. 4.8*d* we obtain

$$g_m = \frac{(6.5 - 5)\text{mA}}{(-1 - (-)1.5)\text{V}} = 3 \cdot 10^{-3} \text{ S}$$

which is a typical value for FETs. The effectiveness of current control by the gate voltage is given by g_m: the larger the g_m, the larger the amplification that a transistor can produce.

4.4.3   Transfer Characteristics

An alternative to drain characteristics for describing the electrical properties of a FET is by transfer characteristics, which are I_d-V_{gs} plots obtained by applying (4.10). They compare variations in drain current to corresponding variations in gate voltage. The slope gives g_m, which is a measure of the transistor's control ability—larger values correspond to larger amplification. Figure 4.9 shows the transfer characteristic of a FET for which the drain characteristics of Fig. 4.8*d* apply. The curve is parabolic, reflecting the square-law nature of (4.10). Note that the transfer curve corresponds to the saturation region (or the on-region) of Fig. 4.8*d*.

4.4.4   Other Types of FETs

There are two different field effect transistors. One is the *junction FET*, usually denoted as JFET, which is the type that we have been considering and which always has a negative gate voltage if the FET is *n*-channel. The other is the *metal-oxide-semiconductor FET* (MOSFET), of which there are two types: *depletion-mode* (DE MOSFET) and *enhancement-mode*. The DE MOSFET is electrically similar to the JFET, except that the gate voltage V_{gs} may be of either polarity. The enhancement-mode MOSFET operates generally with a positive V_{gs} if it is *n*-channel. Besides the flexibility in gate voltage polarity offered by the different FETs, perhaps the greatest advantage of the MOSFET is its practically infinite input resistance (it can be as high as $10^{15}\Omega$), allowing only a trickle of electrons to activate the device, i.e., requiring practically no power from the input signal.

As observed for the BJT where *npn*'s predominate, most FETs in use are *n*-type, as the lighter electrons respond faster than holes, allowing faster switching in digital systems and higher frequency response in analog amplifiers. Furthermore, FETs have an advantage over BJTs in integrated circuits where high element density, low-power requirements, and FETs functioning as capacitors and resistors are important.

4.5   THE TRANSISTOR AS AMPLIFIER

In the introduction to the first chapter, we stated that an amplifier is a device which consists of interconnected transistors, resistors, inductors, and capacitors. Having studied RLC circuits and active elements such as transistors, we are ready to integrate the active and passive elements into an amplifying device. In electronics, the primary purpose of an amplifier is to amplify a signal to useful levels (say 1–10 V), considering that the amplifier input is usually from feeble devices such as transducers, microphones, antennas, etc., which normally produce signals in the micro- or millivolt range. Signals in the μV range, besides needing amplification, are also easily interfered with by noise. Once the signal is in the volt range, it can be considered immune to noise and other disturbing signals and is ideally suited to control other devices such as waveshaping circuits and power amplifiers. In the case of power amplifiers which are able to deliver hundreds or thousands of watts, a large controlling voltage is needed at the input.

If we are treating an amplifier as a fundamental unit, it is not by accident. Once we have an amplifier, we can make many other devices with it. For example, an oscillator is simply an amplifier with feedback.

4.5.1   Elements of an Amplifier

Figure 4.10 shows the three basic elements of an amplifier: an *active element* (transistor), a *resistor*, and a *DC power supply* such as a battery. A signal to be amplified is placed at the control electrode and the amplified output is taken across the series combination of resistor and battery. The battery, which is the source of energy for the amplifier

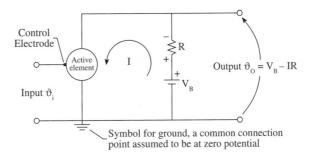

FIGURE 4.10   The components of a basic amplifier are an active element which has a control electrode, a resistor, and a battery.

(and the amplified signal), causes a current I to flow in the output loop. The input voltage which controls current I will add variations to the current I. We have now a varying voltage across the resistor which mimics the varying input voltage—what remains to be shown is that it is an amplified trace of the input voltage. Note that the polarity of the IR voltage drop across the resistor is opposite to that of the battery voltage V_B. Hence, the output voltage v_o, which is the difference between the varying voltage across R and the constant battery voltage, can range only between 0 and V_B volts. A resistor in series with a battery is a combination that is frequently used in electronic circuits. It is a convenient way to develop an output voltage and at the same time supply power to the circuit.

A typical amplifier is made up of several amplifying sections that are needed to produce a large gain that one section alone cannot provide. Figure 4.11 shows the input and output voltages for a two-stage amplifier. We start with a small AC signal at the input and realize a greatly amplified copy of it at the output. Before we can simply

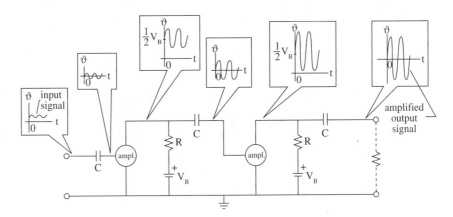

FIGURE 4.11   A small AC signal at the input of a two-stage amplifier. The amplified signal at selected points in the amplifier is shown.

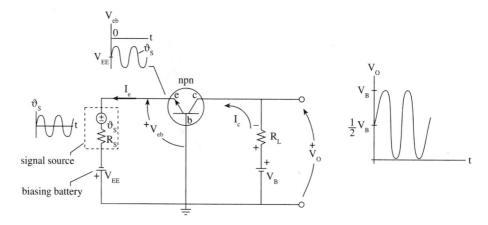

FIGURE 4.12   A grounded-base amplifier, showing a signal voltage at input and amplified output voltage.

cascade two amplifiers of the type shown in Fig. 4.10, we need a DC blocking capacitor C between amplifier stages and at the output, as shown in Fig. 4.11. The purpose of C is to strip off the DC component in a signal containing AC and DC, which is important when only the information-carrying part (AC) is desired at the output of each stage. But another important function of C is to block the large DC supply voltage from reaching the control electrode. For example, in the beginning stages of an amplifier, transistors can be used that are very sensitive, meaning that input voltages can only be in the millivolt range. Since V_B is usually much larger, it could saturate the following stage and keep it from operating properly (it could even damage or destroy it). In that sense, the blocking capacitor acts as a high-pass filter. The disadvantage of using a high-pass filter between stages is that in addition to blocking DC, low frequencies are also discriminated against—hence very low frequencies of the original signal will not be present in the amplified signal. Design engineers go to great length to avoid low-frequency loss, often resorting to the more expensive direct-coupled stages, thus avoiding interstage capacitors.

4.5.2   Basic Design Considerations

If we replace the active element in Fig. 4.10 by a *npn* transistor and connect to the input an AC signal v_s with a biasing battery V_{EE} in series, we have the circuit of Fig. 4.12. We call this a *grounded-base* amplifier. The purpose of the battery V_{EE} is to forward-bias the transistor. The input voltage to the transistor is

$$V_{eb} = -V_{EE} + v_s \tag{4.12}$$

To find what the correct voltage of the biasing battery ought to be, we reduce the signal voltage v_s to zero, which leaves only V_{EE} as the input voltage. A DC current I_e still flows in the output circuit, resulting in an output voltage

$$V_o = V_B - I_e R_L \tag{4.13}$$

A good design criterion is to choose a voltage for V_{EE} that sets the output voltage at one-half of the battery voltage V_B when $v_s = 0$. With an input signal present again, the output voltage can now have equally large swings about $\frac{1}{2} V_B$ in the positive as well as the negative direction, thus allowing the largest input signal to be amplified before distortion of the amplified signal takes place. To see this, let us examine the variations in the output signal as the input signal varies. As v_s swings positively, V_{eb} becomes less negative (see Fig. 4.12), which means that the transistor is less forward-biased; less forward bias reduces the output current which means that the output voltage V_o swings positive (V_o can go as high as V_B). Now as v_s swings negatively, the transistor is more forward-biased than with just voltage V_{EE} alone and the output voltage swings negatively as the output current increases under the increased forward bias (V_o can go as low as zero). The output voltage is shown in Fig. 4.12; it is amplified and is in phase with the input voltage. It should be clear by now that setting the output voltage at $\frac{1}{2} V_B$ with the biasing battery allows v_s to have the largest amplitude before distortion in V_o sets in. For example, if we increase the biasing voltage V_{EE} such that the amplifier sits at $\frac{1}{4} V_B$, the output voltage could only decrease by $\frac{1}{4} V_B$ on the downward swing, but could increase by $\frac{3}{4} V_B$ on the upward swing. Hence, for a symmetric input signal, the output will be undistorted if the output voltage swings are confined to $\pm \frac{1}{4} V_B$. The voltage range between $\frac{1}{2} V_B$ and V_B is thus not utilized, which is not efficient.

Common-base transistors, which have a low input impedance and good voltage gain but no current gain, are used as special-purpose amplifiers. They are applicable when the driving source has an inherently low impedance and maximum power transfer is desired. For example, in mobile use, carbon microphones which are low-impedance devices ($\approx 200\ \Omega$) have frequently a common-base amplifier build right into the microphone housing.

The biasing considerations outlined above apply equally well to FET amplifiers.

4.5.3   The BJT as Amplifier

The common-emitter configuration is the most widely used for amplifiers as it combines high gain with a moderately high input impedance. Figure 4.13 shows a simple common-emitter amplifier and the transistor characteristics.

As shown by the output characteristics (I_c-V_{ce}) in Fig. 4.13b, a transistor is a highly nonlinear device (although normal operation is in the flat portion of the curves, the so-called linear region). On the other hand, the external circuit elements which are connected to the transistor, like the load resistor and the battery, are linear elements. Interconnecting linear and nonlinear elements form a circuit that can be analyzed by Kirchhoff's laws. For example, in the output circuit of the common-emitter amplifier, summation of voltages around the loop gives

$$V_{CC} - I_c R_L - V_{ce} = 0 \tag{4.14}$$

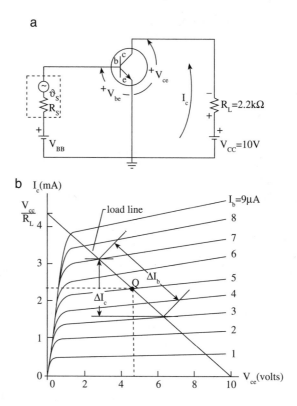

FIGURE 4.13   (a) A grounded-emitter amplifier. (b) Typical collector characteristics for a *npn* transistor. A load line obtained from the output circuit of (a) is superimposed.

Rearranging this equation so it can be plotted on the output characteristics of Fig. 4.13*b* which have I_c and V_{ce} coordinates, we obtain

$$I_c = V_{CC}/R_L - (1/R_L)V_{ce} \tag{4.15}$$

This is an equation of a straight line[11] like that we plotted on Fig. 4.13*b* and we will refer to it from now on as a load line with slope $-1/R_L$. Superimposing the load line on the output characteristics in effect gives us a graphical solution to two simultaneous equations: one equation, belonging to the transistor, is nonlinear and is represented by the family of I_c-V_{ce} graphs (too complicated to be represented analytically), and the other equation, belonging to the output circuit, is linear and is represented by (4.15). Where the family of transistor curves I_c-V_{ce} intersect the load line determines the possible I_c, V_{ce} values that may exist in the output circuit. Hence, transistor current and voltage may only have values which are on the load line. Obviously, an amplified output

---

[11]Remember from analytical geometry that the equation of a straight line is given by $y = b + mx$, with b for the *y*-axis intercept and m for the slope of the line.

voltage can have values anywhere along the load line, from $V_{ce} = 1$ V to 10 V, and an output current from $I_c = 4$ mA to 0. The magnitude of the base current I_b will now determine where on the load line the amplifier will "sit" when there is no input signal, i.e., when it is not amplifying. We call this point the quiescent point, operating point, or simply Q-point. The voltage of the biasing battery V_{BB} will determine this point, and from our previous discussion it should be about halfway on the load line.[12]

At this time it is appropriate to consider what battery voltage, load resistor, load line, and Q-point should be chosen for good transistor operation. In the previous section on design considerations, we started to address some of these issues. If we have no other constraints, good transistor operation can be defined as optimal use of the operating region, which is approximately the area of the flat curves, i.e., the area between $1 < V_{ce} < 10$ and $0 < I_c < 4.5$ in Fig. 4.13b. To use this area optimally, we first pick a battery with voltage of $V_{CC} = 10$ V; this pick determines one endpoint of the load line. The load resistor R_L should then be chosen such that the load line goes through the "knee" of the uppermost I_b curve and intersects the I_c-axis at the point V_{CC}/R_L. This intersection point determines R_L as V_{CC} is known. A proper load line therefore divides the operating region approximately in half. The choice of the Q-point is now obvious: it should be in the middle of the load line (in Fig. 4.13b at $I_c = 2.4$ mA and $V_{ce} = 5$ V, or roughly where the load line intersects the $I_b = 5 \ \mu$A curve).

Choosing a load line and picking a Q-point is referred to as DC design of an amplifier. It is an important part of amplifier design as it sets the stage for proper amplification of a varying input signal. Unless there are other considerations than those outlined above for good design, a poorly DC-designed amplifier can waste power and can lead to distortion. For example, a load line chosen to lie in the low part of an operating region indicates that a less powerful and less costly transistor would do—one for which the load line goes through the entire operating region.

Graphical AC analysis of the above amplifier gives a current gain as the ratio of output and input current

$$G = \frac{i_o}{i_i} = \frac{\Delta I_c}{\Delta I_b} = \frac{(3.2 - 1.5)\text{mA}}{(7 - 3)\mu\text{A}} = 425 \qquad (4.16)$$

where for a sinusoidal input variation, the output current i_o is the sinusoidal variation in collector current I_c; similarly the input current causes the sinusoidal variation in I_b about the Q-point. These variations are sketched in on Fig. 4.13b for the values picked in (4.16). The current gain of a common-emitter amplifier, using representative collector characteristics for the *npn* transistor shown in Fig. 4.13b, is thus substantial.

---

[12]For the amplifier to "sit" at the halfway point on the load line when $v_s = 0$, a biasing current of $I_b = 5 \ \mu$A must be injected into the base. As small voltage changes in the forward bias of 0.7 V cause large changes in I_b (look at Fig. 4.5 near 0.7 V), a signal v_s present at the input is substantially amplified. We will also show in the next example that a self-bias circuit is more practical than one requiring a seperate biasing battery V_{BB}.

4.5.4   DC Self-Bias Design and Thermal Runaway Protection

Consider the commonly used circuit shown in Fig. 4.14a. We have omitted the battery and only shown a 12 V terminal; it is assumed that a 12 V battery is connected between the terminal and ground. This is customary in electronic circuits as only one battery or one power supply is used to power all circuits in an electronic device. As shown in the figure, the biasing voltage is obtained from the common 12 V bus, thus eliminating the need for a separate biasing battery. The voltage divider circuit of R_1 and R_2 provides the voltage for forward bias of the *npn* transistor. For forward bias, the base voltage must be more positive than the emitter voltage by 0.7 V. Therefore R_E is part of the biasing circuit as it raises the emitter voltage when collector current flows. As long as we can maintain the base 0.7 V more positive than the emitter, the transistor is properly biased. Furthermore, this ingenious arrangement of three biasing resistors also provides thermal runaway protection for the transistor: we know that as the temperature of silicon material increases, its resistance decreases, which in turn causes the current in the material to increase; the cycle repeats until the device is destroyed. Should this happen in the circuit of Fig. 4.14a, the additional voltage drop due to any increased current through R_E decreases the forward-bias voltage, thus reducing the current through the transistor and stabilizing the circuit before any damage is done.

EXAMPLE 4.5   Before a signal can be amplified the DC design must be completed, which we will now show for the grounded-emitter amplifier of Fig. 4.14a. The load line to be superimposed on the collector characteristics is obtained by applying Kirchoff's voltage law to the output loop in Fig. 4.14a, which gives the equation

$$I_c = \frac{V_{CC}}{R_E + R_L} - \frac{1}{R_E + R_L} V_{ce} \tag{4.17}$$

As the values for V_{CC}, R_E, and R_L are given in the circuit of Fig. 4.14a, we can sketch in the load line on the output characteristics. All that remains now is to set the Q-point (operating point), which should be chosen to be in the middle of the load line, that is, $I_b \approx 17 \ \mu A$. To design a biasing circuit which will give that value, we proceed by first finding Thevenin's equivalent circuit for the voltage-divider biasing circuit. This is straightforward once we redraw the voltage divider as shown in Fig. 4.14b. Replacing the voltage divider in Fig. 4.14a by Thevenin's equivalent, we obtain the circuit in Fig. 4.14c, which is easier to analyze. For the input loop, we can now write

$$V_{Th} = R_{Th} I_b + V_{be} + R_E I_c \tag{4.18}$$

where we have approximated the emitter current that flows through R_E by the collector current, i.e., $I_e \approx I_c$. The above equation, which has two unknown currents, can be further simplified by using design equation (4.8) for a BJT, which relates the two unknowns as $I_c = \beta I_b$ to give the biasing base current as

$$I_b = \frac{V_{Th} - V_{be}}{\beta R_E + R_{Th}} \tag{4.19}$$

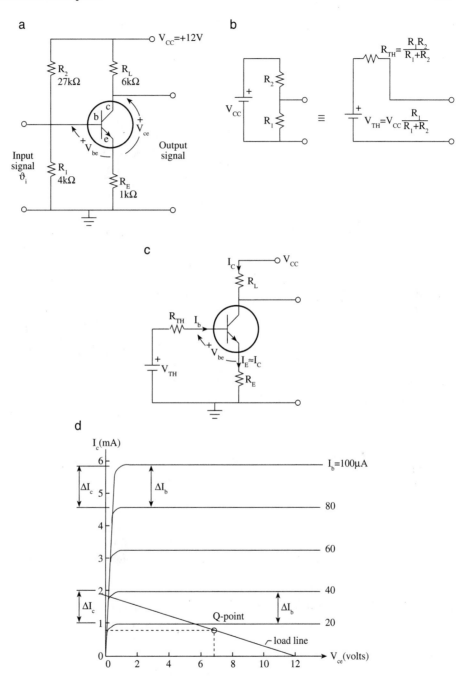

FIGURE 4.14   (a) Self-biasing transistor amplifier. (b) Thevenin's equivalent voltage divider. (c) DC equivalent circuit for setting the DC operating point. (d) Transistor collector characteristics. Load line and Q-point for the circuit of (a) are shown.

This is the design equation for I_b to be used to fix the Q-point. R_{Th} and V_{Th} can be chosen by choosing R_1 and R_2, $V_{be} \approx 0.7$ V for silicon-based transistors, and β can be obtained from the collector characteristics. Once I_b is determined, I_c and V_{ce} at the operating point can be found from βI_b and (4.17) or simply by reading off I_c, V_{ce} values from the collector characteristics at the operating point. (Note that there is continuity between the amplifier of the previous section, Fig. 4.13a and Fig. 4.14c. Except for R_E, they are the same if $V_{Th} = V_{BB}$ and $R_{Th} = R_s$.)

Let us now check the values given in Fig. 4.14a and verify that they do give the desired Q-point. $V_{Th} = (4/(4 + 27)) \cdot 12 = 1.55$ V and $R_{Th} = 4.27/(4 + 27) = 3.5$ kΩ. The gain factor β is determined by finding the ratio $\Delta I_c/\Delta I_b$ using the collector characteristics graphs. Thus near the top, $\beta = \Delta I_c/\Delta I_b = (5.9 - 4.6)\text{mA}/(100 - 80)\mu A = 65$, while near the bottom of the graphs $\beta = (2 - 1)/(0.04 - 0.02) = 50$. Since the operating point is in the bottom of the graphs, we use $\beta = 50$ and obtain for $I_b = (1.55 - 0.7)\text{V}/(50.1 + 3.5)k\Omega = 16$ μA, which gives us a Q-point near the middle of the load line. The collector current at the Q-point is $I_c = \beta I_b = 50 \cdot 16 = 0.8$ mA and the collector–emitter voltage using (4.17) is $V_{ce} = V_{CC} - I_c(R_E + R_L) = 12 - 0.8(1 + 6) = 6.4$ V. These values check with the values obtained by reading off I_c, V_{ce} from the collector characteristics at the Q-point. ∎

We can also check that the transistor has the correct forward bias. Using Fig. 14.4c, Kirchoff's voltage law for the input loop gives us $V_{be} = V_{Th} - I_c R_E - I_b R_{Th} = 1.55 - 0.8 \cdot 1 - 0.016 \cdot 3.5 = 0.75$ V, which is sufficiently close to the correct biasing voltage. Note that the voltage drop across R_{Th} is negligible as the base current is so small; hence the biasing voltage is determined primarily by the voltage-divider action of $R_1 R_2$ and the voltage drop across the thermal-runaway protection resistor R_E. The self-bias circuit therefore injects the correct biasing current into the base.

Now that the amplifier is properly designed, we can amplify a signal. If an AC signal at the input causes the input current I_b to vary about the Q-point between the top of the load line (37 μA) and the bottom of the load line (0), the output current I_c will vary between 1.7 mA and 0. The amplifier has thus a current gain of 1.7 mA/37 μA = 46, which is smaller than $\beta = 50$ as $\beta = 50$ was calculated a bit higher in the active area. This amplifier also has power and voltage gain. To calculate the voltage gain one would have to first relate the voltage of an input source to the input current I_b.

Note that for this example, we chose a load line which is not "optimum" according to our previous discussion as it goes only through the bottom of the active area. A better load line would be steeper (smaller load resistor R_L) and go through the knee of the $I_b = 100$ μA curve. It should be clear that such a load line would require use of a different β—a value closer to the value 65 obtained for the top of the active region.

4.5.5   Fixed-Current Bias

The self-bias circuit of the previous example provides a stable Q-point even as the parameters of the transistor vary with temperature or vary among mass-produced transistors. As the temperature increases, for example, I_c and β increase almost linearly with

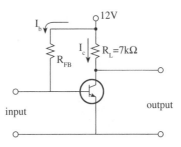

FIGURE 4.15 Resistor R_{FB} injects a bias current for setting the Q-point.

temperature. If stabilization and drift of the Q-point are not of primary importance, a much simpler biasing circuit that injects the correct amount of base current into the transistor for a desired Q-point may suffice, and is considered in the following example.

EXAMPLE 4.6 Let us consider the common-emitter amplifier of Fig. 4.14a and re-place the biasing circuit with that shown in Fig. 4.15, where the single resistor R_{FB} provides the fixed-current bias. To obtain the same load line on the drain character-istics of Fig. 4.14d, we have changed the load resistor to $R_L = 7$ kΩ in Fig. 4.15. To obtain the same Q-point on the load line, we have to inject the same 16 μA base current. As the base–emitter input junction must be forward-biased, the base–emitter voltage will be about 0.7 V. Therefore, the value of the biasing resistor must be $R_{FB} = (V_{CC} - 0.7)/I_{b,Q} = (12 - 0.7)V/16 \mu A = 706$ kΩ. ∎

4.5.6 The FET as Amplifier

The design of a FET amplifier is similar to that for the BJT. After picking a transistor with the desired characteristics, the DC design is carried out next. This involves choos-ing a suitable battery or DC supply voltage, choosing a suitable load resistance which will determine the load line, and finally designing a biasing circuit to give a suitable Q-point (DC operating point). This procedure leaves a good deal of freedom in the choice of parameters and is often as much an art as it is engineering. Only after the DC design is completed is the amplifier ready to amplify a signal.

A basic FET amplifier is that of Fig. 4.13a with the substitution of a FET for the BJT. The load line, which depends on the supply voltage and R_L, can then be drawn on the output (drain characteristics) of the FET. As for the BJT amplifier, to avoid using a separate biasing battery we substitute the self-bias circuit of Fig. 4.14a with its stabilization features and transistor thermal-runaway protection. Figure 4.16a shows a practical FET amplifier using the 2N5459, an n-channel JFET transistor.

One aspect is different for a FET. The characteristics are more nonlinear for a FET than a BJT (the curves for the FET are not as evenly spaced as for the BJT).[13] Recall

---

[13] The nonlinearities of the FET graphs make the FET less suitable as a large-signal amplifier. Distortion

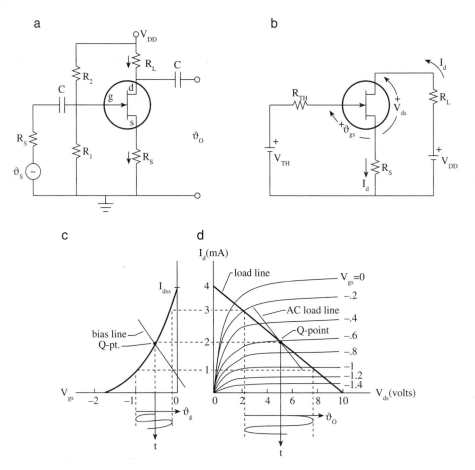

FIGURE 4.16 (a) A FET amplifier with self-bias. (b) DC equivalent circuit using Thevenin's voltage and resistance for the input circuit (see Fig. 4.14*b*). (c) The transfer characteristics and (d) the output characteristics of an *n*-channel JFET 2N5459 transistor.

that for the BJT we used, in addition to the output characteristics, the linear relationship $I_c = \beta I_b$ to set the Q-point, whereas for the FET the comparable relationship is (4.10), which is nonlinear. To avoid complicated algebra due to the nonlinearity of (4.10), we can use the transfer characteristics which are a plot of (4.10) in addition to the output characteristics to set the Q-point graphically. Such a graphical technique aids in understanding, but a more practical technique is simply an approximate "cut-and-try," which makes the design of the operating point for the FET not much more complicated than that for the BJT.

is reduced when the input signal swings are small, as then the operation is confined to a more linear region. Thus FET amplifiers are more suited as front-end stages (preamps) in an amplifier when the signals are still small.

Before the circuit of Fig. 4.16a can be used, let us complete the DC design which will determine all resistance values. The signal source v_s should have no effect on the biasing of the transistor, because for DC operation v_s is essentially disconnected from the amplifier by the coupling capacitor C (an open circuit at DC). The coupling capacitor C also ensures that the DC biasing current only flows into the base. Without C one could find that the biasing current is flowing back into the signal source instead of into the base. Looking at the drain characteristics (Fig. 4.16d), we see that a good choice for supply voltage is $V_{DD} = 10$ V, which fixes one end of the load line. To have the load line pass through the knee of the uppermost curve, pick $R_L + R_S = V_{DD}/I_{dss} = 10$ V/4 mA $= 2.5$ kΩ. The Q-point will be determined when three unknowns, I_d, V_{ds}, and V_{gs}, for the Q-point are specified. Once the load line is drawn in on the drain characteristics, we can easily see that a desirable Q-point is at $I_d = 2$ mA, $V_{ds} = 5$ V, and $V_{gs} = -0.6$ V. We can now use either the graphical or the approximate methods to find the values of the resistors that will set the desired Q-point.

4.5.7   Graphical Method

The equation for the load line is determined by the DC voltage drops in the output circuit, which by the use of Kirchoff's voltage law are $V_{DD} = I_d R_L + V_{ds} + I_d R_S$. Rearranging this in the form of a load line equation,

$$I_d = \frac{V_{DD}}{R_S + R_L} - \frac{1}{R_S + R_L} V_{ds} \tag{4.20}$$

which is plotted on the output characteristics. To fix a Q-point on the load line we need a relationship between gate voltage and drain current which can be obtained from the input circuit. Summing the voltage drops in the input circuit of Fig. 4.16b gives us $V_{Th} = V_{gs} + I_d R_S$, where the Thevenin's voltage is $V_{Th} = V_{DD} R_1/(R_1 + R_2)$, $R_{Th} = R_1 R_2/(R_1 + R_2)$, and the voltage drop $I_g R_{Th}$ is ignored as the gate current into a FET is insignificant. We can now put this equation in the form of a load line

$$I_d = \frac{V_{Th}}{R_S} - \frac{1}{R_S} V_{gs} \tag{4.21}$$

and plot it on the transfer characteristics. When this is done, the straight load line will intersect the transfer characteristic and thus determine the Q-point as shown in Fig. 4.16c. Hence V_{gs} and I_d are fixed, and projecting the Q-point horizontally until it intersects the load line on the drain characteristics, determines V_{ds}. As can be seen by studying the bias load line equation (4.21), the Q-point is determined by the values of the three resistors R_1, R_2, and R_S, which determine the slope and I_d-axis intersect of the bias line.

4.5.8   Approximate Method for the Q-Point

Once the Q-point, V_{Th}, and R_{Th} are chosen, R_1, R_2, and R_S are easily determined by (see Fig. 4.14b)

$$R_2 = R_{Th} V_{DD}/V_{Th} \quad \text{and} \quad R_1 = R_{Th} R_2/(R_2 - R_{Th}) \qquad (4.22)$$

and R_S from (4.21) as $R_S = (V_{Th} - V_{gs})/I_d$. For a good design choose R_{Th} in the megaohm range: the large resistance will keep the input impedance high and will drain little power from the power supply. Similarly, make V_{Th} large compared with V_{gs}, so when V_{gs} varies from transistor to transistor and with temperature, the effects of the variations are minimized.

EXAMPLE 4.7   Design the DC bias circuit of Fig. 4.16a that would fix the Q-point at $V_{gs} = -0.6$ V, when $V_{DD} = 10$ V and $R_L + R_S = 2.5$ kΩ.

Choosing a value of 1 MΩ for R_{Th} and 1.2 V for V_{Th} yields $R_2 = 1$ MΩ $\cdot$ 10 V/1.2 V = 8.3 MΩ and $R_1 = 1$ MΩ $\cdot$ 8.3 MΩ/(8.3 − 5)MΩ = 2.5 MΩ. The value of the sink resistor is $R_S = (1.2 - (-0.6))/2$ mA = 0.9 kΩ, where 2 mA was used as the value of drain current at the Q-point. The load resistor to be used is therefore $R_L = 2.5$ kΩ − 0.9 kΩ = 1.6 kΩ. These are reasonable values. Had we chosen, for example, 10 MΩ and 3 V for R_{Th} and V_{Th}, we would have obtained 14.3 MΩ, 33.3 MΩ, and 1.8 kΩ for R_1, R_2, and R_S, respectively, which are also reasonable values except for R_S, which would require a load resistor of only 0.6 kΩ, a value too small to give sufficient signal gain.

Even though the voltage divider of $R_1 R_2$ places a positive voltage on the gate, the gate is negatively biased with respect to the source. The reason is that source voltage with respect to ground is more positive than gate voltage with respect to ground, the effect of which is to make the gate more negative with respect to source. To check the circuit of Fig. 4.16a, we find that gate voltage with respect to ground is $V_{Th} = 1.2$ V, source voltage with respect to ground is $I_d R_S = 2$ mA $\cdot$ 0.9 kΩ = 1.8 V, and therefore $V_{gs} = 1.2$ V − 1.8 V = −0.6 V, which is the correct bias.

Now that the DC design is completed, the amplifier is ready to amplify an input signal. If v_s produces an AC signal at the gate as shown on the transfer characteristics, the output will appear as v_o on the drain characteristics for a gain of $G = v_o/v_g = (7.5 - 2.2)/(-1 - (-0.2)) = -6.6$. The signal is thus magnified by 6.6 with a 180° phase reversal implied by the minus sign (as input voltage increases, output voltage decreases). We are being somewhat optimistic here: on closer examination we realize that only the voltage that is developed across R_L is available as output voltage; the voltage across R_S is not part of the output voltage. This becomes apparent when we consider the output loop in Fig. 4.16b: v_o is the constant voltage V_{DD} minus the variable voltage across R_L. We can obtain the correct gain by drawing an AC load line which has a slope given by $-1/R_L$ and passes through the Q-point. The correct gain using the AC load line is only $G = (7 - 3)/-0.8 = -5$. ∎

4.5.9 Biasing of MOSFETs

The properties of a DE MOSFET allow the use of a particularly simple biasing circuit which uses a single resistor R_g connected between gate and ground, as shown in Fig. 4.17a. Recall that the transfer characteristics of a DE MOSFET like that shown in Fig. 4.17b allow operation with positive and negative gate voltages, which implies that the Q-point can be set at $V_{gs} = 0$. The gate resistor R_g ties gate g effectively to source v_s such that $V_{gs} = 0$. Recall that there is practically no current flow into a DE MOSFET gate; hence no potential is developed across R_g. On the other hand, should a charge build up on the gate, R_g will allow it to bleed off to ground before any damage is done to the gate. The gain that this amplifier produces, when $V_{DD} = 30$ V and $R_L = 30$ V/4 mA = 7.5 kΩ, can be obtained graphically and is $G = v_o/v_g \approx -6$. Note that most of the source voltage v_s appears across the gate, i.e., $v_s \approx v_g = v_{gs}$ (because of

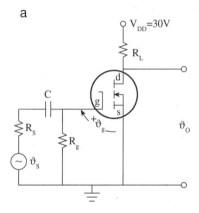

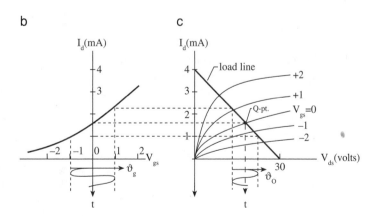

FIGURE 4.17 (a) An amplifier using a DE MOSFET transistor. The (b) transfer characteristic and (c) drain characteristics showing the load line and an amplified AC signal.

the extremely high input impedance of a DE MOSFET and because resistance for R_g is in the megaohm range, we can assume that $R_g \gg R_s$).

4.5.10   Loss of Gain Due to Biasing Resistor

The BJT biasing resistor R_E and the FET biasing resistor R_S stabilize the Q-point and reduce the effects of transistor parameter variations due to temperature changes (transistor operation can be severely affected by temperature changes—thermal stress is the most common cause of electronic component failure). However, biasing resistors also reduce the gain of the amplifier as follows: the output current in Figs. 4.14a and 4.16a flows through the load resistor R_L as well as the biasing resistor; the amplified signal appears in part across each of these resistors, even though the usable output voltage is only the voltage across R_L. The part of the output voltage across the biasing resistor, voltage V_B, is out of phase with the input voltage v_i and thus reduces the voltage to the transistor. For example, in Fig. 4.14a, the voltage to the transistor is

$$V_{be} = v_i - V_B \tag{4.23}$$

where $V_B = I_c R_E$. Hence, the negative feedback[14] due to part of the output voltage into the input loop reduces the gain of the amplifier in comparison with the gain of an amplifier for which R_E is zero. To avoid such negative feedback, we can connect a large capacitor across R_E or R_S, effectively providing a low-impedance path to ground for AC signals. Thus for AC signals the emitter of a BJT or the source of a FET is placed at ground, while for DC operation the biasing resistor is still effective since the shunting capacitor for DC is an open circuit.

Figure 4.18a shows a BJT amplifier connected to a source v_s, whose signal needs to be amplified, and an external load resistor R'_L, across which the amplified signal is developed. For proper DC operation, R_E is necessary, but if no signal loss is to occur due to the presence of R_E, R_E must be zero. Hence a bypass capacitor C_E is placed across R_E. C_E provides a direct path to ground for signal currents, but since C_E acts as an open circuit for DC, the DC operation of biasing is not affected. The DC–AC filtering action of capacitors was covered in Section 2.3 and again in Fig. 4.11. Here we are making the assumption that frequencies of the input signal and the capacitances are large enough to replace all capacitors by short circuits. In other words, the capacitive reactance $\frac{1}{\omega C}$ should be much smaller than any resistance at the lowest frequency of

---

[14] Any time part of the output is applied to the input of an amplifier we have feedback. If the fed-back voltage is in phase with the input voltage, we have positive feedback. The gain of the amplifier is then increased, which can have undesirable effects of instability and oscillation. If we want to use an amplifier to perform as an oscillator, then positive feedback is fine. The reduced gain of negative feedback, on the other hand, has many desirable features which are commonly used in the design of amplifiers. For example, with negative feedback we obtain an amplifier which is more stable as it is less affected by temperature variations. Also because of the many small nonlinearities in the amplifier, the output can be moderately distorted. Negative feedback, which applies part of the distorted output voltage to the input, reduces distortion due to nonlinearities.

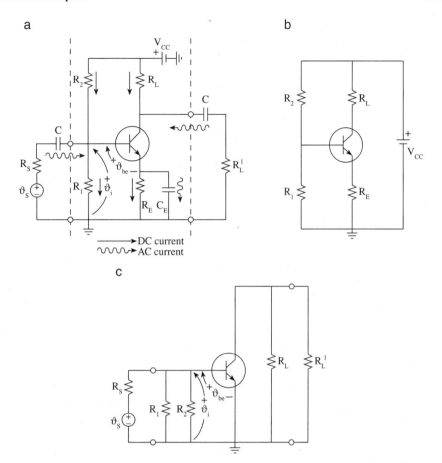

FIGURE 4.18   (a) An amplifier showing AC and DC current paths. (b) A DC equivalent circuit. (c) An AC equivalent circuit.

interest. For example, one rule of thumb is to make

$$\frac{1}{\omega C_E} \leq 0.1 R_E \tag{4.24}$$

at the lowest frequency, ensuring that the signal current will pass through the bypass capacitor. The two coupling capacitors (C) serve a similar function: their purpose is to block DC but to allow passage of signals (AC). Figure 4.18a shows the paths of DC and AC currents, and Figs. 4.18b and c give the equivalent circuits for DC and AC. Thus in the DC circuit we have replaced all capacitors by open circuits and in the AC circuit all capacitors are replaced by short circuits. Furthermore, since the internal resistance of an ideal battery is zero (or a fraction of an ohm for a fully charged, practical battery), it is shown in the DC circuit as a DC voltage V_{CC}, but in the AC circuit it is replaced by a short circuit (a battery has no AC content). Therefore, in Fig. 4.18a, for AC currents, the battery shorts R_2 and R_L to ground and R_2 appears in parallel with R_1 (as was already shown in Figs. 4.14b and c). Note also that the external load resistor R_L' appears in parallel with the internal load resistor R_L.

It should now be mentioned that the gain that was calculated in Example 4.7 using the AC load line is valid for the case of a bypassed source resistor R_S. If R_S is not shunted by a large capacitance, the gain would be less than that calculated using the AC load line in Fig. 4.16d (see Problem 29).

Bypass capacitors can have values in the hundreds or even thousands of microfarads (μF). Such high-capacitance capacitors are bulky and cannot be used in integrated circuits, where even one such capacitor might be larger than the entire chip. Hence in integrated circuits bypass capacitors are omitted and the resulting loss of gain is made up by additional transistor stages.

4.6   SAFETY CONSIDERATIONS AND GROUNDING

In Figs. 4.7 and 4.10 we introduced the ground symbol and stated that it is a common connection point which is assumed to be at ground potential. In more complex circuits involving many transistors it is convenient to tie all circuits to common ground,[15] which can be a larger wire (conducting bus), a metal plate, or a conducting chassis. It is usual practice to refer all voltages to common ground. As ground is considered electrically neutral, the ground points of separate circuits can be connected together without influencing the separate circuits. For example, ground points of the input and output circuits of the two amplifiers in Fig. 4.11 are connected together without interfering with the operation of each circuit.

Another reason to use the ground symbol would be for convenience. Using Fig. 4.11 as an example again, we could have drawn the circuit as shown in Fig. 4.19, omitting the common wire and instead using a number of ground symbols. The omission of a

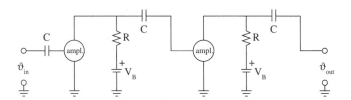

FIGURE 4.19   Alternative representation of the electric circuit of Fig. 4.11 using multiple ground symbols.

---

[15] A good example is provided by the electrical wiring of an automobile. There, common ground is the metal chassis of the automobile, consisting of the metal body, frame, and engine, to which all electric circuits are tied to. The negative terminal of the car battery is tied to common ground. This type of ground is called chassis ground as it is different from earth ground.

common ground wire in a complex circuit schematic could actually make the schematic easier to read.

We should also distinguish between the chassis ground symbol ($\not\!\!\perp$ text-1) and the earth ground symbol ($\not\!\!=$ text-2). For complicated circuits, when such a distinction can be made, it can simplify the tracing of schematics.[16] Chassis ground is a common connection point in electronic devices such as television sets and electrical appliances such as washing machines. Earth ground is provided by a connection to a metal rod driven into the earth or to any other metallic structures that are buried deeply into earth such as metal water pipes. The earth can be considered as a huge capacitor that readily accepts any kind of charge or current flowing into it. A lightning rod is a device to ensure that destructive lightning bolts, which can carry currents of 20,000 A, flow safely to ground instead of through the underlying structure. In electrical equipment it is usually prudent to connect the chassis directly to ground to avoid being shocked when the equipment malfunctions as, for example, when a "hot" wire accidentally shorts to the chassis—a person touching the chassis could be fatally shocked. However, if the chassis is earth-grounded, the current would flow harmlessly to earth (most likely circuit breakers would also trip immediately).[17] Of course in some devices such as automobiles it is impractical to have chassis ground be the same as earth ground because tires provide effective insulation, but then there is really no need for such a linkage. Inside an automobile 12 V does not provide much of a shock hazard; but car mechanics who grope under the dashboard should avoid wearing metallic watches or rings, which could become dangerously hot when accidentally shorted across 12 V. Safety concerns can be further clarified by considering residential wiring.

4.6.1 Residential Wiring

One of the more confusing issues in house wiring is the purpose of the third ground wire in outlets and plugs. Figure 4.20 shows a three-wire system and the incoming line voltages in a typical residence. The neutral wire is grounded with 120 V_{AC}, 60 Hz on either side of it. Line B and line R are 180° out of phase such that $V_B = -V_R$, resulting in 240 V_{AC} if a connection between the two hot wires is made. This system provides for the residence two separate 120 V lines and a single 240 V line for heavier-duty equipment such as air-conditioning and electric ranges. An examination of the individual 120 V outlets raises the question of the need for the extra ground wire running to each outlet. Assuming an appliance is connected to an outlet, why not simply connect the neutral wire to the chassis of the appliance and save one wire? The problem with this is as follows: should the plug be accidentally reversed, the hot wire with 120 V would be connected to the chassis with the possibility of fatal injury if someone were to touch the chassis while standing on ground. The ground wire, on the other hand, if properly connected to the chassis, would allow current to flow harmlessly to ground

---

[16]For relatively simple circuits, most books use only the earth ground.

[17]There is also good reason to connect chassis ground to earth ground in microelectronic systems because experience shows it tends to minimize damage to components due to static electricity.

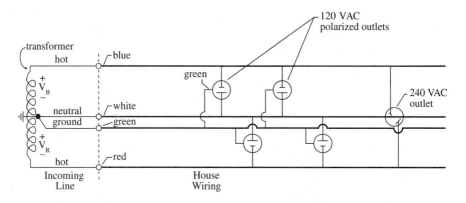

FIGURE 4.20 Typical residential wiring, showing a three-wire single-phase system that provides two circuits of 120 V and one at 240 V. Color convention is white (W) for neutral, red (R) and black (B) for the hot wires, and green (G) for the ground wire.

before a fuse would blow. Hopefully it is clear to the reader that ungrounded chassis or metallic cases that enclose electrical equipment can be lethal devices in the event of insulation failure or accidental hot-wire contact to chassis or case.

To avoid the possibility of accidental shock, many communities mandate GFIs (ground-fault interrupters) in bathrooms, kitchens, and swimming pools. An electric appliance that is plugged into an outlet has currents in the hot and neutral wire that are equal and opposite in a properly working appliance. Unequal current flow could be the result of a malfunctioning appliance causing current to flow through the third ground wire or through a person that has touched the case of a malfunctioning appliance. GFIs, which look like ordinary outlets except for the presence of a trip switch, detect unequal current flow in the hot and neutral wire and instantly disconnect the faulty appliance from the outlet.

4.7 SUMMARY

- This chapter lays the foundation for more complicated electronics such as multi-stage amplifiers, operational amplifiers, integrated circuits, oscillators, and digital and analog electronics.

- We showed that conduction can be by electrons and by holes and that doping of a semiconductor can increase its conductivity substantially. In a *p*-type doped semiconductor, holes are majority carriers and electrons are minority carriers.

- The *pn*-junction was shown to be practically an ideal diode which under forward bias acted as a switch in the on-position and under reverse bias as a switch in the off-position. The rectifier equation showed this behavior mathematically.

- A bipolar junction transistor (BJT) was formed by two diodes back-to-back with the input junction forward-biased and the output junction reverse-biased. Amplification is possible as the current that flows through the low-resistance input junction (emitter-base) is forced to flow also through the high-resistance output (base-collector) junction. The BJT is basically a current amplifier.

- A second type of a transistor, simpler in concept, is the field effect transistor (FET). An input voltage varies the width of a channel in a doped semiconductor through which a current flows, thus controlling the output current. Since the input impedance of a FET is very high, practically no input current flows and the FET can be considered as a voltage amplifier.

- Amplifier action was shown graphically by first deriving the equation of a load line and plotting it on the output characteristics of the transistor. After choosing the Q-point on the load line and designing the DC biasing circuit to establish that Q-point, the amplifier gain was calculated by assuming an input voltage (or current) variation and using the load line to read off the corresponding output voltage (or current) variation.

- An amplifier, in addition to voltage and current gain, also provides power gain. In that sense it is fundamentally different from a device such as a transformer which can also provide voltage or current gain but never power gain. The energy of the amplified signal, which can be much larger than that of the input signal, has its source in the battery or the DC power supply. Since only the input signal should control the output, the voltage from the power supply must be constant so as not to influence the variations of the output signal, if the output is to be a faithful but magnified replica of the input signal. As electric utilities provide AC power only, rectifiers and filters studied in previous chapters are part of a key component of electronic equipment, namely, the DC power supply.

Problems

1. Determine the concentration of electron–hole (e-h) pairs and the resistivity of pure silicon at room temperature.
   Ans: $1.5 \cdot 10^{16}$ e-h pairs/m^3, 2273 $\Omega \cdot$ m.
2. Find the resistance of a 1-m-long conducting wire with a cross-sectional area of 10^{-6} m^2. The material of the wire has a concentration of 10^{21} electrons/m^3 with a mobility of 1 m^2/V·s.
3. A silicon sample is doped with donor impurities at a level of 10^{24}/m^3. For this sample determine the majority and minority carrier concentration and the conductivity.
   Ans: Electron majority concentration $n = N_d = 10^{24}$, minority hole concentration $p = 2.25 \cdot 10^8$/m^3, and $s = 2.16 \cdot 10^4$ S/m.
4. Calculate the forward bias needed at room temperature on a germanium pn-junction to give a current of 10 mA. Use a reverse saturation current of $I_o = 10^{-6}$A.

5. The reverse saturation current for a silicon *pn*-diode at room temperature (293 K) is $I_o = 1$ nA $(= 10^{-9}$A). If the diode carries a forward current of 100 mA at room temperature, calculate the reverse saturation current and the forward current if the temperature is raised by 50°C.

   Ans: $I_o(70°C) = 32$ nA; $I(70°C) = 216$ mA.

6. In the rectifier circuit of Fig. 3.2*a*, assuming room temperature and a reverse saturation current $I_o = 1$ μA, find current i when (a) $v = 0.2$ V and $R_L = 0$, (b) $v = -4$ V and $R_L = 100$ Ω, and (c) $v = +4$ V and $R_L = 100$ Ω.

7. Plot the input–output characteristics v_o/v_i for the circuit shown in Fig. 4.21. Assume the diode is ideal (on–off switch), the input voltage varies in the range -10 V$< v_i < +10$ V, and $R_1 = 10R_2$.

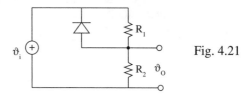

Fig. 4.21

8. Three diodes have i-v characteristics given by the a, b, and c graphs in Fig. 4.22. For each diode sketch a circuit model which can include an ideal diode, a battery, and a resistor.

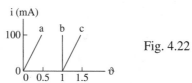

Fig. 4.22

9. If the diodes of Problem 4-8 are used in the half-wave rectifier circuit of Fig. 3.2*a*, find the output peak voltage V_p when the input voltage is 10 V$_{AC}$ (rms) and $R_L = 30$ Ω.

   Ans: (a) 12.12 V, (b) 13.14 V, and (c) 11.26 V.

10. Determine if the ideal diode in the circuit shown in Fig. 4.23 is conducting by finding either the voltage V_d or the current I_d in the diode.

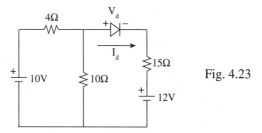

Fig. 4.23

11. Find the current I_B through the 11 V battery in Fig. 4.24. Assume the diode is ideal.

Ans: $I_B = 0$.

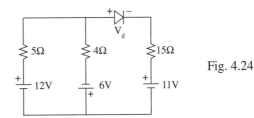

Fig. 4.24

12. If β for a BJT is given as 150, find the emitter current I_e if the collector current I_c is given as 4 mA.

13. Find β and α for a BJT transistor whose collector characteristics are shown in Fig. 4.13b.
    Ans: $\beta \approx 500$, $\alpha \approx 1$.

14. Plot the transfer characteristics for a FET which has drain characteristics given by Fig. 4.8d. Compare to Fig. 4.9, which is the transfer characteristic obtained by applying (4.10).

15. Find the transconductance g_m for a FET which has drain characteristics given by Fig. (16d.) 4.16d   (30 JUNE 2015)

16. In the amplifier of Fig. 4.12, the voltage of the biasing battery V_{EE} is increased until the output voltage has a DC value of $\frac{1}{4}V_B$ when the input voltage $v_s = 0$. Sketch the output voltage V_o when v_s varies with the same sinusoidal amplitude as implied in Fig. 4.12. Repeat for the case when the output voltage has a DC value of $\frac{3}{4}V_B$. To give an undistorted output, how must the amplitude of v_s change?

17. Using the grounded-emitter circuit of Fig. 4.13a,

    (a) Find the value of resistor R_s for a Q-point at $I_b = 3$ μA, when the biasing battery has a voltage of $V_{BB} = 2$ V.
    (b) Calculate I_c, I_e, and V_{ce} if $R_L = 2.2$ kΩ, $V_{CC} = 8$ V, and $\beta = 500$.
        Ans: (a) $R_{BB} = 433$ kΩ; (b) $I_c = 1.5$ mA, $I_e = 1.503$ mA, $V_{ce} = 4.7$ V.

18. Using the grounded-emitter circuit of Fig. 4.13a with a change to $V_{CC} = 8$ V and Q-point $I_b = 3$ μA,

    (a) Find the current gain G of the amplifier using a graphical method; that is, find G from the load line.
    (b) How do the values of I_c and V_{ce} at the Q-point compare to those calculated in Problem 17?

19. Redesign the self-biasing transistor amplifier of Fig. 4.14a to operate in most of the active area (shown by the collector characteristics of Fig. 4.14d); that is, use a load line which has one end at the battery voltage $V_{ce} = 12$ V and the other end at the knee of the $I_b = 100$ μA curve in Fig. 4.14d. To narrow the design, use $V_{CC} = 12$ V, $R_E = 0.5$ kΩ, and $R_1 = 10$ kΩ. Find R_L, R_2, and the current gain $G = \Delta I_c / \Delta I_b$.

20. A silicon BJT uses fixed-current bias as shown in Fig. 4.15. If $V_{CC} = 9$ V, $R_L = 3$ kΩ, $\beta = 100$, and $I_c = 1$ mA at the Q-point, find R_{FB}, I_b, and V_{ce}.
    Ans: $V_{ce} = 6$ V, $I_b = 10$ μA, and $R_{FB} = 1.13$ MΩ.

21. Design a grounded-emitter amplifier that will amplify signals with the largest possible amplitudes. Use a silicon transistor whose collector characteristics are shown in Fig. 4.7b in a fixed-bias circuit of the type in Fig. 4.15. Specify the battery voltage V_{CC}, the load resistor R_L, the DC operating point (Q-point), and the biasing resistor R_{FB} to give that Q-point.

22. Determine the Q-point of the self-bias, germanium transistor amplifier of Fig. 4.14a, given that $R_L = 5$ kΩ, $R_E = 2$ kΩ, $R_1 = 30$ kΩ, $R_2 = 120$ kΩ, $V_{CC} = 12$ V, and $\beta = 100$.
    Ans: $I_{b,Q} = 9.8$ μA, $I_{c,Q} = 0.98$ mA, and $V_{ce,Q} = 5.1$ V.

23. A JFET whose drain characteristics are shown in Fig. 4.8d is in the grounded-source circuit of Fig. 4.16b. Assume that $R_E = R_{Th} = 0$, $V_{DD} = 15$ V, and $R_L = 3$ kΩ.

    (a) Determine V_{Th} for a drain current of $I_d = 2.8$ mA.
    (b) Determine V_{Th} for a drain-source voltage $V_{ds} = 10$ V.

24. In the grounded-source n-channel FET amplifier of Fig. 4.16a, $R_1 = 3.3$ MΩ, $R_2 = 15$ MΩ, $R_{rmL} = 1$ kΩ, $R_s = 1$ kΩ, and $V_{DD} = 15$ V. If the drain characteristics of Fig. 4.8d apply, determine the Q-point of the amplifier. Ignore the input source (at DC, $X_c = \infty$). *Hint*: either construct a transfer characteristics graph by use of (4.10) or by use of the saturation (constant current) region of the drain characteristics and then plot the bias line, or use a trial-and-error method to find the $I_{d,Q}, V_{gs,Q}$ point on the load line.
    Ans: $V_{gs,Q} = -1.8$ V, $I_{d,Q} = 4.5$ mA, and $V_{ds,Q} = 6$ V.

25. Design the self-biasing circuit for the grounded-source amplifier of Fig. 4.16a if $V_{CC} = 8$ V, $R_L = 2.5$ kΩ, $R_s = 1.5$ kΩ, and the Q-point is in the middle of the load line. Use the characteristics of Fig. 4.16d.

26. An enhancement MOSFET whose transfer and drain characteristics are shown in Figs. 4.25a and b is to be used as a basic amplifier in the circuit shown in Fig. 4.25c. Recall, that an n-channel enhancement-mode MOSFET operates with positive gate–source voltages between a threshold voltage V_T (typically between 2 and 4 V) and a maximum voltage $V_{gs,on}$. For $V_{DD} = 15$ V, $V_{GG} = 7$ V, and $R_L = 3$ kΩ, find the Q-point.
    Ans: $V_{gs,Q} = 7$ V, $I_{d,Q} = 2.5$ mA, and $V_{ds,Q} = 8$ V.

27. Self-bias can also be used for an enhancement-mode MOSFET. Consider the circuit of Fig. 4.16a with R_s shorted as it is not needed because R_1 and R_2 alone can provide the positive biasing voltage needed for the enhancement-mode MOSFET. If a MOSFET transistor whose characteristics are shown in Fig. 4.25 is used in the self-bias circuit of Fig. 4.16a with $R_1 = 400$ kΩ, $R_2 = 600$ kΩ, $R_L = 3$ kΩ, $R_s = 0$, and $V_{DD} = 15$ V, specify V_{gs}, I_d, and V_{ds} at the Q-point. *Hint*: Assume that $I_g = 0$.

28. To avoid negative feedback and the accompanying reduction of gain in the transistor amplifier of Fig. 4.16*a*, calculate the value of a bypass capacitor C_E that needs to be placed in parallel with R_S if the amplifier is to amplify signals down to 30 Hz. Use the value for R_S calculated in Example 4.7.

29. Using the AC equivalent circuit of Fig. 4.16*a*, which is shown in Fig. 4.25.

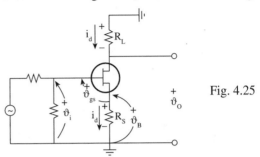

Fig. 4.25

(a) Calculate the gain $G = v_{out}/v_{in}$ of the amplifier (this is the gain with negative feedback).

(b) Calculate the gain G, assuming a large bypass capacitor C_E parallels R_S, which in effect reduces R_S to zero for AC signals.

(c) Compare the two gains and state which is the larger. *Hint*: use $i_d = g_m v_{gs}$, derived in (4.11), to relate the varying part i_d of the drain current to the varying part v_{gs} of the gate–source voltage.

Ans: (a) $G = -g_m R_L/(1 + g_m R_L)$, (b) $G = -g_m R_L$, and (c) gain with $R_S = 0$ is larger.

CHAPTER 5

Practical Amplifier Circuits

5.1 INTRODUCTION

The previous chapter covered the fundamentals of a single-stage amplifier, primarily the DC design which after choosing the battery voltage and the load resistor, determined the biasing needed for an optimum Q-point on the load line. After designing the biasing circuit, the amplifier is ready to amplify a weak signal to useful levels. Signals from sensors or transducers such as those from an antenna, microphone, or tapehead typically are weak signals—frequently so faint that they are at the ambient noise level. Other weak signal examples are

(1) car radio reception that becomes weaker and weaker with distance from the radio station, prompting the driver to eventually switch to a stronger station to avoid the overriding crackling noise.

(2) television reception with a great deal of "snow," indicating that atmospheric noise is becoming comparable in strength to the desired signal.

Our study in the previous chapter was confined to single-stage amplifiers. A practical amplifier, on the other hand, consists of several stages which are cascaded to produce the high gain before a weak input signal can be considered sufficiently large to drive, for example, a power amplifier. Typically input signals, including those mentioned above, are on the order of microvolts (μV), whereas usable signals should be in the volt range. Once the signal is in the volt range, it can be considered immune from interference by noise and other disturbing signals and is ready to be acted on by other devices such as waveshaping circuits and power amplifiers. To drive a power amplifier, which can deliver hundreds or thousands of watts, a large, noise-free voltage signal is needed—typically 1 to 10 V, which means that the voltage amplifier section of an amplifier must have a signal gain as high as 10^6. Obviously such a large gain cannot be obtained by a single amplifier stage. Several stages are needed, each one with a gain of 10 to a 1000. For example, to produce an overall gain of 10^6, which is required to amplify a signal

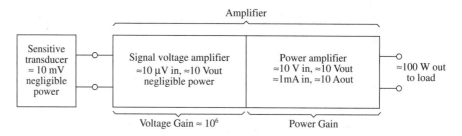

FIGURE 5.1 Typical amplifier showing the separate signal and power amplifier sections. A sensitive pickup device which generates microvolt signals that have practically no power is shown as driving the input of the amplifier.

near the ambient noise level, three stages, each with a gain of 100, are needed.

As a general guide, we can represent in block diagram form an amplifier such as that in a radio receiver or in a television set as consisting of a voltage amplifier followed by a power amplifier.

Figure 5.1 shows such an amplifier with an overall gain of 10^6. This gain is achieved in the voltage amplifier section of the amplifier. The power amplifier section does not contribute to voltage gain—it is basically a current amplifier. Another way of looking at it is that the voltage section is a signal amplifier with no significant power at its output. It is the task of the power amplifier to produce substantial power at the output, which it does by amplifying the current to levels that are typically 1 to 100 A, while voltage swings are in the tens of volts. It should also be mentioned that it is the signal amplifier section which has more components and is more complicated, but since no power is involved and heat dissipation is small, an entire voltage amplifier can be manufactured as an integrated circuit in chip form.[1]

The objective of this chapter is to master the components that compose a multistage, high-gain amplifier, as well as the characteristics of such an amplifier. Although modern implementation of such an amplifier is frequently an integrated circuit in chip form, a study of interconnected, discrete devices is necessary for a complete understanding. For example, the beginning and the end stages of an amplifier are different because the functions performed are different. The first stage receives signals which are usually very small, making the FET ideally suited as a front-end stage—the high input impedance requires little power from the input device and even though the FET's characteristics are more nonlinear than those of a BJT, for small amplitudes signals this is of no consequence (any nonlinear curve is linear over a sufficiently small interval).

---

[1] An entire power amplifier can be similarly manufactured as an integrated circuit. But because of the large amount of heat that needs to be dissipated, the chip is much larger, usually on the order of inches, and is mounted on a substantial heat sink such as a large metal plate with fins.

5.2   THE IDEAL AMPLIFIER

An amplifier is a device that takes an input signal and magnifies it by a factor[2] A as shown in Fig. 5.2a, where $v_{\text{out}} = Av_{\text{in}}$. If one could build an ideal amplifier, what would its characteristics be? In a nutshell, the gain should be infinite, the frequency response should be flat from DC to the highest frequencies, the input impedance should be infinite, and the output impedance should be zero. Let us arrive at these conclusions by using a typical amplifier circuit, like that shown in Fig. 5.2b. Here we have used Thevenin's circuit to represent the source, the input, and the output of the amplifier as well as the load that is connected to the output. Recall that in Section 1.6, we showed that Thevenin's theorem guarantees that "looking" into two terminals of a complex circuit, the complex circuit can be represented by Thevenin's equivalent circuit at the two terminals. Hence, the input terminals of an amplifier "see" the input source as a resistance R_s in series with an ideal voltage source v_s. At the same time, the input source "sees" the amplifier as a load resistance R_i. Similarly, the output of an amplifier acts as a source to the load resistance R_L. The load resistance represents loads such as speakers, printers, motors, display devices, etc. The amplification factor A_r of the real amplifier can now be stated as

$$A_r = \frac{v_o}{v_s} = \frac{i_o R_L}{v_s} = \frac{Av_i/(R_o + R_L)}{v_s} R_L \tag{5.1}$$

$$= A \frac{R_i}{R_s + R_i} \frac{R_L}{R_o + R_L}$$

where the voltage v_i at the input terminals of the amplifier is related to the source voltage v_s by $v_i = v_s \frac{R_i}{R_s+R_i}$. The last expression of (5.1) clearly states that overall gain A_r is less than intrinsic or open-loop gain A of the amplifier. However, it also suggests changes that can be made in the parameters of the amplifier for maximizing the overall gain A_r, which was our initial goal. Thus for an ideal amplifier:

   (a) $R_i \to \infty$, so the entire source voltage v_s is developed across R_i (in other words,

a   b

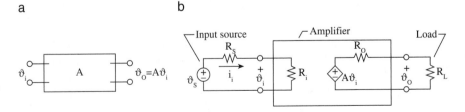

FIGURE 5.2   (a) An ideal amplifier. (b) Thevenin's equivalent of an amplifier with signal source connected to input and a load impedance connected to the output.

---

[2] Also known as *open-loop* or *open-circuit* gain. G is also commonly used for gain.

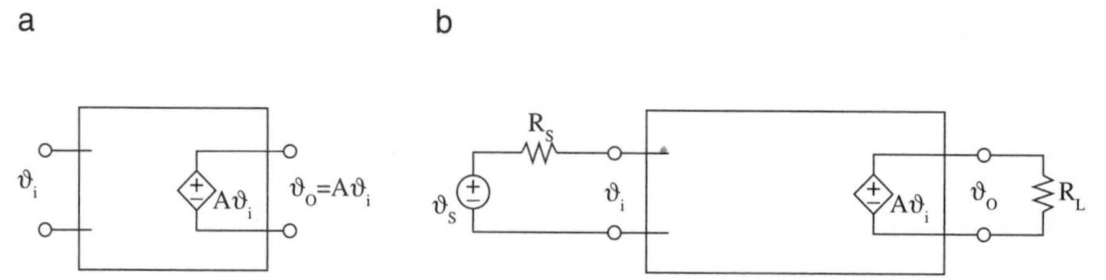

a   b

FIGURE 5.3   (a) An ideal amplifier is represented by input terminals which are open-circuited ($R_i = \infty$) and output terminals which connect to a controlled voltage source. (b) Source and load are connected to an ideal amplifier.

all of v_s is placed across the amplifier input) and the input source v_s does not have to develop any power ($i_i = 0$ when $R_i = \infty$).

(b) $R_o \to 0$, so all of the available voltage Av_i is developed across R_L and none of it is lost internally. Also if it were possible to have R_o be equal to zero, the amplifier would then be a source of infinite power; hence, the smaller R_o in a real amplifier, the less effect a load has on the amplifier (in other words, R_L does not "load" the amplifier).

(c) $A \to \infty$ (for obvious reasons) and A should be constant with frequency, that is, amplify all frequencies equally.

In conclusion, we can state that

$$\text{ideal } A_r = \lim_{\substack{R_i \to \infty \\ R_o \to 0}} A_r = A \qquad (5.2)$$

An ideal amplifier can be represented as in Fig. 5.3a or as in Fig. 5.3b, which shows source and load connected to the input and output terminals.[3] When designing a practical amplifier, we should be aware of these characteristics and use them as a guide in the design. For example, the first stage of a multistage, high-gain amplifier should have an input impedance as high as possible to accommodate feeble input sources that, by their very nature, are high-impedance sources.

EXAMPLE 5.1   Temperature variations are to be recorded on a chart recorder. The output of a temperature transducer (a device that converts temperature to low-level voltages) must be sufficiently amplified so it can drive the chart recorder (a transducer that converts voltage levels to pen positions). At maximum temperature the transducer puts out 10 mV. The chart recorder requires 1V for maximum pen position. If we were to use an amplifier such as that in Fig. 5.2b with an open-loop gain $A = 1\text{V}/10 \text{ mV} = 100$,

---

[3]We will shortly show that an operational amplifier (op amp), which is an integrated circuit chip of a multistage, high-gain voltage amplifier, comes close to the specifications of an ideal amplifier. As such, it finds wide use in industry.

what would the pen position be at maximum temperature? The temperature transducer has an internal impedance of 600 Ω, the chart recorder has an internal impedance of 1200 Ω, the amplifier input impedance is 3000 Ω, and the amplifier output impedance is 200 Ω.

The open-circuit voltage of the temperature transducer at maximum temperature is 10mV, of which only $10\text{mV}\cdot(3000/(600 + 3000)) = 8.33$ mV is available for amplification because of voltage division between source and amplifier input resistance. The open-circuit voltage of the amplifier is therefore $100 \cdot 8.33 = 833$ mV, of which only $833\cdot(1200/(200+1200)) = 741$ mV is available to the chart recorder because of voltage division. Hence the overall gain A_r of the amplifier is $A_r = v_o/v_i = 741/10 = 74.1$ (this result could have also been obtained from (5.1)). The chart recorder will therefore read 71.4% of maximum at maximum temperature. ∎

Let us now proceed with the analysis of the beginning stage of a multistage amplifier, and after that progress through the amplifier to the larger-voltage amplifier stage, and finally to the power amplifier stage.

5.3 SMALL-SIGNAL AMPLIFIERS

We showed in the previous chapter that gain of an amplifier could be obtained by a graphical method, simply by comparing output and input voltage swings. This method, however, becomes useless when millivolt input signals are plotted on characteristic curves that have an axis in the volt range. Thus for input stages of an amplifier, the input signal variation becomes a dot on the characteristic curves and gain cannot be read off. This problem, however, can be converted to our advantage as follows: even though the transistor characteristic curves are nonlinear, using a small portion of the curve allows us to approximate the curve by a straight line. Thus for small input signals, we can linearize the transistor, which until now was a mysterious, three-terminal, nonlinear device. It can now be replaced by resistors and a controlled source. A transistor circuit can then be treated as an ordinary circuit, which is a big advantage when analyzing transistor amplifiers.

5.3.1 Small-Signal Model (FET)

If we look at FET characteristics of, say, Fig. 4.8*d* or 4.16*d*, we observe that the current I_d depends on the gate voltage V_{gs} as well as the drain voltage V_{ds}, i.e.,

$$I_d = I_d(V_{gs}, V_{ds}) \tag{5.3}$$

For small signals which have small excursions Δ about the Q-point (operating point), we have from elementary calculus

$$\Delta I_d = \frac{\partial I_d}{\partial V_{gs}}\Delta V_{gs} + \frac{\partial I_d}{\partial V_{ds}}\Delta V_{ds} \tag{5.4}$$

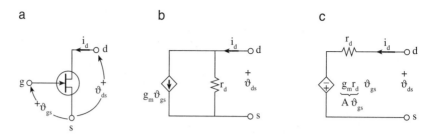

FIGURE 5.4 (a) A FET and its (b) small-signal model. (c) The equivalent voltage-source model for a FET.

We can identify the Δ's with the varying or the AC part of the total signal. For example $I_d = I_{d,Q} + \Delta I_d = I_{d,Q} + i_d$ (see Fig. 4.13b), where small-case letters are used to represent the AC part of the signal. Equation (5.4) can then be written as

$$i_d = g_m v_{gs} + \frac{1}{r_d} v_{ds} \tag{5.5}$$

where $g_m = \Delta I_d / \Delta V_{gs}$ and is called the *transconductance* and for most FETs has values in the range of 1000 to 10000 μS. It is a measure of the effectiveness of drain current control by the gate voltage. Transconductance is obtained experimentally by holding the drain–source voltage constant and taking the ratio of changes of drain current to gate voltage. It is frequently used as a measure of quality of a FET transistor. The other parameter, a measure of the small upward tilt of the output characteristics in Fig. 4.16d, is the drain resistance $r_d = \Delta V_{ds} / \Delta I_d$, obtained by holding the gate voltage constant. A typical value is 50 kΩ.

Expression (5.5) is an equation for current summation at a single node; the circuit corresponding to this equation is a current source in parallel with a resistor as shown in Fig. 5.4b. Thus the AC equivalent circuit for the FET transistor at its output terminals is a practical source. It was shown in Section 1.6 that a practical source can have the form of Norton's equivalent circuit (current source in parallel with a resistor) or that of Thevenin's equivalent circuit (voltage source in series with a resistor). Thus an alternate way to represent a FET is by Thevenin's equivalent circuit, shown in Fig. 5.4c. We can go back and forth between these two circuits by noting that as long as the impedance and the open-circuit voltage at the two terminals is the same, the circuits are equivalent. Therefore, the magnitude of the voltage source in Thevenin's equivalent is $g_m r_d v_{gs}$ or simply $A v_{gs}$, where $g_m r_d = A$ is a dimensionless factor which is the voltage gain of the transistor.[4]

---

[4]We should now distinguish between dependent and independent sources. An independent voltage (denoted by ⊕) or current (circle with an arrow ↑) source delivers an unchanging magnitude of voltage or current (examples are a battery and an outlet with 120V$_{AC}$). A dependent source, on the other hand, has its magnitude controlled by some other quantity, usually a voltage or current that can be anywhere in the circuit. For example, the dependent sources of Figs. 5.3 and 5.4 are in the output circuit but are controlled by a voltage

Let us now see how a typical amplifier circuit, such as that in Fig. 5.5, simplifies by use of the small-signal model. Figure 5.5a shows a FET amplifier with a bypass capacitor, coupling capacitors, a source v_i connected to the input, and the output terminated in a load resistor. Figure 5.5b shows the same circuit but with the transistor replaced by the small-signal equivalent circuit. For small-amplitude signals we now have an ordinary circuit (*ordinary* means that the mysterious transistor has been replaced by a current source and resistor). We can simplify it further by noting that in an AC circuit,[5] large capacitances can be replaced by short circuits, giving us the circuit in Fig. 5.5c. The input impedance is seen to be determined by the parallel combination of the biasing resistors R_1 and R_2. If they are of high resistance, which they usually are so as to avoid loading the input source, we can approximate $R_1 \parallel R_2$ by infinite resistance. Furthermore, if we use Thevenin's equivalent in place of Norton's we arrive at the final form of the equivalent circuit[6] shown in Fig. 5.5d, which depicts the FET to be a voltage-controlled amplifier (the controlling voltage is the input voltage). The input to the amplifier is shown as an open circuit, which is a reasonable approximation for FETs as their input impedance is very high, typically 10^{14} Ω.

We have now succeeded in reducing a typical FET transistor amplifier (Fig. 5.5a) to the elementary amplifier form (Fig. 5.2b) that we considered at the beginning of the chapter. In Fig. 5.2b we used Thevenin's theorem to show that the essential components of an amplifier are a resistance at the input terminals and a practical source at the output terminals. Comparing the equivalent circuit of Fig. 5.5c or d to that of Fig. 5.2b, we see obvious similarities.

The gain $A_r = v_{\text{out}}/v_{\text{in}}$ of the amplifier can now be readily obtained by first finding the output voltage

$$v_{\text{out}} = -\frac{A v_{gs}}{r_d + R_{\text{L}}} R_{\text{L}} = -g_m v_{gs} \frac{r_d R_{\text{L}}}{r_d + R_{\text{L}}} \tag{5.6}$$

The real signal gain of the FET transistor amplifier is then

$$A_r = \frac{v_{\text{out}}}{v_{\text{in}}} = \frac{v_o}{v_{gs}} = -g_m \frac{r_d R_{\text{L}}}{r_d + R_{\text{L}}} \approx -g_m R_{\text{L}}|_{r_d \gg R_{\text{L}}} \tag{5.7}$$

where the approximation $r_d \gg R_{\text{L}}$, valid for most transistors in practical situations, was made. The last expression in (5.7) is very useful: it states that for amplifier gain *the important parameters are the transconductance g_m of the transistor and the external load resistance R_{L}*. Often, the easiest way to increase gain is simply to increase the external load resistance. Should this not be practical, then either a transistor with greater transconductance or an additional amplifier stage should be used.

---

at the input. A controlled or dependent voltage source is denoted by a diamond with a plus–minus sign and a dependent current source by a diamond with an arrow.

[5]For audio amplifiers, high-value capacitors should begin to act as shorts ($1/\omega C \approx 0$) for frequencies larger than 20–30Hz (note that "short" is an abbreviation for "short circuit").

[6]From a cursory examination, the equivalent circuits of Figs. 5.5b, c, and d appear to have two independent halves. Obviously this is not the case as the two halves are linked by the input voltage v_{gs}, which appears as the controlling voltage in the output half.

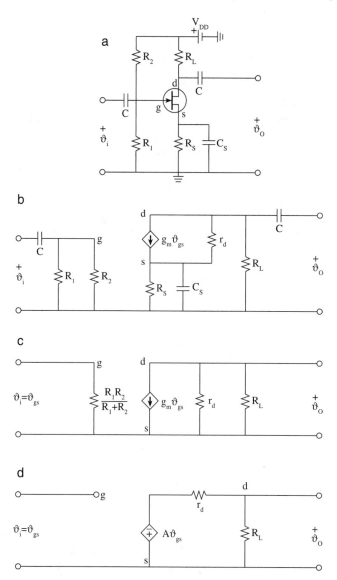

FIGURE 5.5   (a) A FET amplifier. (b) The transistor is replaced by the small-signal model. (c) AC equivalent circuit. (d) Same circuit as (c), except that $R_1 \parallel R_2$ is approximated by infinity.

EXAMPLE 5.2   Apply the gain formula (5.7) to find the gain of the FET amplifier shown in Fig. 5.5. Use the resistance values and the output characteristics graph that were used in Example 4.7 to obtain the gain graphically for the amplifier in Fig. 4.16*a*.

In order to calculate the gain mathematically using (5.7), we need to first calculate the transconductance g_m and output resistance r_d from the graph in Fig. 4.16*d*. Using an area centered around the Q-point, we obtain the transconductance (while holding V_{ds} constant, or equivalently $v_{ds} = 0$) as

$$g_m = \frac{\Delta I_d}{\Delta V_{gs}} = \frac{(2.3 - 1.7) \text{ mA}}{(-0.5 - (-0.7)) \text{ V}} = \frac{0.6}{0.2}$$

$$= 3 \text{ mS} = 3 \cdot 10^{-3} \text{ S}.$$

Similarly for the output resistance, which is the slope of the output characteristic curves near the Q-point, we obtain (holding V_{gs} constant or what amounts to the same thing, $v_{gs} = 0$) $r_d = \frac{\Delta V_{ds}}{\Delta I_d} = (10 - 0) \text{ V}/(2.1 - 1.9) \text{ mA} = 10/0.2 = 50 \text{ k}\Omega$. In Example 4.7 the load resistance was given as $R_L = 1.6 \text{ k}\Omega$. Thus r_d is much larger than R_L, which justifies use of the last expression in (5.7). Hence, the gain for the FET amplifier is $A_r = -g_m R_L = (-3 \cdot 10^{-3})(1.6 \cdot 10^3) = -4.8$. This result compares favorably with the gain of -5 obtained graphically in Example 4.7. ∎

5.3.2 Small-Signal Model (BJT)

Similarly to the previous section, we would now like to develop a linear circuit model for BJT transistors which is valid for small input signals. As in the previous section, we will equate the small signals with the AC signals that need amplifying and that normally ride on top of the DC voltages and currents at the Q-point. If we examine typical BJT characteristics, Figs. 4.7, 4.13, or 4.14, we find them highly nonlinear, but if the excursions about a point (like the Q-point) along one of these curves are small, the nonlinear curves can be approximated by straight lines at that point. Proceeding as in the case of the FET, we note that the collector current I_c in the above-mentioned figures depends on the base current and the collector voltage, i.e.,

$$I_c = I_c (I_b, V_{ce}) \tag{5.8}$$

Differentiating, using the calculus chain rule, gives us an expression in which the Δ's can be associated with small variations in total voltage or current (generally about the Q-point of a properly biased transistor):

$$\Delta I_c = \frac{\partial I_c}{\partial I_b} \Delta I_b + \frac{\partial I_c}{\partial V_{ce}} \Delta V_{ce} \tag{5.9}$$

As before, we identify the small variations with the AC part of the signal,[7] i.e., $i_c = \Delta I_c$, $i_b = \Delta I_b$, and $v_{ce} = \Delta V_{ce}$, and $\partial I_c/\partial I_b$ with the current gain β (evaluated at the Q-point with V_{ce} constant, β is also known as h_f) and $\partial I_c/\partial V_{ce}$ with the slope of the collector characteristics (with I_b constant), usually called the collector conductance

---

[7]The convention that each voltage and current is a superposition of a DC component (the Q-point current or voltage) and a small AC component is used (see Fig. 4.11 or Fig. 4.13b). DC components are denoted by uppercase letters and AC components by lowercase letters (as, for example, $I_b = I_{b,Q} + \Delta I_b = I_{b,Q} + i_b$).

h_o or collector resistance $r_c = 1/h_o$. Thus, the output of the small-signal model is characterized by

$$i_c = \beta i_b + \frac{1}{r_c} v_{ce} \tag{5.10}$$

for which the equivalent circuit is given in Fig. 5.6b. As in the case of the FET (or Fig. 5.4b), at the output terminals (which are the collector–emitter terminals), the BJT is represented by a controlled-current source in parallel with a resistance.

What about the input terminals of a BJT amplifier—how do we characterize them? In the case of the FET, the input terminals were simply represented by an open circuit as shown in Fig. 5.5d, which is a valid representation as the input resistance of FETs is very high, typically 10^{14} Ω. For the BJT, on the other hand, this would not be a valid approximation, as the BJT by its very nature is a current-controlled amplifier, whereas the FET is a voltage-controlled amplifier.[8] To find the input resistance, we must remember that a BJT operates properly only if the input junction is forward-biased; in other words, the base–emitter junction must be at the turn-on voltage $V_{be} \approx 0.7$ V. If it is less[9] than 0.7 V, the transistor is reverse-biased and no emitter or collector current will flow, i.e., the transistor is cut off. Therefore, for DC biasing voltages, which are on the order of a volt, the input junction acts as a forward-biased diode and can be modeled as such. The situation is different, though, when a small signal voltage on top of the 0.7 V_{DC} is applied to the input (base–emitter junction), as in Fig. 5.6. Then the small changes in V_{be} (denoted as v_{be}), which are in response to a weak input signal, will cause i_b to change and i_c to change greatly but in unison ($i_c = \beta i_b$). The input resistance that a signal source will "see" and which we are trying to find can be determined from the characteristics of a forward-biased diode, which are given in Fig. 4.5, with a vertical axis I_b and horizontal axis V_{be}. Hence, using Equation (4.5) under forward bias, when the -1 term is negligible in comparison with the exponential term, we have

$$I_b = I_o \exp(eV_{be}/kT) \tag{5.11}$$

To obtain the resistance to small changes in current, we take the total derivative of (5.11)

$$\Delta I_b = (\partial I_b / \partial V_{be}) \Delta V_{be} \tag{5.12}$$

and (using Ohm's law) identify the partial derivative term with the input resistance r_i. The steps in detail are $r_i = 1/(\partial I_b / \partial V_{be}) = 1/(I_b e / kT)$, which at room temperature ($T = 20^\circ C = 293$ K) gives

$$r_i = \frac{0.025}{I_b} = \beta \frac{0.025}{I_c} \tag{5.13}$$

---

[8]Current-controlled devices must have a low input resistance in order to have adequate current flow into the device, whereas voltage-controlled devices must have high input resistance in order to have adequate voltage developed across their terminals.

[9]It cannot be more, as an attempt to increase the voltage above 0.7V will only increase the current drastically, as shown in Fig. 4.5, without increasing the voltage.

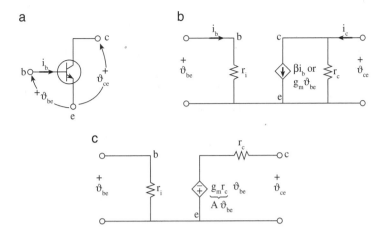

FIGURE 5.6 (a) The equivalent circuit for the small-signal model of a BJT transistor. (b) Current-source model. (c) Voltage-source model.

where $kT/e = 0.025$ V, $k = 1.38 \cdot 10^{-23}$ J/K, $e = 1.6 \cdot 10^{-19}$ C, and I_b and I_c are the total currents at the Q-point. Equation (5.12) can now be stated as

$$i_b = \frac{1}{r_i} v_{be} \tag{5.14}$$

To summarize, we can say that for small signals (v_{be} and i_b) the base–emitter junction acts as resistor and not as a diode (the junction does act as a diode for larger voltages). Typical common-emitter values for r_i are 1–3 kΩ, for β are 50–150, and for r_c are 10^5 Ω (in the literature r_c is also known as $1/h_o$, r_i is known as r_π or as h_i, and β as h_f). Equations (5.14) and (5.10) define the small-signal model of the BJT amplifier for which the equivalent circuit is shown in Fig. 5.6. The current source is seen to be controlled by the input current i_b. The input voltage v_{be} is equally suitable to control the source, because input current and voltage are related as $i_b = v_{be}/r_i$. Hence $\beta i_b = \beta v_{be}/r_i = g_m v_{be}$, which also gives the transconductance of a BJT as $g_m = \beta/r_i$. For direct comparison with the equivalent circuit for the FET and for deriving the voltage-source equivalent circuit for the BJT (Fig. 5.6c), a knowledge of g_m for a BJT is valuable. However, for BJTs, the parameter most often used is the current gain β and not g_m, which finds primary use in FET circuits.

The BJT equivalent circuit is very similar to that of the FET (Fig. 5.4), except for the infinite input resistance of the FET, whereas the BJT has a finite input resistance which is rather small—r_i is typically 1 kΩ.

Let us now see how a typical BJT amplifier circuit, such as that in Figs. 4.14, 4.18, and 5.7, simplifies by use of the small-signal model. Figure 5.7a shows a BJT amplifier with a bypass capacitor, coupling capacitors, a source v_i connected to the input, and the output terminated in a load resistor. In Fig. 5.7b, we have replaced the BJT by its AC equivalent circuit and shorted the battery (as the AC voltage across the battery is zero, it

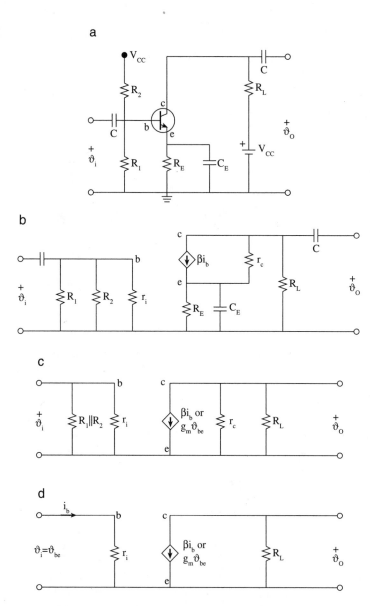

FIGURE 5.7   (a) A BJT amplifier. (b) The transistor is replaced by the small-signal model. (c) AC equivalent circuit. (d) Simplified AC circuit, assuming that $R_1 \parallel R_2 \gg r_i$ and $r_c \gg R_L$.

acts as a short for AC and hence all power supply or battery nodes are AC ground). For sufficiently high frequencies (typically larger than 20 Hz), the capacitors act as shorts, which gives us the circuit of Fig. 5.7c. To ensure that most input power flows into the r_i of the amplifier and not into the biasing resistors R_1 and R_2 (where it would be wasted), the parallel combination of R_1 and R_2 is usually larger than r_i.[10] Also, the collector resistance r_c (typically 50–100 kΩ) is usually much larger than the load resistance R_L (typically 10 kΩ). Neglecting the biasing and collector resistances gives the simple equivalent circuit of Fig. 5.7d, for which the real voltage gain is readily obtained as

$$A_v = \frac{v_{\text{out}}}{v_{\text{in}}} = \frac{v_o}{v_{be}} = -\frac{\beta i_b R_L}{i_b r_i} = -\frac{\beta R_L}{r_i} = -g_m R_L \qquad (5.15)$$

Similarly, the current gain of the amplifier is

$$A_i = \frac{i_{\text{out}}}{i_{\text{in}}} = \frac{i_o}{i_b} = \frac{\beta i_b}{i_b} = \beta \qquad (5.16)$$

The current gain of the amplifier is thus simply the current gain factor β of the transistor (provided the current in r_c can be neglected in comparison with that in R_L). For the BJT, the power gain can also be stated as

$$A_p = \frac{P_{\text{out}}}{P_{\text{in}}} = \frac{v_o i_o}{v_i i_i} = A_v A_i = \frac{(\beta i_b)^2 R_L}{i_b^2 r_i} = \frac{\beta^2 R_L}{r_i} \qquad (5.17)$$

where we have neglected the minus sign.

Once again, as in the case of the FET amplifier, we have shown that we can reduce the BJT amplifier to its fundamental form as shown in Fig. 5.2b. Comparing the circuit of Fig. 5.7d to that of Fig. 5.2b, we see the obvious similarities (the comparison is even better if in the circuit of Fig. 5.7d we convert the current source to a Thevenin-type voltage source).

EXAMPLE 5.3 Design an amplifier of the type shown in Fig. 5.7a which is to drive another amplifier whose input impedance is 10 kΩ; i.e., assume the amplifier of Fig. 5.7a has a load resistance of $R_L' = 10$kΩ connected across its output terminals. The requirement is to develop 1 V across the input of the second amplifier when a 100 Hz voltage source ($v_s = 10$ mV in series with a resistance R_s) is connected across the input of the amplifier under design. The circuit of the amplifier under design looks like that of Fig. 4.18a. In the design use the BJT collector characteristics of Fig. 4.14d, have the load line go through the middle of the active area, and have the Q-point be in the middle of the load line. For the power supply voltage use $V_{CC} = 12$ V. Specify R_L, R_E, R_1, R_2, input impedance Z_i and output impedance Z_o (as seen by the input of the second amplifier), and the value of R_s to give the desired 1 V output. ∎

---

[10]To keep the Q-point from wandering along the load line and to stabilize the β of the transistor, a good design procedure is to have the current through the biasing resistors R_1 and R_2 be approximately 10 times as large as the current into the base I_b.

DC DESIGN The DC equivalent circuit is given in Fig. 4.18b. If the load line goes through the two points $V_{ce} = 12$ V and $I_c = 6$ mA in Fig. 4.14d, we have for $R_L + R_E = 12$ V/6 mA $= 2$ kΩ. A Q-point in the middle of the load line gives $V_{ce,Q} \approx 6.2$ V, $I_{c,Q} \approx 2.8$ mA, and $I_{b,Q} \approx 50$ μA. We must now choose R_E: a larger R_E will stabilize the circuit better but will also reduce the gain as the available amplified voltage is developed only across R_L. Let us choose $R_E \approx 0.1 R_L$ as a reasonable compromise—we can always make it larger should excessive environmental temperature variations necessitate it. This gives $R_E = 0.2$ kΩ and $R_L = 1.8$ kΩ. The voltage drop across R_E is therefore 0.2k$\Omega \cdot 2.8$ mA $= 0.56$ V, which places the emitter voltage 0.56V above ground. Since the base-emitter voltage must be 0.7 V for the transistor to be on, the voltage at the base must be $0.56 + 0.7 = 1.26$ V above ground, which the biasing voltage divider must provide. Thus, $R_1 = 1.26$V/I_1, except that we have not chosen a value for I_1. Since we desire a stable current flowing into the base, which is provided by the voltage divider R_1 and R_2, a good engineering practice is for the current in the biasing resistors to be about $10 I_b$. Of course, the larger the current through R_1 and R_2, the more stable I_b is when the temperature fluctuates. However, the power consumption in the biasing resistors will increase, the input impedance will decrease, and the current gain of the amplifier is reduced, all undesirable affects. As a compromise, choose $I_1 \approx 10 i_b = 0.5$ mA, which gives for $R_1 = 1.26/0.5 = 2.56$ kΩ. To find the other biasing resistor, we note that $R_1 + R_2 = V_{CC}/I_1 = 12$V/0.5mA $= 24$ kΩ, which gives $R_2 = 24$ k$\Omega - R_1 = 21.44$ kΩ. This completes the DC design: the Q-point as well as the β of the transistor which can vary with temperature variations should be well stabilized in this circuit.

AC DESIGN The equivalent AC circuit of the amplifier is given by Fig. 4.18c. From the output characteristics in Fig. 4.14d, we find $\beta \approx 60$, r_c (which is difficult to estimate because of the small slope) is ≈ 60kΩ, and r_i, using (5.13), is $0.025 \cdot 60/2.8 = 536$ Ω. Now that we have these values, we can calculate the gain and the impedances. The voltage gain, using (5.15), is $A_v = -\beta(R_L \parallel R'_L)/r_i = -60 \cdot (1.8$ k$\Omega \parallel 10$k$\Omega)/0.536$ k$\Omega = -171$. This gain is too large as it would result in a 10mV$\cdot$171 $= 1.71$V at the input of the second amplifier when only 1 V is desired. To reduce this voltage, we can reduce the input voltage from 10mV to 5.8 mV by increasing the source resistance R_s such that the voltage divider action of R_s and r_i gives 10 mV$\cdot r_i/(r_i + R_s) = 5.8$ mV at the input of the amplifier under design. Since $r_i = 536$ Ω, we obtain for $R_s = 407\Omega$. Hence an input source of $v_s = 10$ mV which is in series with $R_s = 407$ Ω will produce, after amplification, 1 V at the input of the second amplifier. The input impedance which the source sees is $Z_i = r_i = 536$ Ω, and the output impedance of the amplifier under design and what the input of the second amplifier sees is $Z_o = R_L = 1.8$ kΩ (if a more accurate answer for Z_o is needed use $R_L \parallel r_c$).

5.3.3   Comparison of Amplifiers

We have already observed that the FET is basically a voltage amplifier (strictly speaking, a voltage-controlled current source) and the BJT a current amplifier (strictly speaking, a current-controlled current source). Only a trickle of electrons is needed to activate a FET, implying that its input impedance is very high, which in turn makes the FET ideally suited as a front-end stage in amplifiers where the input signals are very weak (low power). Such an amplifier would not load a weak input source (it would not draw any appreciable current from the source), allowing all of the voltage of the source— often in the microvolt range—to appear across the input of the amplifier. Because of the high input impedance,[11] the power gain of FETs can be very large, which is of little importance unless we are considering the final stages of an amplifier (see Fig. 5.1). Throughout most of the amplifier we need signal gain, which means voltage gain. Being fundamentally a current amplifier implies that the BJT has a low input impedance, typically a thousand ohms. Except for that shortcoming, it generally has superior voltage gain when compared to the FET, and hence is an ideal amplifier to follow a first-stage FET amplifier. It goes without saying that the high gain of practical amplifiers is obtained by cascading many amplifier stages, each with different properties.

The output impedance Z_o of FET and BJT amplifiers, which is the parallel combination of r_d or r_c with a load resistance R_L, is typically 5kΩ. As the internal load impedance R_L (in contrast to the external load impedance R'_L shown in Fig. 4.18a) is ordinarily much smaller than r_d or r_c, we can state that $Z_o \approx R_L$. An output impedance of 5kΩ is a rather high value and is frequently not suitable for driving a following stage. What we need is a buffer that could be inserted between two amplifier stages. The buffer should have a high input impedance (in the megaohms) and a low output impedance (about a hundred ohms). Such a device is called an *emitter follower* (a grounded or common collector BJT) or a *source follower* (a grounded or common drain FET); it has a voltage gain of 1 but transforms the output impedance to about 100Ω. Thus a common source FET followed by an emitter follower or a source follower is an excellent two-stage amplifier with the desirable properties of very high input impedance and very low output impedance. In integrated form it is very small and makes an excellent input amplifier, commonly used in instruments such as the oscilloscope.

Another possible, though infrequently used, configuration is the grounded or common base BJT and the grounded or common gate FET. This configuration is characterized by an unusually low input impedance (as low as 20Ω), which is generally undesirable.

To summarize other differences, we can state that the BJT is a bipolar (two *pn*-junctions with flow of majority and minority current) device, whereas the FET is unipolar (one junction with majority current only). The BJT is a more linear device—the output characteristic curves are straighter and more evenly spaced than those of FETs. The FET has lower power consumption and can be made smaller and cheaper but the BJT is more rugged, can handle higher power loads, has a wider frequency response,

---

[11]Convention is to refer to input and output impedance, even though in most practical cases it is a resistance.

and, because it has a larger transconductance (typically the g_m for BJTs is 50000 μS and that for FETs is 2000 μS), can have a substantially higher voltage gain.

5.4   DECIBEL NOTATION FOR GAIN

For proper operation of a system, such as a communication network, we find that amplifiers and other devices such as filters, signal processing circuits, transmission lines, etc., are cascaded (output of one device connected to the input of the next). For example, even a typical amplifier is a cascading of a preamplifer, main amplifier, and a power amplifier, often in one enclosure. When calculating the gain of such a chain, it is more convenient to add the logarithm of the gain than to multiply the gain of the individual stages. Specifying power gain A in decibels (dB) of cascaded gains $A_1, A_2, A_3, \ldots$, that is, when $A = A_1 \cdot A_2 \cdot A_3 \cdots$, we have[12]

$$
\begin{aligned}
A_{dB} &= 10 \log A = 10 \log A_1, A_2, A_3, \ldots & (5.18) \\
&= 10 \log A_1 + 10 \log A_2 + 10 \log A_3 + \cdots \\
&= A_{1,dB} + A_{2,dB} + A_{3,dB} + \cdots
\end{aligned}
$$

As an example, a system consisting of a transmission line which has a power loss of 2dB, a filter with a power loss of 3dB, and an amplifier with a power gain of 20dB would have an overall system gain of 15dB (-2dB $-$ 3dB $+$ 20dB $=$ 15dB).

"Decibel" has meaning only when the ratio of two numbers is involved. For a device with input and output terminals such as an amplifier, the power gain in dB is related to the log of the power ratio $A = P_o/P_i$ as[13]

$$
A_{dB} = 10 \log A = 10 \log \frac{P_{out}}{P_{in}} \tag{5.19}
$$

where the logarithm is to base 10. If the output of an audio amplifier is 3W (watts) but can be changed to 6W, then dB $= 10 \log(6/3) = 3.01$. A change in power of 2:1 increases the power level by 3 dB. Since 3dB is a minimum change in power level that an average person can perceive (1 dB is the minimum change in level detectable by the human ear), it is surprising to many that doubling the power produces only a slightly noticeable change in sound. Increases in power level by 10, 100, and 1000 correspond to increases in dB by 10, 20, and 30. Similarly, decreases in power level by the same amount, correspond to -10 dB, -20 dB, and -30 dB.

---

[12]For simplicity, we are omitting the subscript p in the symbol for power gain A_p. Also note, that if current gain is $A_i = I_2/I_1$ and voltage gain is $A_v = V_2/V_1$, then power gain is $A_p = |A_i A_v|$.

[13]Even though the decibel is a measure of relative power, we can also use it to state absolute power by defining a reference power level, which in engineering practice is taken as 1 milliwatt (mW) for P_i. Consequently, an amplifier with a 15 W output is $10 \log(15/0.001) = 41.8$ dB above the reference level of 1mW. The symbol "dBm" is used to indicate a reference level of 1mW. Hence the above amplifier would be referenced as 41.8 dBm.

Since power is proportional to the square of voltage or current, (5.19) may also be expressed as $A_{dB} = 10 \log(V_o^2/R_o)/(V_i^2/R_i)$, and if $R_o = R_i$, as $20 \log V_o/V_i = 20 \log I_o/I_i$. It is common practice to use this last expression even when $R_o \neq R_i$. Furthermore, the dB is such a convenient measure that voltage gains of amplifiers are calculated in this manner. Thus an amplifier with a signal gain of 1000, indicating that its output voltage V_o is 1000 times larger than its input voltage V_i, has a voltage gain of 60 dB. In general, when comparing two signals, if one signal has twice the amplitude of a second, we say it is +6 dB relative to the second. A signal 10 times as large is +20 dB, a signal 1/10 as large is −20 dB, and a signal 1/1000 as large is −60 dB.

5.5   FREQUENCY RESPONSE OF AMPLIFIERS

The gain A that we calculated in the previous sections was assumed to be constant, independent of the magnitude or the frequency of the input signal. In practice, A can remain constant only over a finite range of frequencies, known as the *midband*. The gain in this range is known as *midband gain*. For frequencies lower and higher than the midband, the gain of the amplifier decreases and continues to decrease until the gain reduces to zero. The causes of this gain reduction are the coupling capacitors at low frequencies and shunting capacitances at the high frequencies. Let us consider the low frequencies first.

5.5.1   Loss of Gain at Low Frequencies

Figure 5.8a shows a typical two-stage, RC-coupled amplifier. The coupling capacitors between amplifier stages and any resistance before or after the capacitor form a high-pass filter of the type shown in Fig. 2.7a. The purpose of C is to prevent DC from reaching the input of the following amplifier (where it could saturate the base or gate) but not to prevent signal frequencies from reaching the input. As pointed out in Section 2.3, this action is not perfect: in addition to blocking DC, the low frequencies are also attenuated. Attenuation can be calculated by considering the small-signal equivalent circuit, shown in Fig. 5.8b, where the biasing resistors R_1 and R_2 have been neglected as they are usually much larger than the input resistance r_i of the following amplifier (we can always include them by considering the parallel combination $R_1 \parallel R_2 \parallel r_i$). The midband gain (the C's have negligible reactance ($1/\omega C$) at midband and are assumed to be short circuits) of the first amplifier, using (5.15), is $A_v = v_{b2}/v_{b1} = -g_m(R_L \parallel r_i)$, where $\parallel$ denotes the parallel combination of R_L and r_i. At lower frequencies, when the reactance of C increases to where it is comparable to r_i and R_L, we have for gain

$$A_{v,l} = \frac{v_{b2}}{v_{b1}} = -\frac{i_2 r_i}{v_{b1}} = -g_m \frac{r_i R_L}{r_i + R_L} \frac{j\omega C(r_i + R_L)}{1 + j\omega C(r_i + R_L)} \tag{5.20}$$

a

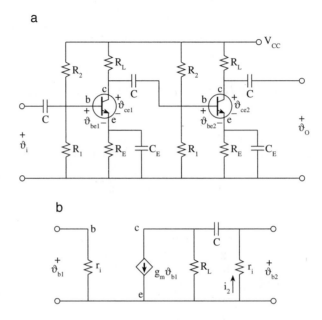

b

c

$$\text{gain} \, |A_\vartheta(f)| = |A_\vartheta| \cdot \frac{1}{\sqrt{1+(f_l/f)^2}} \cdot \frac{1}{\sqrt{1+(f/f_h)^2}}$$

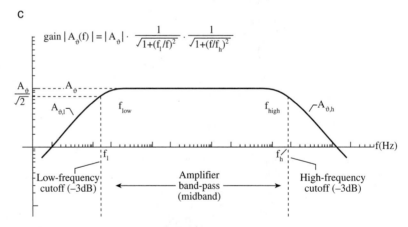

FIGURE 5.8   (a) A two-stage amplifier. (b) Small-signal equivalent circuit of the first stage at low frequencies. The input resistance r_i acts as an external load resistance. (c) Band-pass of the amplifier, showing attenuation at low and high frequencies.

where i_2 is obtained (by current division) from Fig. 5.8b as

$$i_2 = g_m v_{b1} \frac{R_L}{R_L + r_i + 1/j\omega C} \tag{5.21}$$

The low-frequency gain is thus seen to be equal to the midband gain multiplied by a filter function, that is, $A_{v,l} = A_v \cdot j\omega C(r_i + R_L)/(1 + j\omega C(r_i + R_L))$. As the frequency ω increases toward midband and higher, the filter term multiplying A_v becomes unity, that is, $\lim_{\omega \to \infty} A_{v,l} = A_v$. Hence, coupling capacitors do not affect the gain at higher frequencies. If we define a corner or half-power frequency as

$$f_l = \frac{\omega_l}{2\pi} = \frac{1}{2\pi C(r_i + R_L)} \tag{5.22}$$

we can write the low-frequency gain as

$$A_{v,l} = A_v \frac{jf/f_l}{1 + jf/f_l} = A_v \frac{1}{1 - jf_l/f} \tag{5.23}$$

Frequently only the magnitude of the gain is desired. The absolute value (which eliminates imaginary quantities) is then

$$\left| A_{v,l} \right| = |A_v| \frac{1}{\sqrt{1 + (f_l/f)^2}} \tag{5.24}$$

and is plotted in Fig. 5.8c as the left-hand curve.[14] Coupling capacitors thus limit the low-frequency performance of amplifiers. For more details on high-pass filters, see Fig. 2.7. The *corner* (also known as *cutoff* or *half-power) frequency* f_l is the frequency at which the gain is $1/\sqrt{2}$ of midband gain, or equivalently is reduced by 3 dB from midband gain. The gain is seen to decrease by 20 dB every 10-fold decrease in frequency (the slope is 20 dB per decade). Hi-fi audio amplifiers, which need to reproduce low frequencies well, should have $f_l = 20$ Hz or lower; otherwise the result is a tinny sound. To have good low-frequency response in an amplifier, we need (see (5.22)) large values for r_i, R_L, and C, or we should entirely eliminate coupling capacitors and use only direct-coupled amplifiers. Direct-coupled amplifiers are more difficult to design, are more critical, and are less flexible, but in integrated circuits their use is common since capacitors take up too much space. Similarly, because of the large area required, bypass capacitors (which increase the gain of an amplifier stage) for the biasing resistors R_E and R_D are impractical in integrated circuits. Nevertheless, cascaded direct-coupled amplifiers are a very useful form in miniaturized circuitry and, because they have no capacitive coupling reactance, they have another important feature: they have no low-frequency cutoff and will amplify down to $f_l = 0$, in other words, down to DC.

---

[14]The midband gain A_v is thus seen to be multiplied by a high-pass filter function which is of the same form as the high-pass RC filter expression of (2.15).

EXAMPLE 5.4   In the circuit shown in Fig. 5.8b, we have $R_L = 5$ kΩ and $r_i = 1$ kΩ. Find the required value of the coupling capacitor C for $f_l = 20$ Hz. Repeat for a FET amplifier for which $R_g = 1$ MΩ.

For the BJT transistor, using (5.22), we obtain

$$f_l = 20 \text{ Hz} = \frac{1}{2\pi C(1000 + 5000)}$$

$$C = \frac{1}{2\pi \cdot 20(1000 + 5000)} = 1.33 \ \mu\text{F}$$

For the FET transistor, the equivalent FET input impedance r_i is very high and can be approximated by an open circuit. The input impedance to the amplifier is then determined by the biasing resistors, which we refer to as R_g, and which can be equal to R_{Th} of the FET amplifier circuit of Fig. 4.16. Therefore, for the FET

$$f_l = 20 \text{ Hz} = \frac{1}{2\pi C(10^6 + 5000)}$$

$$C = \frac{1}{2\pi \cdot 20 \cdot 1.005 \cdot 10^6} = 0.008 \ \mu\text{F}$$

Because the input impedance of FETs is much higher than that of BJTs, much smaller coupling capacitors can be used for FET amplifiers. This is a great advantage in miniaturized circuits as lower-capacitance capacitors are typically smaller in size. As stated before, in integrated circuits, capacitors are omitted because of their bulkiness. Use is made of direct-coupled amplifiers which have no coupling or bypass capacitors even though direct-coupled amplifiers have lower gain than RC-coupled amplifiers.   ∎

5.5.2   Loss of Gain at High Frequencies

As the frequency increases above midband, we again find that gain begins to decrease below the midband gain value of A_v. We will also again find that a cutoff or half-power frequency f_h exists, at which the gain is reduced by 3dB and above which the gain continues to decrease at a rate of 20 dB per decade. The gain characteristics of a typical amplifier are shown in Fig. 5.8c: constant gain around midband (or midrange frequencies) with gain falling off on either side of midband. This resembles band-pass characteristics, with a bandwidth defined by $B = f_h - f_l$, where subscripts h and l stand for high and low.

What causes the limitation to gain at high frequencies? The answer is that it is due to the small shunting (or parasitic) capacitance that exists between any two terminals. We know that any two conductors, when separated by a small distance, form an effective capacitor with capacitance given by the formula $C = \varepsilon A/d$, where A is the *effective area* of each conductor and d is the *separation* (see Eq. (1.15)). As shown in Fig. 5.9a, small stray capacitances C_{ce}, C_{be}, and C_{cb} exist between any two terminals or leads of

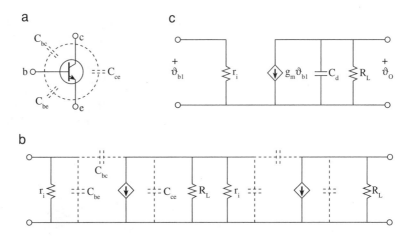

FIGURE 5.9   (a) Stray capacitances between terminals of a transistor. (b) High-frequency equivalent circuit of a two-stage amplifier. (c) Equivalent circuit of a single stage showing only the largest stray capacitance.

a transistor. These capacitances, although small (in the picofarad range), nevertheless become effective shunts at sufficiently large frequencies, with the effect that part of the signal current will flow through the shunt, thus decreasing the available signal current to any device that is connected after the shunt.

Figure 5.9b gives the high-frequency equivalent circuit of the two-stage amplifier of Fig. 5.8. At midband frequencies and greater, the coupling and bypass capacitors C and C_E have a very low reactance and can be replaced by shorts, but the shunting capacitances now become important and are shown as dashed capacitors. To find the gain at frequencies above midband, we consider a single stage, shown in Fig. 5.9c, and only the largest capacitance C_d, which is the combination of C_{ce} and the shunting capacitance of the following stage. The voltage gain of this single stage can be stated as

$$A_{v,h} = \frac{v_o}{v_{b1}} = \frac{-g_m v_{b1}(R_L \parallel C_d)}{v_{b1}} = A_v \frac{1}{1 + j\omega R_L C_d} \tag{5.25}$$

where v_o is given by the current $g_m v_{b1}$ that flows through the parallel impedance $R_L \parallel C_d = (R_L/j\omega C_d)/(R_L + 1/j\omega C_d)$ and where the midband gain is given by $A_v = -g_m R_L$. Taking the absolute value, we obtain[15]

$$\left| A_{v,h} \right| = |A_v| \frac{1}{\sqrt{1 + (f/f_h)^2}} \tag{5.26}$$

where the high-frequency cutoff or corner frequency is given by

$$f_h = 1/2\pi R_L C_d \tag{5.27}$$

---

[15]The expression multiplying A_v in (5.26) is a low-pass filter function considered previously as (2.14).

Formula (5.26) is similar to (5.24), except for the corner frequency f_h being in the denominator, whereas in (5.24), the corner frequency f_l is in the numerator. We see that for frequencies smaller than f_h, the square root term becomes unity and the gain is the midband gain A_v. For frequencies higher than f_h, the square root term can be approximated by f/f_h, and $|A_{v,h}| \approx |A_v| f_h/f$. The gain decreases as $1/f$, or decreases with a slope of -20 dB per decade (a range of frequencies where the highest frequency is 10 times the lowest; hence dB $= 10 \log(1/10) = -20$). In Fig. 5.8c, we show on the right-hand side of the figure the decrease in frequency response due to shunting capacitances. Thus, it is the shunting stray capacitances that decrease the high-frequency response of an amplifier. In large bandwidth amplifiers, designers go to great length to decrease any stray or parasitic capacitances to guarantee the highest possible f_h.

EXAMPLE 5.5   A transistor in a multistage amplifier has parasitic capacitances $C_{be} = 40$ pF, $C_{bc} = 5$ pF, and C_{ce}. The combined output capacitance C_d, which is made up of C_{ce} and the input capacitance of the following stage, is $C_d = 300$ pF. Hence, given the size of C_d, we can ignore all stray capacitances except for C_d. If the load resistance is $R_L = 10$ kΩ, then using the high-frequency equivalent circuit of Fig. 5.9c, we obtain for the high-frequency cutoff $f_h = 1/2\pi R_L C_d = 1/6.28 \cdot 10^4 \cdot 300 \cdot 10^{-12} = 53$ kHz. This is a comparatively low cutoff frequency, and therefore this amplifier is only good for amplifying audio signals. For example, to amplify television signals, the upper half-power frequency f_h must be in the megahertz range. A better transistor, for which $C_d = 5$ pF, would extend this to 5.3 MHz. Reducing R_L would also increase the high-frequency cutoff, but the voltage gain of the amplifier, which at midband is given by $A_v = -g_m R_L$, would then be reduced.

The high-frequency equivalent circuit of Fig. 5.9c is also valid for FET amplifiers because the AC equivalent circuits for FETs and for BJTs are similar.   ■

5.5.3   Combined Frequency Response

The complete response of an amplifier can now be stated by combining the gain formulas for low and high frequencies, (5.24) and (5.26), as

$$|A_v(f)| = |A_v| \frac{1}{\sqrt{1 + (f_l/f)^2}} \frac{1}{\sqrt{1 + (f/f_h)^2}} \tag{5.28}$$

where $A_v(f)$ stands for frequency-dependent voltage gain. For frequencies lower than f_l, the first square-root factor dominates while the second square-root factor is unity (this is the range of frequencies for which (5.24) was derived). As the frequency increases to where $f \gg f_l$ but still $f \ll f_h$, both square-root factors can be approximated by unity and the gain is given by $A_v(f) = A_v$, that is, we are at midband and the range of frequencies is called midband or midrange frequencies. Continuing to increase frequency until $f > f_h$, we find that the second square-root factor begins to dominate while the first remains at unity (we are now in the range of frequencies for which (5.26) was derived). This function is plotted as Fig. 5.8c and shows the band-pass characteristics of an amplifier, being limited at the low frequencies by coupling capacitors and at high frequencies by stray shunting capacitances. Midband gain A_v occurs only over a finite range of frequencies, called the bandwidth $B = f_h - f_l$, where f_h and f_l are called the high and low half-power, corner or cutoff frequencies at which the gain is reduced to $1/\sqrt{2} = 0.707$ of midband voltage gain or equivalently is reduced by -3 dB. The gain continues to decrease at either end with a slope of -20 dB per decade.

Amplification is usually done in a number of stages to achieve the necessary gain. When amplifiers with identical band-pass characteristics are cascaded, that is, connected in series for additional gain, the band-pass characteristics of the cascaded amplifier are inferior to those of a single stage. Recall that we define band-pass as the frequency interval between the high- and low-frequency points where the gain falls to $1/\sqrt{2}$ of midband gain. Say we cascade two amplifier stages with identical band-pass characteristics. At the corner frequencies where the gain of each stage was down by $1/\sqrt{2}$ (or -3 dB), when cascaded, the gain will be down by $1/\sqrt{2} \cdot 1/\sqrt{2} = 1/2$ (or -6 dB), because the gain of the two stages is multiplied. As bandwidth is defined between the $1/\sqrt{2}$ points, we readily see that the bandwidth of the cascaded amplifier is smaller than the bandwidth of a single stage. Bandwidth deterioration is a serious restriction when cascading stages.

As an example, if we have two stages with similar band-pass characteristics but with a gain of 100 for the first and a gain of 30 for the second stage, the gain at midband for the cascaded stages will be $100 \cdot 30 = 3000$. The midband gain, which is the gain over the flat portion of the band-pass curve as shown in Fig. 5.8c, is not as wide as that for a single stage and also drops off more sharply at a rate of -40 dB per decade, whereas for a single stage the drop-off is more gentle and is at a rate of -20 dB per decade.

5.5.4 Cascading of Amplifier Circuits

As pointed out above, we resort to cascading of amplifier stages when substantial gain is needed. Figure 5.8a shows a two-stage amplifier. In this configuration, the output of the first stage becomes the input of the second stage. This can be continued until the desired gain is achieved, i.e., the desired signal voltage level is obtained at the output of the cascaded amplifier.

In the two-stage amplifier of Fig. 5.8a, the two individual sections are connected by the coupling capacitor C. This capacitor passes the amplified AC signal from the first transistor to the second, while blocking the DC collector voltage of the first transistor from reaching the base of the second transistor. The DC voltage V_{CC}, which is usually much larger than the AC signal voltage, would saturate the second transistor if it were applied to it, making it inoperative or even destroying it due to excessive heat buildup (in addition, it would upset the proper DC bias of the second transistor provided by resistors R_1 and R_2). The first C and the last C in the circuit of Fig. 5.8a serve a similar function of isolating input and output from DC potentials.

At midband, when the reactances of the coupling capacitors C are small and can be approximated by short circuits, the output voltage of the two-stage amplifier can be written as

$$v_o = (v_{o1})A_2 = (v_i A_1)A_2 = v_i A \qquad (5.29)$$

where $v_{o1} = A_1 v_i$ is the signal voltage at the output of the first stage in Fig. 5.8a (and which becomes the input voltage to the second stage) and A_1 and A_2 are the gains of the first stage and second stage, respectively. Hence the overall gain of the two-stage amplifier is $A = A_1 A_2$ or simply the product of the gains of the individual stages. To repeat, the above expression for midband gain, (5.29), is an approximation valid for signal frequencies which are neither so low that the coupling reactances cannot be ignored nor so high that shunting transistor stray capacitances reduce the gain.

5.6 TIME RESPONSE AND PULSE AMPLIFIERS

In the preceding section the inputs to an amplifier were sinusoidal voltages of a single frequency. We then showed that all signals with frequencies which fall in an amplifier's midband are amplified equally, but frequencies outside the midband are attenuated. Imagine a complex signal such as a speech pattern that is composed of many frequencies is now applied to the amplifier. In order for the amplifier to faithfully magnify this signal it should be obvious that the speech signal not have frequencies outside midband. Such frequencies would be amplified less, with the result that the magnified signal would be distorted. Clearly the frequency content of the signal must fit within the midband of the amplifier for a faithful magnified replica of the input signal. How can we tell what the frequency content of a complex signal is? Fourier analysis is a simple mathematical method that breaks a signal into its sinusoidal components.

5.6.1 Fourier Series

Periodic waveforms may be represented by summing sine waves of different frequencies and amplitudes. Consider, for example, the square wave in Fig. 5.10a. If we take only two sinusoids, a fundamental and a third harmonic as shown in Fig. 5.10b by the dashed curves, and add them we obtain the solid curve, which is beginning to look like the square wave. If we add more harmonics (of the right frequency and amplitude), we can come as close as we like to a square wave. Fourier series analysis (which is beyond the scope of this book) gives us a prescription of how to decompose a complex signal into its Fourier components. For example, the Fourier series of a square wave of peak amplitude V_p and period T is given by

$$v(t) = \frac{4V_p}{\pi} \left(\sin \omega t + \frac{1}{3} \sin 3\omega t + \frac{1}{5} \sin 5\omega t + \frac{1}{7} \sin 7\omega t + \cdots \right) \qquad (5.30)$$

where the period T is related to the angular frequency ω and frequency f by $T = 1/f = 2\pi/\omega$. A practical interpretation of (5.30) is as follows: if we take a number of sinusoidal generators (one with frequency ω and amplitude $4V_p/\pi$, another with frequency

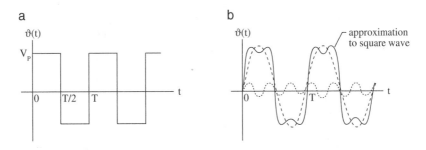

FIGURE 5.10   (a) A square wave of period T and amplitude V. (b) Approximation to a square wave, obtained by summing only two sinusoids.

3ω and amplitude $4V_p/3\pi$, etc.) and string them together in series and use the string to drive a circuit, the circuit will respond in the same manner as if driven by a square wave of period T and amplitude V_p. Other periodic waveshapes such as sawtooth, triangular, half-wave rectified, etc., can equally be well represented by a Fourier series.[16] It is interesting to note that sharp-cornered signals such as square waves and sawtooths can be represented by rounded signals such as sine waves. This being the case, it is equally obvious that it is the high frequencies in the Fourier series which give rise to the sharp corners in a periodic signal. Hence, for an amplifier to faithfully magnify a sharp-cornered, periodic signal, the amplifier must have a large bandwidth—in other words, the high-frequency cutoff f_h must be large. We have now related qualitatively bandwidth of an amplifier to its ability to amplify sharp-cornered, periodic signals.

5.6.2   Pulse Amplifiers

In addition to amplifying a single frequency or a narrow band of frequencies, amplifiers are also needed to amplify rapidly changing periodic signals (such as square waves), rapidly changing nonperiodic signals (such as speech), and single pulses. Such signals have a wide band of frequencies and if their signal bandwidth exceeds the bandwidth of the amplifier, signal distortion will occur in the output of the amplifier. Such a distortion is sometimes desired as in waveshaping circuits but mostly we are interested in an undistorted but magnified signal.

An alternative to bandwidth as a criterion of amplifier fidelity is the response of the amplifier to a square-wave input. Square-wave generators are readily available and square-wave testing of amplifiers is common. We will show that a relationship between the leading edge of a square wave and the amplifier high-frequency response f_h (see (5.27)) exists. Also, a relationship between the flat portion of a square wave and the low-frequency response f_l (see (5.22)) exists. This should not be too surprising as f_h characterizes the amplifier's ability to magnify the rapid variations in a signal, whereas

---

[16]Computer programs usually perform the Fourier analysis. Spectrum analyzers can also be used to display the Fourier coefficients when a periodic signal such as a square wave is applied to the analyzer.

f_l characterizes the ability to magnify slow variations in a signal. A square wave is therefore ideally suited as it possesses both an abrupt change in voltage and no change (the flat portion of the square wave).

5.6.3   Rise Time

Assume a square wave, Fig. 5.10a, is applied to an amplifier. At the output it is magnified but distorted as in Fig. 5.11, which shows the first pulse of the square wave. The distortion in the vertical part of the square wave is caused by the shunting capacitances that are present in the amplifier. The amplified pulse does not rise instantaneously to its maximum value but is delayed by time t_r, which we call the rise time. To show the effects of shunting capacitances on a sharply rising signal such as a square wave, we use the same $R_L C_d$ circuit in Fig. 5.9c that was used to derive f_h as $f_h = 1/2\pi R_L C_d$. The $R_L C_d$ combination in that figure is driven by a current source (Norton's equivalent). To relate it to a voltage step, we change the circuit to a Thevenin's equivalent, which gives us a circuit of the type shown in Fig.1.25a, the voltage response of which is given by (1.51) as $v_C = V(1 - e^{-t/\tau})$, where $\tau = R_L C_d$. The convention is to define rise time as the time required for the pulse voltage to go from 0.1 V to 0.9 V as shown in Fig. 5.11. It is a measure of how fast an amplifier can respond to step discontinuities in the input voltage. If we call t_1 the time that it takes for the capacitor voltage to reach $1/10$ of its final value, we have

$$0.1\text{V} = V(1 - e^{-t_1/\tau}) \tag{5.31}$$

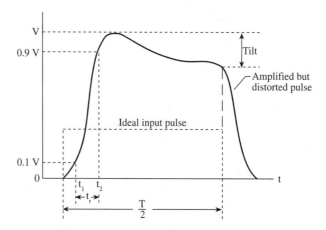

FIGURE 5.11   An ideal square input pulse is distorted after amplification. The distortion is a finite rise time and a tilt.

which gives $t_1 = 0.1\tau = 0.1 R_L C_d$. Similarly, if t_2 is the time to reach $9/10$ of the final voltage, we obtain $t_2 = 2.3\tau$. The rise time is then

$$t_r = t_2 - t_1 = 2.2\tau = 2.2 R_L C_d = \frac{2.2}{2\pi f_h} \approx \frac{1}{3 f_h} \tag{5.32}$$

where $f_h = 1/2\pi R_L C_d$, given by (5.27), was used. We have now shown that the rise time is related to the inverse of the cutoff frequency f_h of an amplifier. Accordingly, a pulse will be faithfully reproduced only if the amplifier has a good high-frequency response; i.e., the rise time t_r of an amplified step voltage will be small only if f_h of the amplifier is large. For example, for a pulse rise time less than 1μs, the amplifier bandwidth must be larger than 340 kHz. Conversely, for an amplifier with a 1 MHz band-pass, $t_r = 0.33\ \mu$s.

The amplified pulse in Fig. 5.11 shows a trailing edge as well as a leading edge. Both of these are due to the presence of shunting (parallel) capacitance in the amplifier. The leading edge is the result of charging the shunting capacitance, whereas the trailing edge is due to discharge of the same capacitance. Although we have shown the shunting capacitance as a lumped quantity C_d, it usually is distributed throughout the amplifier. For convenience and for ease of calculation, we decided to represent it by a lumped quantity. In the design of an amplifier care must be taken to reduce any stray capacitance, such as input or output leads too close together, which would add to the total parallel capacitance and thereby decrease the high-frequency performance of the amplifier. (See also Problem 5.29).

5.6.4   Tilt

The second type of distortion in the amplified pulse is the *sag* or *tilt* in the DC portion of the square pulse, which is caused by the presence of DC coupling capacitors in the amplifier (Fig. 5.8a, for example, shows three coupling capacitors C). DC coupling capacitors form the type of high-pass filter considered in Fig. 2.7a. To perfectly reproduce the horizontal portion of the square pulse would require a DC coupled amplifier for which $f_l = 0$. We can relate the distortion (tilt) in the DC portion of the square wave to the low-frequency cutoff f_l of the amplifier by using the circuit in Fig. 2.7a or in Fig. 1.25a. To mimic a pulse, let us assume that a DC voltage V is suddenly applied to this circuit. As shown in Fig. 1.25a, the voltage across R will jump to V (assuming the capacitor is initially uncharged) and then decay depending on the time constant RC. That is, using Eq. (1.50), $v_R = Ri(t) = V \exp(-t/RC)$. For minimum decay, the time constant must be very large. If we expect the decay or tilt at the end of the pulse (given by $t = T/2$) to be small, we can approximate the exponential voltage across R by only two terms, that is,

$$v_R(t = T/2) = V e^{-(T/2)/RC} \approx V(1 - T/2RC) \tag{5.33}$$

and the percentage decay, sag, or tilt P in Fig. 5.11 is then given by

$$P = \frac{V - v_R(t = T/2)}{V} = \frac{T}{2RC} \tag{5.34}$$

If we now introduce from (5.22) the low-frequency cutoff $f_l = 1/2\pi RC$, we have

$$P = 2\pi (T/2) f_l \qquad\qquad (5.35)$$

which shows that for a given pulse length $T/2$, the distortion in the amplified pulse as measured by tilt P is proportional to f_ℓ of the amplifier. Long pulses therefore require amplifiers with excellent low-frequency response (i.e., f_l should be as small as possible). For example, if a 1ms pulse is to be passed by an amplifier with less than 10 % sag, f_l must not exceed 16 Hz, where $f_l = P/2\pi (T/2) = 0.1/(6.28)(0.001) = 15.9$ Hz.

5.6.5   Square-Wave Testing

Band-pass characteristics of an amplifier can be determined by applying a single-frequency voltage to the input and measuring the output voltage. If this is repeated for many frequencies, eventually a plot of the type shown in Fig. 5.8c is obtained. An easier, less laborious procedure is provided by square-wave testing in which a square wave is applied to the input and the output is observed on an oscilloscope. First we reduce the frequency of the square wave until tilt P is measurable in the output waveform, which allows us to determine the low-frequency cutoff f_l using (5.35), i.e., $f_l = P/\pi T$. Frequency f_l is usually identified with the onset of low-frequency distortion. Analogously, we increase the frequency of the square wave until the rise time of the amplified pulses becomes observable on the oscilloscope. As we increase the frequency of the square wave, the sweep of the oscilloscope must also be increased so only a few pulses are displayed, which then allows us to measure the rise time accurately (of course the tilt will not be visible at the high frequencies at which leading edge distortion is observable; correspondingly, at low frequencies and at slow sweeps, tilt is visible but not leading edge rise time). Once the rise time t_r is measured, (5.32) can be used to obtain the high-frequency cutoff f_h, i.e., $f_h = 1/3t_r$. Frequency f_h is usually identified with the onset of high-frequency distortion.

Summarizing, we can state that two square-wave frequencies establish the band-pass characteristics of an amplifier. Square-wave testing, which determines f_l and f_h, is especially useful when changes in the amplifier circuitry need to be made so as to achieve specified or desired characteristics. It is advantageous to be able to make changes and simultaneously observe the amplifier output waveform. It is perhaps redundant to state that for accurate square-wave testing, high-quality oscilloscopes and square-wave generators are necessary.

5.7   POWER AMPLIFIERS

As discussed in the Introduction, a power amplifier is usually the last stage in an amplifier. The preceding voltage sections have taken a feeble signal voltage and amplified it to volt levels that can easily control a power amplifier whose purpose is to give the signal muscle. Figure 5.1 shows this process in block diagram form. The power amplifier, in essence, is a current amplifier. It steps up signal current, which at the input of the power amplifier is small, to substantial values at the output. Even though voltage levels at input and output of the power amplifier are about the same, the power is considerably increased at the output of the power amplifier stage.

5.7.1   Transformer-Coupled Class A Amplifier

Because of their high cost and bulkiness, transformers are usually avoided in electronic circuitry. Nevertheless, they provide an ideal way to couple a power amplifier to a load. Using resistors in the collector or drain circuits when large powers are involved would lead to excessive DC losses, that is, the I^2R losses due to the high currents associated with high powers would be excessive. The efficiency of power transfer from amplifier to load is less than 25% in resistive circuits, whereas in transformer-coupled amplifiers it can be 50% in Class A operation.[17] The reason is the DC resistance of transformer windings, which is typically very low. Furthermore, the impedance of many loads might differ substantially from that needed for optimum operation of the amplifier. As shown in (2.49), a transformers can provide impedance matching; that is, if a load with impedance Z_L is connected to the secondary of a transformer, and the primary is connected in the collector or drain circuit of an amplifier, the amplifier would "see" an impedance equal to $Z = (N_1/N_2)^2 Z_L$, where N_1 and N_2 are the winding turns of the primary and secondary, respectively. By using this impedance-changing ability of transformers we can provide any amplifier with the optimum load impedance. An additional feature of a transformer is that it isolates the load which is connected to the secondary from DC currents that flow in the primary (a transformer only transforms AC; DC currents in the primary and secondary do not affect each other). This can be important as many loads, such as speakers, cannot tolerate direct current.

A power amplifier, which by its very nature is a large-signal amplifier, can only be analyzed graphically. Hence, we cannot find an equivalent circuit for it, as we did for the small-signal case in the previous sections. A typical, common-emitter power amplifier is shown in Fig. 5.12. Let us first consider the DC design. As the load is the primary of the transformer, which for DC can have practically zero resistance but for AC can have a very large impedance, we need to consider two separate load lines, a DC load line and a AC load line. For the DC load line: to set the operating point (Q-point) for proper DC bias, we first obtain the DC voltage equation for the output loop which is

$$V_{CC} = V_{CE} + R_E I_C \qquad (5.36)$$
$$10 = V_{CE} + 4I_C$$

where we have neglected the small DC resistance of the primary winding and have

---

[17] In Class A amplifiers the Q-point is chosen to lie in the center of the linear portion of the transistor output characteristics, as, for example, in Fig. 4.13b. Amplification is essentially linear, the output signal being an amplified replica of the input signal. Output current I_c (or drain current I_d if it is a FET amplifier) flows at all times. Class B operation, on the other hand, is nonlinear, with output current flowing only half the time.

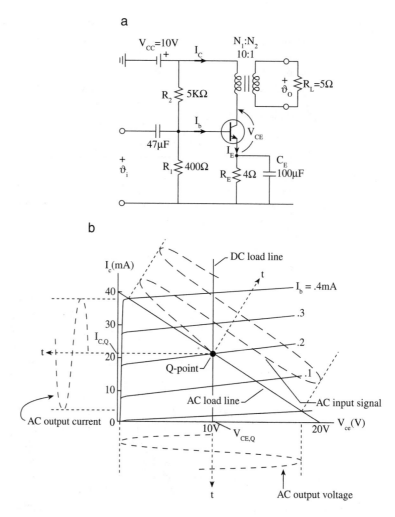

FIGURE 5.12 (a) A transformer-coupled Class A power amplifier. (b) DC and AC load lines. The excursions of I_b (dashed sinusoid) due to an AC input signal are shown swinging up and down the AC load line about the Q-point. These excursions cause corresponding excursions in collector voltage v_{ce} (between 0 and 20 V) and collector current i_c (between 0 and 40mA).

approximated $I_E \cong I_C$. Using (5.36) to plot the DC load line on the collector character-istic curves of Fig. 5.12b, we obtain essentially a vertical line (the slope is $dV/dI = -4$ V/A). The Q-point should be in the middle of the active area, i.e., at $I_C = 20$ mA and $I_b = 0.2$ mA. DC biasing is set by the three resistors (5 kΩ, 400 Ω, and 4 Ω) which were calculated by procedures outlined in Example 4.7. For good bias stability, the cur-rent through the 400 Ω bias resistor should be about 10 times larger than the 0.2 mA DC base current at the Q-point ($I_{400} \cong 10 \text{ V}/(5000 \text{ Ω} + 400 \text{ Ω}) = 1.9$ mA, which is approximately $10I_b$) and the base–emitter voltage should be between 0.6 and 0.7 V (it is approximately 10 V·400 Ω/5400 Ω − 4 Ω · 20 mA= 0.74 − 0.08 = 0.66 V).

An AC signal applied to the input will see a different load line because the reflected impedance $R'_L = (N_1/N_2)^2 R_L = 10^2 \cdot 5 = 500 \ \Omega$ in the primary of the transformer is much larger than the DC winding resistance. The AC load line passes through the Q-point and is determined by

$$0 = v_{ce} + R'_L i_c \tag{5.37}$$

Unlike in (5.36), R_E is not present in the above equation as R_E is bypassed by C_E; in other words, AC signals at the emitter are shorted to ground by C_E. V_{CC} is similarly not present as a battery cannot have an AC voltage across it. The slope of the AC load line is then $di/dv = -1/R'_L = -1/500 \ \Omega$, which is used to draw the load line in Fig. 5.12b. As the operation of the transistor must be confined to the area of the characteristic curves of Fig. 5.12b,[18] we locate the operating point in the middle of that area which will allow the largest AC signal to be applied at the input in order to have the largest undistorted power output.

Power amplifier efficiency is given by the ratio of AC signal power to DC power supplied either by a battery or by a power supply. The average DC power is the product of voltage and current at the Q-point, or $V_{CE,Q} \cdot I_{C,Q}$. This power is taken from the battery regardless if the amplifier is amplifying or not. It is also the power that must be dissipated by the transistor and hence an adequate heat sink must be provided to prevent the transistor from overheating. For a Q-point located in the middle of the active area, the AC signal can swing about the Q-point from zero to twice the value at the Q-point, as shown in Fig. 5.12b. Hence the peak value of an undistorted output signal voltage is $V_p = V_{CE,Q}$ and similarly for the current. The *efficiency* of the amplifier can now be stated as

$$\text{Efficiency} = \frac{\text{AC power}}{\text{DC power}} = \frac{\frac{1}{2} I_C V_C}{I_C V_C} = \frac{1}{2} = \text{or } 50\% \tag{5.38}$$

where for simplicity we have omitted the additional subscripts (E and Q) on collector voltage and current at the Q-point and where effective values for voltage and current were used to express AC power. Thus in an optimally designed Class A amplifier (Q-point in the middle of the AC load line, allowing equal swings of the signal for the largest undistorted amplification), the ideal efficiency is 50%. In practical situations, the signal is rarely at maximum amplitude at all times, yielding an average efficiency substantially less than 50%.

This concludes the design of a simple power amplifier, which finds common use as an audio amplifier. All signal power developed in the transformer primary circuit is transferred practically without loss to the 5 Ω load. Even though transformer-coupled amplifiers, because of their weight, bulkiness, and expense of iron-core transformers,

---

[18]If collector voltage exceeds 20 V, it can break down the collector junction; exceeding 0.4 mA base current can saturate the collector junction; and, in general, the power dissipation in the collector junction is limited by allowable temperature rise in the junction.

are not as popular as in the days of discrete components and are completely unsuited for integrated circuits, the principles involved in the design of these amplifiers are important and applicable in any amplifier design.

5.7.2   Class B Push–Pull Amplifiers

Class A amplifiers have desirable linear features that cause little distortion, but their low efficiency generates high levels of heat which must be dissipated in the transistor. Class B operation is more efficient and thus is more attractive for high-power amplifier stages. The large distortion that Class B operation introduces (current flows only half the time, which makes the output look like that of a half-wave rectifier) is avoided by using two transistors in a push–pull arrangement, as shown in Fig. 5.13*a*. Two *pnp* transistors are biased near the turn-on voltage of approximately 0.6–0.7 V by battery

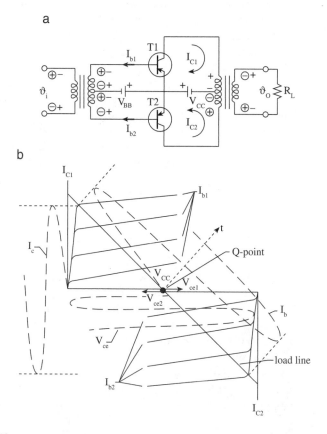

FIGURE 5.13   (a) A push–pull power amplifier circuit, showing polarities of voltages throughout the circuit. (b) The AC load line of the push–pull amplifier, showing that the amplifier does not draw any current at the *Q*-point.

V_{BB} (this means that should the input voltage go positive, T2 will conduct or go negative when T1 will conduct) and are connected by a center-tapped secondary of an input transformer. The collector outputs are also connected by a center-tapped primary of an output transformer, the secondary of which is connected to a load such as a speaker. We use two sets of plus–minus signs to illustrate the operation. Assume the AC input signal is a simple sine wave. The circled set denotes the polarity of the input signal when it swings positively. The bottom transistor T2 becomes forward-biased, which causes an output current I_{c2} to flow through T2, the battery V_{CC}, and the bottom half of the output primary, while the top half of the primary is idle because transistor T1 is cut off, i.e., $I_{c1} = 0$. During the positive half of the input sinusoid, the circled signs give the polarities of all voltages. We observe that the output voltage at the secondary of the output transformer is 180° out of phase with the input voltage.

During the second half of the input sinusoid, the input voltage is negative, which is denoted by the uncircled signs. The top transistor T1 conducts now while T2 is cut off. This produces current I_{c1} in the direction shown and now the top half of the primary produces the entire output voltage. If the two transistors are matched, a push–pull amplifier produces a nearly undistorted sinusoidal voltage at the output, even though each transistor delivers only one-half of a sine wave to the output transformer.

Push–pull, Class B operation can be further clarified by showing the AC load line of the push–pull amplifier. In Fig. 5.13b we have taken the output characteristics of each amplifier and put them back-to-back, given that the total collector current in the primary is $I_c = I_{c1} - I_{c2}$. This yields a composite load line which is made up of the load lines for T1 and T2. The operating point is shown in the middle of the figure, that is, at the cutoff point on the load line for each amplifier. As pointed out before, each transistor conducts or carries current for only half the time. The combined output current and output voltage, corresponding to a sinusoidal input voltage, is sketched in also and shows that the peak value of the AC output voltage is the battery voltage V_{CC}. Similarly, the peak AC current is the maximum I_c, giving for the efficiency

$$\text{Efficiency} = \frac{\text{AC power}}{\text{DC power}} = \frac{\frac{1}{2} V_{CC} I_c}{V_{CC}(2I_c/\pi)} = \frac{\pi}{4} = 0.785 \text{ or } 78\% \qquad (5.39)$$

where the DC power supplied by the battery is obtained as follows: from Fig. 5.13a we see that the pulsating current from each transistor flows through the battery in the same direction, causing the battery current to pulsate just like the current in the full-wave rectifier shown in Fig. 3.3d. The effective value of such a pulsating current is given by (3.2) as $2I_p/\pi$. A high 78% efficiency is possible because battery current flows only when a signal is present. In the absence of a signal, the amplifier sits at the Q-point and does not draw any current, which implies that the battery is not supplying any DC power. It is now apparent that the push–pull Class B amplifier is significantly more efficient than the Class A amplifier, which means that more power is delivered to the load and less is wasted in the transistors. This makes the push–pull amplifier the configuration of choice when large powers are needed.

5.7.3   Class B Complementary Amplifiers

An intriguing combination of a *npn* and *pnp* transistor is the complementary-symmetry circuit shown in Fig. 5.14*a*. It is suited for integrated circuits (ICs) as it is direct coupled, eliminating coupling capacitors or bulky and expensive transformers. One drawback is that two batteries or two power supplies with opposite polarities are needed.

The operation is as follows: in the absence of an input signal, the base bias on both transistors is zero so both transistors are cut off. Furthermore, both transistors remain off for input signals in the range between -0.7 V and 0.7 V. Since neither transistor conducts, the output voltage v_o is zero. As v_i increases to higher than 0.7 V, the *npn* transistor T1 turns on and provides current to the load R_L, while T2 remains off. Similarly, when v_i decreases to less than -0.7V, the *pnp* transistor T2 turns on and supplies current to the load, while T1 is off. The output voltage across the load is therefore given by[19]

$$
\begin{aligned}
v_o &= v_i - 0.7 \text{ V} \quad \text{when } v_i > 0.7 \text{ V} \\
v_o &= v_i + 0.7 \text{ V} \quad \text{when } v_i < -0.7 \text{ V}
\end{aligned}
\tag{5.40}
$$

and is plotted versus input voltage in Fig. 5.14*b*. Such a curve is called a transfer characteristic. In this case it shows that the voltage gain of the amplifier (the slope of the curve) is unity, except for the flat center region where the amplifier has zero gain. It is in the flat region that conduction is shifting from one transistor to the other. This nonlinearity in the transfer characteristics of the amplifier causes distortion, which is referred to as crossover distortion. Even though such an amplifier has no voltage gain, it can have substantial current gain and therefore substantial power gain.

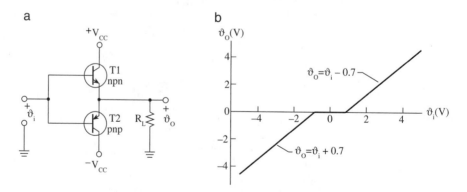

FIGURE 5.14   (a) A transformerless push–pull amplifier suitable for ICs. (b) Transfer characteristics of the amplifier, showing severe crossover distortion.

---

[19]Because the output voltage seems to follow the input voltage except for a small constant term, such an amplifier is also referred to as an emitter follower.

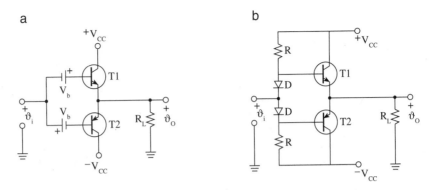

FIGURE 5.15 (a) The addition of biasing batteries reduces crossover distortion. (b) Replacing the batteries with diodes gives biasing voltages which automatically compensate for temperature variations.

Figure 5.15*a* shows how we can modify the complementary amplifier to eliminate crossover distortion. By adding batteries which have biasing voltages of about $V_b = 0.5$ V to 0.6 V, both transistors will be on the verge of turning on when there is no input signal, i.e., $v_i = 0$. Even a small positive input voltage will now cause T1 to conduct and similarly a small negative voltage will cause T2 to conduct, thereby eliminating most of the crossover distortion. A distortion-free amplifier would have a transfer characteristic of a straight line in Fig. 5.14*b*.

The amplifier circuit of Fig. 5.15*a*, besides having biasing batteries V_b which are awkward and difficult to provide for in an integrated circuit, has a more serious flaw, which can lead to the destruction of the transistors as the temperature increases even moderately. Recall that silicon devices are very sensitive to temperature increases. In Example 4.3 we showed that the reverse current in a diode increases with temperature rises. In Fig. 4.14, we showed that adding an emitter resistance to the biasing circuit will protect a transistor against thermal runaway destruction, which is caused by the decreasing resistance of silicon material with increasing temperature. Allowing the temperature to rise can rapidly lead to a runaway process as the increasing current causes increased $I^2 R$ losses which further increase the heat and temperature in a silicon device. Correspondingly, we can state that if the turn-on voltage for room-temperature silicon is 0.7 V, then warmer silicon will have a smaller turn-on voltage; typically V_{be} will decrease by 2.5 mV for every 1°C rise. Therefore, maintaining a constant biasing voltage on a transistor as the temperature increases in effect increases the forward bias on the transistor, speeding the runaway process until large currents destroy the transistor. This effect is critical in power amplifiers which carry significant currents. To protect against heat damage, power amplifiers have efficient and frequently large heat sinks—usually thick aluminum plates directly attached to power transistors.

To avoid this type of destruction, we modify the circuit of Fig. 5.15*a* to that of Fig. 5.15*b* by replacing the bias batteries with diodes D whose forward voltages will track the base–emitter voltages of the transistors. In practice, the diodes are mounted on the same heat sink as the transistors, guaranteeing the same temperature changes. Now as the temperature rises, the biasing voltage V_{be} will decrease automatically because the forward voltage drop of the diode decreases. The current through the transistor is reduced and the circuit is stabilized. Frequently the diodes are replaced by thermistors, which are a type of resistor whose resistance decreases as the temperature rises.

We now have an amplifier circuit that is ideally suited for integrated circuits. It is efficient because it is Class B and it can be produced at low cost since coupling capacitors and transformers are absent. It is Class B because the base bias on both transistors is so adjusted that in the absence of a signal the transistors are cut off. Therefore, current flows in each transistor only when the input signal biases its emitter–base junction in the forward direction. Because of the opposite polarities of the transistors, this happens on alternate half-cycles of the input voltage. Hence one transistor delivers current to the load while the other is cut off (as shown in Fig. 5.13b, which applies to this case). The output signal v_o is a replica of the input signal v_i, even though each transistor operates only half the time. The high efficiency of the complementary circuit is due to small I^2R losses because the DC current in the load resistor is zero at all times.

5.8   AM RADIO RECEIVER

Now that we know the characteristics of electrical components, let us see how they are used in a system such as an AM radio (AM stands for *amplitude modulation*). Figure 5.16 shows a *superheterodyne* receiver for the 550kHz–1.6MHz broadcast band. The objective of this system is to receive a signal containing desired information, in this case music or speech signals that were placed on a carrier at the broadcast station, separate the signal from the carrier at the receiver, amplify it, and reproduce it by a speaker for our listening pleasure.

The need for such a system is obvious: it is desired to provide entertainment and information for many people, including those that are far away from the source of entertainment. Considering that sound from music or speech carries only over short distances, we need a carrier that could deliver the desired information over long distances. We find such a carrier in electromagnetic (EM) waves, which have the ability to travel over long distances at the speed of light. All that remains is to mount the information on these carriers. This is done at the radio station and is referred to as *modulation*. Modulation is nothing other than superimposing two voltages—information signal plus carrier. The combined signal is then sent out (by transmitter and antenna) as an electromagnetic wave. EM waves are therefore a desirable medium for transmission of a modulated signal. EM waves have different properties depending on their frequency.[20] AM radio waves in the broadcast band (550 kHz–1.6 MHz) are generated at radio stations with kilowatts (kW) of power and carry information signals well for distances up to hun-

---

[20]As a rule of thumb, the higher the frequency of a wave, the more information it can carry but its ability to go long distances decreases, and as frequency continues to increase, waves begin to mimic light waves.

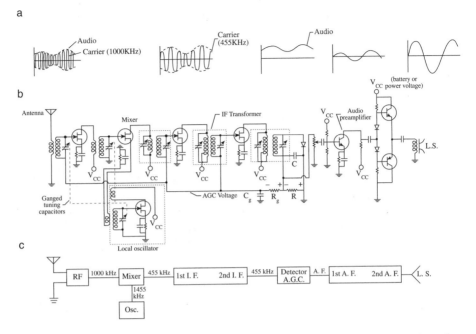

FIGURE 5.16 (a) Signal voltages throughout the receiver. (b) Circuit of the complete superheterodyne receiver. (c) Block diagram showing arrangement of the component parts.

dreds of miles. Exceeding these distances, they are subject to atmospheric noise, and because amplitude modulation is used, any voltage spikes due to noise or lightning can become objectionable. AM radio, using only 10 kHz of bandwidth per station, serves the purpose of distributing information to masses of people extremely well.

Yet if music is the primary product, AM radio has its faults. The limited bandwidth and amplitude modulation, with its susceptibility to all types of noise spikes, are limitations. FM radio, on the other hand, is well suited for music. The higher frequencies of the FM band (88–108 MHz) allow the use of a wider bandwidth (200 kHz) per station and the use of frequency modulation (FM), because of its noise-limiting ability, guarantees almost noise-free reception. Even though FM transmission is theoretically limited to reception along a line-of-sight path, it is at present preferred for high-quality music. Still higher frequencies such as UHF or microwaves could be used, but because they are strictly limited to line-of-sight, only an area near the transmitter could be covered. Also, perhaps more importantly, the frequency spectrum is extremely crowded and all available frequencies are already assigned for other purposes.

Having briefly described the properties of the carrier, which is a single-frequency signal that is broadcast by the radio station and is observable during moments when the station is not broadcasting speech or music, let us now study the AM radio receiver. The block diagram in Fig. 5.16 displays the major components. To learn what each one does, it is helpful to sketch the signal voltages throughout the receiver which are shown as the top figures.

5.8.1   RF Stage

The antenna receives a wide range of broadcast signals, which are weak signals typically in the microvolt (μV) range, and feeds these to the RF (radio frequency) stage. Before any amplification takes place, a single signal representing a broadcast station is first selected by the LC parallel resonant circuit which is at the input of the RF stage. The variable capacitor of the LC circuit is tuned to the selected frequency by the user (a variable capacitor is denoted by an arrow across the capacitor symbol). At the input of the RF transistor, there is now a single broadcast signal for amplification—all other signals in the broadcast band are attenuated because their frequencies are not at the peak of the resonance curve. This incoming signal is sketched in the first top figure. It is a carrier frequency of a broadcast station (in this case a 1000 kHz signal) that is modulated by a single-frequency audio tone; that is, the peaks of the carrier voltage rise and fall with the audio tone, which is referred to as amplitude modulation. To keep it simple, we have chosen a single audio tone for broadcasting. At the output of the RF stage, we now have an amplified broadcast signal, denoted as 1000 kHz in the block diagram.

5.8.2   Mixer

In order to reduce the number of tunable resonance circuits that have to be synchronized in a radio receiver whenever a new station is tuned in, we use the principle of *super-heterodyne*. Using heterodyne circuitry, we can decrease the number of tunable stages and replace them with fixed-tuned stages which stay tuned to the same frequency, called the *intermediate* frequency (IF), as the receiver is tuned to a variety of different stations. The term *heterodyne* implies the use of a heterodyne frequency (more commonly known as a *beat* frequency), which is the difference between two combining frequencies. Furthermore, this frequency is chosen so it can be amplified with higher gain and selectivity than the incoming, higher broadcast frequency. In AM radio the IF stages are typically tuned to 455 kHz. In order to make this possible, we need a tunable local oscillator which produces single-frequency signals. When this local signal is mixed with the incoming signal, beat (heterodyne) frequencies[21] are produced which are difference and sum frequencies of the incoming and local oscillator signals. The mixing is done in a mixer stage, which uses the nonlinear characteristics of a transistor to produce the beat

---

[21] In order to understand how the frequency of the incoming signal is changed to the IF frequency, let us first consider sound and learn about the phenomenon of beats. Strike middle C on the piano. The sound you hear has a frequency of 256 cycles per second (256Hz). Now strike the note before it, B, on the piano. This note has a frequency of 240 Hz. Now strike both keys together. The sound you hear is neither B nor C but a mixture of the two. If you listen closely you will notice that this new sound rises and falls in loudness or intensity. If you can time the rise and fall of sound you will notice that it occurs 16 times per second, the exact difference between the frequencies of B and C. We call this rise and fall the *beat* note. Its frequency is equal to the difference between the frequencies of the notes producing it. Similarly, when radio waves of different frequencies are mixed, beat (heterodyne) frequencies are produced.

frequencies. To produce a beat frequency of 455 kHz when the incoming signal is 1000 kHz, for example, the oscillator output must be 1455 kHz (or 545 kHz). The output of the mixer stage is now fed into an IF amplifier, but since a LC resonant circuit which is tuned to 455 kHz is placed between the mixer and the first IF stage, only the 455 kHz signal is selected for amplification.

Local Oscillator

As a listener tunes across the broadcast band, the local oscillator is also tuned in synchronism so as to precisely produce a single-frequency signal (essentially an unmodulated carrier signal) which is higher in frequency by 455 kHz in comparison to the incoming signal. For example, if a 800 kHz station is received, the oscillator produces a 1255 kHz signal which again results in a 455 kHz beat signal at the mixer output. An oscillator is an amplifier with positive feedback from output to input. An LC circuit determines the frequency at which this unstable circuit will oscillate. Therefore, in the superhet receiver shown, the tunable capacitors at the antenna input of the RF amplifier and at the oscillator must be synchronized, either by a mechanical linkage or by electronic means. The linkage is indicated by dashed lines.

IF Amplifiers

Much more gain in a receiver is needed (up to 10^6) than a single RF stage can provide. In a superheterodyne receiver, the additional gain is produced at a single, predetermined frequency by the IF stages. In other words, after selecting a particular radio station, we change the frequency of the currents flowing in our receiver to the predetermined frequency of 455 kHz (the intermediate frequency) and then feed it into amplifiers tuned to that frequency. The amplitudes of the predetermined frequency are modulated in the same way as was the incoming signal. Hence, in our second top figure, the modulated signal is shown with a lower carrier frequency but the same audio tone. Summarizing, we can state that superheterodyne circuitry is simpler, less costly, and less critical. In addition, the IF frequency is chosen so as to optimize the sensitivity and selectivity of the receiver. The selectivity of a superhet receiver is high also because six tuned parallel resonant circuits exist in the IF strip. Pairs of such LC circuits are coupled magnetically; that is, the primary magnetic field links the secondary coil, thus transmitting AC but preventing DC from reaching the following transistor where a high DC level could damage or saturate the following transistor. Each pair is shown inside a dashed square and is referred to as an IF transformer. Such IF transformers commonly occur at the input and output of each IF amplifier stage.

Detector or Demodulator Stage

Our signal at the output of the last IF stage is now sufficiently amplified, so it is about 1–10 V. We are now ready to "strip off" the information that is riding on the IF carrier. In our case, for simplicity, we have assumed that the information transmitted from the broadcasting station to our antenna is a simple single-frequency audio tone. How do we recover this audio tone, amplify it further, and finally reproduce it in a speaker so we can listen to it? First, the signal at the output of the second IF amplifier is fed into a diode (which is a nonlinear element). The diode removes the bottom half of the signal, which then passes through a low-pass RC filter. The resulting signal after demodulation looks like that in Fig. 5.16a above the diode: it is the audio tone riding on a DC level. To ensure that the low-pass filter (see Fig. 2.6) passes the audio frequency but not the carrier frequency, the values of R and C are chosen so the cutoff frequency $f_o = 1/2\pi RC$ is, for example, 15 kHz. Then DC and audio frequencies up to 15 kHz will be passed, but the 455 kHz carrier frequency will not.

To show explicitly demodulation action, we redraw the detector circuit in Fig. 5.17a. At the secondary of the last IF transformer in Fig. 5.16b, the signal voltage is v_{in} and is shown in Fig. 5.17b. After passing through the diode, which removes the bottom half of the signal, the rectified signal is smoothed by the low-pass RC filter and is shown in Fig. 5.17c. Typical values of R and C when f_o is 15 kHz are $R = 10$ kΩ and $C = 0.001$ μF. Except for the values of R and C, a demodulator circuit is identical to the power supply rectifier filter of Fig. 3.4. Thus, in Fig. 5.17c, we observe that the audio tone is reproduced, because the voltage across the capacitor cannot follow the rapid variations of the carrier signal, but can follow the much slower audio signal. The wiggles in the audio signal are the exponential decays of the capacitor voltage which have a time constant RC which is much longer than the period T ($= 1/455$ kHz) of the IF signal. These high-frequency 455 kHz wiggles shown in Fig. 5.17c are exaggerated and are evened out by the low-pass filter so the resulting signal looks like the smooth signal in Fig. 5.16. A detector of this type is also referred to as a peak detector, as it follows the peaks of the modulated carrier signal. But even if the RC filter would not smooth the wiggly signal in Fig. 5.17c, the ear would because the ear does not respond to the high frequencies which the wiggles represent. In that sense, the ear acts as its own low-pass filter.[22]

---

[22]To gain additional insight into demodulation, we can look at it from a different viewpoint. We can state that the audio signal (assume it to be a 1000 Hz audio tone, i.e., $f_a = 1$ kHz) is not explicitly present in the IF strip, and only three frequencies are present: f_c, which is the 455 kHz carrier; $f_c + f_a$, which is a 456 kHz frequency; and $f_c - f_a$, which is a 454 kHz frequency. These three signals, in addition to others, were produced by the mixer stage. The IF transformers were tuned to the 455kHz center frequency and thus passed only the three signals. How do we recapture the f_a audio tone? The three signals, after amplification in the IF strip, are fed into a diode (see Fig. 5.17a). The diode is a nonlinear element with an exponential relationship between voltage and current as given by (4.6). If we expand the exponential in a Taylor series as $e^v = 1 + v + v^2/2 + \cdots$, we see that the v^2 term will result in

$$\cos \omega_c t \cos(\omega_c - \omega_a)t = \frac{1}{2}\{\cos(\omega_c + (\omega_c - \omega_a))t + \cos(\omega_c - (\omega_c - \omega_a))t\}$$

$$= \frac{1}{2}\{\cos(2\omega_c - \omega_a)t + \cos \omega_a t\}$$

where $\omega = 2\pi f$. Hence, the diode current contains the audio tone $\cos \omega_a t$, which means that it is recovered and can now be further amplified in the audio stages of the receiver. For the v^2 term to be effective, the voltage v cannot be small, otherwise we obtain a linear approximation of the exponential $e^v = 1 + v$, which is usually desirable in most circumstances. However, here we need the nonlinear term v^2, so when the frequencies f_c and $f_c - f_a$ are multiplied in the diode, i.e., $(\cos \omega_c t + \cos(\omega_c - \omega_a)t)^2$ we obtain the above results. The signal voltages must therefore be adequately amplified in the IF strip, before the v^2 term becomes effective. This is part of the demodulator circuit design. Also, the low-pass action of the RC filter will eliminate all higher frequencies, except for the audio tone.

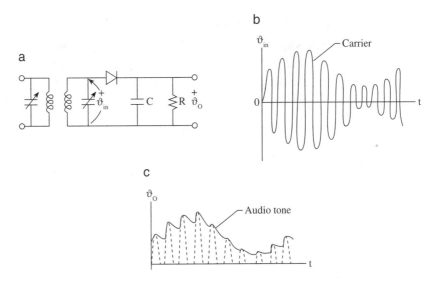

FIGURE 5.17 (a) The demodulating circuit. (b) The signal voltage v_{in} at the last IF stage, showing a carrier signal whose voltage peaks rise and fall in rhythm with the audio tone. (c) The audio tone is demodulated by the diode and the low-pass RC filter.

Automatic Gain Control (AGC)

The audio volume could vary significantly (and annoyingly) as one tunes across the broadcast band and receives near and distant stations were it not for automatic gain control (AGC). A clever circuit is used to keep the gain and hence the audio volume constant from station to station. Recall that the function of the low-pass RC filter in the demodulating circuit was to stop the 455 kHz carrier frequencies but to pass the audio frequencies. The RC filter time-averages out the carrier variations but not the audio variations. Now, if we could also time-average out the audio variations, we would be left with a DC signal which is proportional to received signal strength. This voltage, customarily referred to as *AGC*, is applied as a negative bias to the preceding FET amplifiers. A strong station will now produce a larger negative bias on the gates, reducing the gain, and a weak station will produce a lesser bias, increasing the gain of the amplifiers. Such a feedback technique causes both weak and strong stations to have roughly equal loudness.

We accomplish this by passing the audio signal through another low-pass $R_g C_g$ filter, as indicated in Fig. 5.16, for which the cutoff frequency $f_o = 1/2\pi R_g C_g$ is chosen very low, say 1 Hz (a typical combination would then be $R_g = 15$ kΩ and $C_g = 10$ μF). All variations are thus smoothed out, leaving a DC voltage which is proportional to the received signal strength of a station. Automatic gain control, sometimes referred to as automatic volume control, is used in virtually all receivers.

5.8.3   Audio Frequency Amplification

The demodulated audio signal is applied through an RC coupling circuit to the first audio frequency amplifier. The resistor R is shown as a variable R and is used as a volume control, while the capacitor blocks DC from getting to the first audio stage and also from interfering with the DC biasing of that stage. The signal at that stage is shown in Fig. 5.16 as an AC signal with the DC level removed. After amplification by the first stage, this signal is now of sufficient strength to drive a complementary-symmetry power amplifier. The power amplifier has a sufficiently low output impedance so it can drive a speaker (4–16 Ω) directly. Hence, an impedance-changing transformer to ensure maximum power transfer from power amplifier to speaker is not needed (see Fig. 2.18).

5.8A   Summary

A radio receiver presents a good example of applications of analog circuitry presented thus far, even though a radio receiver is rarely assembled out of discrete components in the age of microchips. As integrated circuits become more common, various components of a receiver became available in chip form. First, preamplifiers came as chips, then the entire IF strip was made as an integrated circuit, and now practically the entire AM receiver is made in chip form. High-power receivers are more modular, with the receiving section as one module, the power supply as another, and the audio section as one chip, usually mounted on a heat sink.

5.9   SUMMARY

In this chapter we explored the use of amplifiers in practical circuits.

- We started by stating the characteristics of ideal amplifiers (infinite input resistance, zero output resistance, very large but constant gain) which are frequently used as design goals for practical amplifiers. When the input signals to the transistor are small, we are able to replace the transistor by a linear model, the advantage of which is that the transistor could be viewed not as a mysterious three-terminal device, but as made up of ordinary circuit elements, namely, a resistor and a controlled source. Replacing the output terminals of a transistor by Thevenin's or Norton's equivalent allows us to treat a circuit with transistors as an ordinary circuit, that is, a circuit without transistor symbols. Furthermore, the gain of a small-signal amplifier can be expressed in terms of circuit parameters. For example, the gain of a FET amplifier is expressed as $A = -g_m R_L$.

- When a numerical value for the gain of an amplifier is given, it is understood to be applicable only for a frequency range called the midband. For frequencies below midband, the gain decreases because coupling capacitors are involved, and for frequencies above midband, the gain decreases because shunting capacitances, either internal to the transistor or external due to the circuit, are present. Midband is customarily defined as the frequency range between f_l and f_h for which the gain A does not fall below -3dB. Typical bandwidth for audio amplifiers is 25 kHZ, for a TV set 6 MHz, for an 80-column monitor 15 MHz, and for an oscilloscope amplifier as much as 100 MHz.

- Cascading a number N of identical amplifiers will decrease the bandwidth in comparison to a single stage as the gain of the cascaded amplifiers will be down by $-3N$ dB at the -3 dB frequencies of the single stage. Thus, even though the gain of cascaded amplifiers is $A_{\text{cas}} = NA$, it is limited to a narrower bandwidth.

- In integrated circuits capacitors are avoided as they take up too much space. It is possible to design direct-coupled amplifiers in which the collector potential of one stage is equal to the base potential of the succeeding stage. Eliminating the coupling capacitor also eliminates the drop in gain at low frequencies. Therefore, in integrated circuits the bandwidth is given by the upper half-power frequency, f_h.

- When pulses or rapidly changing signals need to be amplified, we showed that the more rapidly the signal changes, the larger the bandwidth of the amplifier must be for undistorted amplification of the signal. The rise time of a pulse t_r can be related to the bandwidth by $t_r = 1/3 f_h$, where f_h is the high-cutoff frequency. Thus, larger-bandwidth amplifiers can more faithfully magnify sharper pulses.

- If a square pulse of length T_p is used as the input to an amplifier, what must the high-frequency cutoff f_h of the amplifier be so excessive distortion is avoided? An acceptable ballpark answer to this question is that f_h should be chosen such that the reciprocal of the pulse length T_p is equal to $f_h = 1/T_p$. However, the rise time t_r of the amplified pulse (5.32) gives a more accurate answer.

- After a signal is sufficiently amplified in voltage, which frequently is the end goal in some applications, power capabilities must be also added to the signal. This is achieved by feeding the amplified signal into a power amplifier, which typically is a current amplifier. At the output of a power amplifier we now have large voltages (on the order of power supply voltages) and a large current capability as depicted in Fig. 5.1. Transformer coupling of audio stages is convenient, but because of the cost and bulkiness of transformers, their use is confined to special situations. In solid-state devices, we can use complementary *npn* and *pnp* transistors as Class B push–pull-type power amplifiers which are very efficient and have a low output impedance, allowing low-impedance speakers to be directly coupled to them. Such power amplifiers, capable of delivering 50–200 W of audio power, come typically as flat chips, not much larger than 2×3 in.

- An AM receiver served as an example to tie together seemingly disparate devices into a practical system. An FM receiver, TV set, or some other electronic device could have served that purpose just as well. Even though many components of the receiver shown are implemented in chip form (low-power and miniaturized receivers practically come as single chips), a discrete layout of a receiver allowed us to study the superheterodyne principle, demodulation, and automatic gain control. It should be pointed out that the schematic in Fig. 5.16 appears much simpler than a schematic of a commercial receiver. The reason is that in a commercial version of an AM radio, numerous transistors and diodes are used for all sorts of things such as stabilizing and protecting the circuitry against overvoltages, excessive currents, temperature fluctuations, etc. Furthermore, adding to the apparent complexity are circuits for the convenience of the consumer, such as treble, bass, loudness, etc. These kinds of peripheral circuits were left out of the schematic in Fig. 5.16 as not contributing to the fundamentals of an AM receiver. We can refer to the receiver in Fig. 5.16 as a bare-essentials receiver.

Problems

1. A transducer produces a 1 μW signal at 100 μV, which is applied to an amplifier. If the output of the amplifier is to be 100 W at 1 V, find the voltage gain, current gain, and the power gain of the amplifier.
   Ans: $A_V = 10^4$; $A_I = 10^4$; $A_P = A_V A_I = 10^8$.

2. The (open-loop) gain of an amplifier which has an input/output resistance of 10^5 $\Omega/10^3$ Ω is given as 10^4. If a transducer which has an internal resistance of 100 kΩ and produces a 10 μV signal is connected to the input of the amplifier, find the real gain of the amplifier when a load of 1 kΩ is connected to the output of the amplifier.

3. Using the real amplifier outlined in Problem 2, how would you change the input/output resistance of the amplifier so as to maximize the real gain?

4. Evaluate the small-signal parameters at the operating point of the amplifier of Fig. 4.17 and use these to calculate the gain of the amplifier. Compare your result to the graphical result obtained in the text related to Fig. 4.17 (agreement within 10–15% is good).

5. A MOSFET for which the output characteristics are given by Fig. 4.17c is to be represented by a small-signal model. Find the g_m and r_d parameters near $V_{gs} = -2$V and $V_{ds} = 30$ V and specify the load resistor R_L which would result in a voltage gain of -12.
   Ans: $g_m \approx 0.75$ mS; $r_d \approx 60$ kΩ; $R_L = 16$ kΩ.

6. Calculate the small-signal gain of the amplifier of Fig. 4.16a. Compare the results using (5.6) and (5.7).

7. The source resistor R_s in the amplifier of Fig. 4.16a is not bypassed with a capacitor. Derive an expression for the small-signal gain of the amplifier with R_s not ignored and calculate the gain using values obtained in Examples 4.7 and 5.2.
   Ans: $A_r = -g_m r_d R_L/(r_d + R_s + R_L)$; -4.57.

8. The collector characteristics for BJTs are given by Figs. 4.7, 4.13, and 4.14. In each case calculate a representative value for current gain β (also known as h_f).
   Ans: $\beta = I_c/I_b = 12 \cdot 10^{-3}/100 \cdot 10^{-6} = 120$; 500; 55.

9. A common-emitter amplifier uses a BJT with $\beta = 60$, operating at a Q-point for which $I_c = 1$ mA. Use the small-signal equivalent circuit of Fig. 5.6b to find the voltage and current gain of the amplifier if the amplifier is to deliver an AC signal current of 0.3mA rms to a load of $R_L = 5$ kΩ. Assume $r_c \gg R_L$.

   Ans: $i_o/i_i = i_c/i_b = 60$; $v_o/v_i = v_{ce}/v_{be} = -200$.

10. Using the amplifier specified in Problem 9, calculate the transconductance g_m and use it to find the voltage gain of the amplifier.

11. A device (another amplifier, another load resistance, etc.) is connected to the output terminals of the amplifier in Fig. 5.7a. Find the impedance which the device "sees"; that is, find the output impedance of the amplifier. Use two methods.

    Method A: Assume a voltage source v_i is connected to the input and find $Z_o = v_{o,\text{open-circuit}}/i_{o,\text{short-circuit}}$.

    Method B: Short out v_i. Connect a voltage source v to the output and calculate the resulting current i; then $Z_o = v/i$.

12. A load resistor R_L' is connected to the output terminals of the amplifier in Fig. 5.7a. Find the current gain of the amplifier.

    Ans: $A_i = \dfrac{i_{R_L'}}{i_b} = \dfrac{\beta(r_c\|R_L)}{(r_c\|R_L)+R_L'}$ using Fig. 5.7c and $A_i = \dfrac{\beta R_L}{R_L+R_L'}$ using Fig. 5.7d.

13. If the input to an amplifier is 1 W, calculate the gain of the amplifier in decibels if the amplifier output is 100 W, 1 W, or 0.1 W.

    Ans: 20 dB; 0 dB; -10 dB.

14. In an audio system, the microphone produces a voltage of 10 mV which is connected by a cable to an amplifier with a voltage gain of 30 dB. The cable introduces a loss of 5 dB. Calculate (a) the gain of the system and (b) the output voltage.

15. Adding a preamplifier to an audio amplifier increases the voltage gain by 60 dB. What is the corresponding factor by which the voltage is increased?

    Ans: 10^3.

16. An amplifier has an open-circuit voltage gain of 1500, an input impedance of 3 kΩ, and an output impedance of 300 Ω. Find the voltage and power gain of the amplifier in decibels when a load impedance of 200 Ω is connected to the output of the amplifier.

    Ans: $A_v = 55.6$ dB; $A_p = 67.3$ dB.

17. Some AC voltmeters provide dBm readings which are based on a 600 Ω reference resistance in addition to the 1mW reference power. If an AC voltmeter reads 16 on its dBm scale, what is the rms value of the voltage that corresponds to this reading?

    Ans: 4.89 V.

18. A microphone of 300Ω impedance has an output at -60 dB below a 1 mW reference level. The microphone provides input to an amplifier which in turn is to provide an output level of 30 dBm when a load of 8 Ω is connected to the output terminals of the amplifier. Find

    (a) The output voltage of the microphone

a b

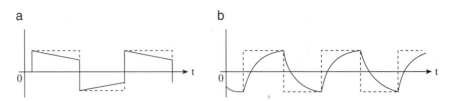

FIGURE 5.18 (a)Tilted square wave caused by insufficient low-frequency content. (b) Smoothed square wave caused by insufficient high-frequency content. See Problem 26.

(b) The gain in dB required of the amplifier
(c) The output power of the amplifier
(d) The load voltage

19. For the amplifier of Fig. 5.8, calculate the value of the coupling capacitor C if the half-power frequency (cutoff or 3dB frequency) for one stage is to be $f_l = 10$ Hz. Assume $R_L = 5$ kΩ and $r_i = 1$ kΩ.

20. For the amplifier of Problem 19, how many decibels (when compared to midband gain) will a two-stage amplifier lose at the one-stage half-power frequency $f_l = 10$ Hz?
    Ans: -6 dB.

21. For cascaded amplifiers in general, at the half-power frequency of one stage, what will the loss in decibels be in comparison to midband for a two-stage amplifier? For a three-stage? For an n-stage?

22. For the amplifier of Problem 19, for which the one-stage half-power frequency was given as $f_l = 10$ Hz, what will the corresponding half-power frequency be for the two-stage amplifier?
    Ans: 15.6 Hz.

23. To calculate the decrease in bandwidth of a n-stage amplifier in comparison to a single-stage one, the -3 dB high and low frequencies are needed. Derive a formula for the low-frequency cutoff of a n-stage amplifier, if the single-stage low-frequency cutoff is f_l hertz.
    Ans: $f_{l,n} = f_l(10^{0.3/n} - 1)^{-1/2}$.

24. The total shunting stray capacitance at the output of an amplifier is given as 100 pF. If this amplifier is working into a load of 10 kΩ, find the high-frequency cutoff f_h.

25. If the -3 dB corner frequency of a low-pass filter is given as f_o, what is the attenuation of this filter at the frequencies $0.1 f_o$ and $10 f_o$?

26. An amplifier's bandwidth is specified as 20 Hz to 20 kHz. What would appropriate square-wave test frequencies be to confirm the specified bandwidth. Sketch the expected output waveforms at both square-wave frequencies.
    Ans: 628 Hz; 6 kHz and Fig. 5.18.

27. A step-voltage (a voltage obtained when a battery is turned on) is to be amplified. If the rise time of the amplified step voltage is to be less than $2\mu s$ (microseconds), find the amplifier bandwidth.

28. A 5ms voltage pulse (like the ideal input pulse shown in Fig. 5.11) is to be amplified. If the tilt in the amplified pulse is to be less than 20%, find the amplifier's low-frequency cutoff.

29. An audio amplifier is to be used to amplify a single square pulse (as shown by the ideal input pulse in Fig. 5.11). If the pulse length is 3 ms ($= T/2$ in Fig. 5.11) and the bandwidth of the audio amplifier is specified by the half-power frequencies of 15 Hz and 20 kHz, estimate the tilt and the rise time of the amplifier output.
    Ans: 0.28; 22 μs.

30. In the transformer-coupled power amplifier of Fig. 5.12, the AC load line was chosen as $R'_L = 500$ Ω. Does this give maximum power output? Would $R'_L = 1000$ Ω or $R'_L = 250$ Ω give more power?

31. Find the transformer turns ratio for maximum output power when the load resistance $R_L = 10$ Ω for the amplifier in Fig. 5.12.

32. In the amplifier of Fig. 5.12, if one of the biasing resistors R_2 is changed to 6 kΩ, what must be the new value of R_1 be so as to maintain the same biasing?
    Ans: 479.5 Ω.

33. The efficiency of the amplifier in Fig. 5.12 is close to 50% with $R_L = 5$ Ω. What is the efficiency of the amplifier if the load resistor is changed to $R_L = 10$ Ω.

34. Design a transformer-coupled power amplifier of the type shown in Fig. 5.12 to deliver power to a 5 Ω speaker. You have available a 5 V battery that can deliver a 10 mA current on the average. Retain R_E and R_2 but

    (a) Find the new value of R_1
    (b) Find the output power and the efficiency of the amplifier
        Ans: $I_{c,Q} = 10$ mA; $V_{ce,Q} = 5$ V; $R_1 = 800$ Ω; $R'_L = 500$ Ω; 25 mW; $\approx 50\%$.

35. For the Class B push–pull amplifier of Fig. 5.13 determine the required reflected load resistance R'_L if the battery voltage $V_{CC} = 9$ V and the output power is 0.5 W.
    Ans: 81 Ω.

36. Determine the maximum current amplitude I_c for the Class B push–pull power amplifier of Fig. 5.13 if the output power is 1 W and $V_{CC} = 12$ V.

37. In the complementary-symmetry amplifier of Fig. 5.15, $R_L = 8$ Ω and $V_{CC} = 12$ V. Assume that when either transistor conducts, $V_{ce,\text{sat.}} \approx 0$; that is, when $v_i > 0$, T1 conducts and T2 is cut off and vice versa. Find

    (a) The maximum power delivered to the load R_L.
    (b) The maximum power dissipated by each transistor. *Note*: for each transistor, the collector dissipation is one-half the total dissipation

38. List the desirable features of a superheterodyne receiver.

39. If an IF frequency of 200 kHz is desired, list the range of frequencies that a local oscillator must produce for AM band reception.

40. In the AM receiver of Fig. 5.16, the low-pass RC filter in the demodulator stage is to be designed so it will pass audio frequencies up to 10 kHz. Find appropriate values for R and C.

    Ans: $R = 10$ kΩ; $C = 1.59$ nF.

41. If the time constant for the AGC signal in an AM radio is chosen as 50 ms, find the value of C_g in Fig. 5.16 if $R_g = 15$ kΩ. What is the cutoff frequency of this filter?

    Hint: A reasonable estimate of the 3 dB frequency f_h of an amplifier used to amplify a pulse of pulse length $T/2$ is to choose f_h equal to the reciprocal of the pulse length, i.e., $f_h = 1/(T/2)$. From (5.32) we then have $t_r = 0.33(T/2)$. Clearly an amplifier with a larger bandwidth than $1/(T/2)$ hertz will reproduce the initial shape of the square pulse more faithfully.

CHAPTER 6

Operational Amplifiers

6.1   INTRODUCTION

Integrated circuit technology made possible the operational amplifier (op amp) on a chip, which is a high-gain, multitransistor voltage amplifier, encapsulated in a small chip and costing less than a dollar. The first popular chip was the 741, a 24-transistor op amp. It came out in the late 1960s and continues to be popular to this day. Op amps were originally used in analog computers where, by changing the external circuitry which is connected to the op amp, they could be used as adders, integrators, and differentiators. Today these versatile amplifiers find applications in all fields, including signal processing, filtering, switching circuitry, instrumentation, and so forth. Their use is only limited by the ingenuity of the designer.

An op amp is designed to handle two inputs at the same time. The output signal will be an amplified version of the difference between the two inputs. That is, if v_+ and v_- are the two inputs, $v_{\text{out}} = A(v_+ - v_-)$, where A is the *voltage gain* of the op amp. If we set one of the inputs to zero (grounded), v_{out} will then just be an amplified version of the nonzero input signal, that is, $v_{\text{out}} = Av_+$ or $v_{\text{out}} = -Av_-$. The ability to amplify the difference between two signals makes the op amp valuable in instrumentation, where unwanted interference signals common to both inputs are automatically canceled out.

6.2   OP AMP—AN ALMOST IDEAL AMPLIFIER

In Section 5.2, we learned that an ideal amplifier is characterized by infinite gain ($A = \infty$), infinite input resistance ($R_i = \infty$), and zero output resistance ($R_o = 0$). To these characteristics we could also add that the bandwidth should be infinite—in other words, the ideal amplifier amplifies all frequencies from DC to the highest frequencies with the same gain—and furthermore that all of the preceding characteristics should remain

a   b

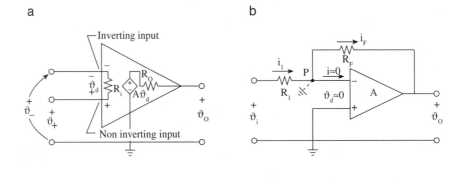

c

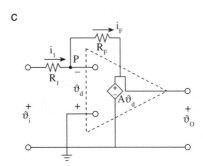

FIGURE 6.1   (a) The equivalent circuit (open-loop) of an op amp showing two input terminals. (b) The inverting amplifier consists of an op amp and external circuitry (closed-loop). For simplicity, power supply connections to the op amp are not shown. (c) The equivalent circuit of the inverting amplifier, using the ideal representation for the op amp.

stable with temperature changes. The equivalent circuit of such an amplifier is given by Fig. 5.3.

The op amp matches these characteristics to a high degree. A popular model is the μA741, a cheap but high-performance operational amplifier for which $A = 10^5$, $R_i = 2\ M\Omega$, and $R_o = 75\ \Omega$. Figure 6.1a gives the equivalent circuit of an op amp enclosed by a triangle, which is the traditional symbol for the op amp. It is basically a differential amplifier as it responds to the difference in the voltages applied to the inverting and noninverting inputs, i.e.,

$$v_o = A(v_+ - v_-) = Av_d \qquad (6.1)$$

where A is the open-loop (no feedback) gain of the op amp (typically 10^5 or larger). Note that the polarity of v_d is given by the $+$ and $-$ signs of the input terminals. For most circuit calculations the ideal representation of an op amp (shown within the dashed triangle in Fig. 6.1c) can be used; that is, in Fig. 6.1a the input resistance R_i is replaced by an open circuit ($R_i = \infty$) and the output resistance R_o is replaced by a short circuit ($R_o = 0$).

An op amp, even though almost ideal in the first three characteristics (A, R_i, and R_o), has a gain that varies substantially as the frequency of the input signal changes, and furthermore its characteristics are not very stable with temperature changes. Therefore, typical use of an op amp is in a closed-loop configuration in which negative feedback is applied which "kills" much of the open-loop gain of the op amp, resulting in a lesser but very stable gain. The loop is closed by using external resistors to feed back some of the output voltage to the input. An example is provided in the following section.

6.2.1   The Inverting Amplifier

The simplest example of a closed-loop configuration is the inverting amplifier shown in Fig. 6.1b, in which an input signal v_i is applied to the inverting (negative) terminal through the R_1 resistor and the noninverting input is grounded, while R_F serves to "feed back" from output to input. The circuit equivalent of this popular amplifier is shown in Fig. 6.1c. Using Kirchoff's voltage law we can readily write for the input loop $v_i = i_1 R_1 - v_d$, and for the output loop $v_o = A v_d = -i_F R_F - v_d$. Note that in the equivalent circuit we show a connection from the controlled voltage source $A v_d$ to ground that is not present in Fig. 6.1b. This can be confusing. However, in Fig. 6.1b the op amp is represented simplistically by a triangle symbol and such a connection would add little meaning. Besides, when analyzing a circuit which has op amps, the equivalent circuit for each op amp should always be used. The voltage gain A_r with feedback is obtained as follows: using the ideal representation ($R_i = \infty$, $R_o = 0$) of the op amp shown in Fig. 6.1c, we conclude that since $R_i = \infty$, $i = 0$. Therefore all current through R_1 passes through R_F

$$i_1 = i_F \qquad (6.2)$$

Furthermore, we note that $v_d = v_o/A \approx 0$ as $A \to \infty$, so the op amp input is in effect shorted and the inverting terminal is effectively grounded.[1] Therefore, the gain can be stated as

$$A_r = \frac{v_o}{v_i} = \frac{-i_F R_F}{i_1 R_1} = -\frac{R_F}{R_1} \qquad (6.3)$$

This is a surprising result, as it says that the gain of the op amp with the external circuitry is equal to a ratio of the resistances of a feedback resistor and an input resistor. The gain is thus determined by external resistors only and is independent of A as long as A is high. Obviously, resistors are much less frequency and temperature sensitive than op amps. Hence, we have obtained an amplifier whose gain is not as high as A of an op amp alone, but it is very stable and has a constant value. The minus sign in (6.3) means that there is a phase reversal in the output voltage.

Let us elaborate on the result obtained in (6.3). A consequence of the output voltage fed back by resistor R_F to the inverting terminal is to make that terminal (point P) a

---

[1]Note that the output voltage v_o is limited by the power supply voltage, which is typically between 5 and 15 V. Hence, the input voltage v_d would be on the order of microvolts, given that $A > 10^5$.

virtual ground (suggested by the dashed ground symbol at P), which means that the voltage at that terminal, v_-, is zero. What makes this result so remarkable is that a real short circuit between ground and point P does not exist as then a substantial current would flow to ground. Point P is at ground potential but no current flows as $i = 0$ for the op amp, and hence the label virtual ground. Point P, for the inverting configuration, remains a virtual ground for any variations in the input signal v_i.

In addition to being a virtual ground, point P, for the inverting configuration, is also referred to as *summing point*. To sum several signals, for example, we can connect several resistors to point P as shown in Fig. 6.4a. The sum of the signal currents in the input resistors equals the current in R_F, since no current to ground exists at point P. Because the addition appears to take place at point P, it is called a summing point.

EXAMPLE 6.1   For the inverting amplifier of Fig. 6.1b, find the gain, input resistance, $v_d, v_i, i,$ and i_F. Assume $A = 5 \cdot 10^5$, $R_i = 10^6 \ \Omega$, $R_o = 0$, $R_F = 100 \ \text{k}\Omega$, $R_1 = 1 \ \text{k}\Omega$, and the power supply voltage is ± 5 V. In order to obtain a better understanding, derive the op amp gain, Eq. (6.3), without initially setting $v_d = 0$.

Since the amplifier input impedance R_i is large, the current i flowing into the op amp is negligible. Therefore, as in (6.2), current in R_1 equals the current in R_F, i.e.,

$$\frac{v_i + v_d}{R_1} = \frac{-v_d - v_o}{R_F}$$

Using (6.1), $v_d = v_o/A$ in the above expression, we obtain

$$v_o \left(1 + \frac{1}{A} + \frac{R_F}{A R_1} \right) = -\frac{R_F}{R_1} v_i$$

Given that the op amp gain is very large, we can take the limit as $A \to \infty$ and obtain

$$v_o = -\frac{R_F}{R_1} v_i$$

which is the desired result of (6.3).

The gain of the op amp with external circuitry depends only on the external resistors that are connected to the op amp. Hence, from (6.3), $A_r = -R_F/R_1 = -100/1 = -100$. The minus sign in the gain expression implies that the amplified output signal is $180°$ out of phase with the input signal.

The input resistance (it is the resistance that a source would see when connected to the v_i input terminals) is simply $R = v_i/i_1 = R_1 = 1 \ \text{k}\Omega$. This is a rather low input impedance (in practice the term *impedance* is used when addressing any kind of input resistance), not suitable when a high-impedance source is to be connected to the input terminals, as then only a small portion of the source voltage would be driving the amplifier. In addition, a feeble source might not be able to provide the large currents which a low input impedance requires. Ideally, a high-impedance voltage source should work into an infinite-impedance voltage amplifier.

The maximum magnitude of the output voltage v_o is limited by the supply or battery voltage (it usually is less by about 2 V). Any figure displaying a load line such as 4.13b

or 4.17c shows that $v_{o,\max} \approx V_{\text{powersupply}}$. Assuming $v_o = -5$ V, we obtain for the differential voltage $v_d = -5 \text{ V}/-A = 5/5 \cdot 10^5 = 10 \ \mu\text{V}$, which is a very small voltage and is usually neglected.

From (6.3), we have that $v_i = (-R_1/R_F)v_o = (-1/100)(-5) = 50$ mV. The current i flowing into the op amp input is given by $i = v_d/R_i = 10 \ \mu\text{V}/1 \ \text{M}\Omega = 0.00001 \ \mu\text{A} = 10$ pA. This current is so small, that the approximation by zero is valid.

The feedback current $i_F = i_1$. Hence, $i_F = v_o/R_F = 5 \text{ V}/100 \ \text{k}\Omega = 50 \ \mu\text{A}$.   ■

In summary we state two basic rules for the analysis of op amp circuits. **Rule 1** is assume the two input terminals are at the same voltage, i.e., the differential voltage $v_d = 0$ (or equivalently $v_+ = v_-$). **Rule 2** is assume no current flows into either input terminal ($i = 0$). Practically any new op amp configuration with feedback can be analyzed this way.

6.2.2   The Noninverting Amplifier

If we apply the input signal to the noninverting terminal and the feedback voltage to the inverting terminal, as shown in Fig. 6.2a, the result is an amplifier with very high input impedance, low output impedance, and no phase reversal. Such an amplifier is ideal as it can be driven by a high-impedance source (it does not load the source) and can drive a low-impedance load (the load does not affect the amplifier).

To realize the noninverting amplifier, we ground R_1 of Fig. 6.1b or 6.1c and apply the input signal v_i at the noninverting terminal. The voltage $i_1 R_1$ is then applied to the inverting terminal as negative feedback voltage v_1. Again, we assume that the input current i is zero and $v_d = 0$, and hence $v_i = v_1$. With op amp input current i equal to zero, the currents through R_F and R_1 are the same, therefore

$$\frac{v_o - v_1}{R_F} = \frac{v_1}{R_1}$$

The real gain of the op amp with external circuitry is then

$$A_r = \frac{v_o}{v_i} = \frac{v_o}{v_1} = \frac{i_1(R_F + R_1)}{i_1 R_1} = 1 + \frac{R_F}{R_1} \tag{6.4}$$

As in the case of the inverting amplifier, gain depends on the ratio of two external resistances.

Whereas the input impedance of the inverting amplifier was low, equal to R_1 in Fig. 6.1b, the input impedance of the noninverting amplifier for practical purposes can be approximated by infinity. Using Fig. 6.2a, the input impedance can be stated as $R_i' = v_i/i$. Kirchoff's voltage law for the input loop gives us

$$-v_i + v_d + i_1 R_1 = 0 \tag{6.5}$$

where $v_d = i R_i$. Solving for i in (6.5), we have for the input impedance

$$R_i' = \frac{v_i}{i} = \frac{v_i}{(v_i - i_1 R_1)/R_i} = \frac{R_i}{1 - i_1 R_1/v_i} \approx \frac{R_i}{1 - 1} = \infty$$

where, using (6.5), we have approximated $i_1 R_1 \approx v_i$. A more careful analysis would show that the input impedance of the noninverting amplifier is $R_i' \gg R_i$, and since R_i is on the order of several megaohms for op amps, the approximation of R_i' by infinity is very useful. The noninverting amplifier is thus ideally suited to amplify signals from feeble (high-impedance) sources.[2]

6.3   VOLTAGE FOLLOWERS AND THE UNIT GAIN BUFFER

A special case of the noninverting amplifier is the useful configuration known as the voltage follower, shown in Fig. 6.2b. It is obtained by letting $R_F = 0$ (short circuit) and $R_1 = \infty$ (open circuit) in the circuit of Fig. 6.2a. The gain of this configuration is then

$$A_r = \frac{v_o}{v_i} = 1 + \frac{R_F}{R_1} = 1 \qquad (6.6)$$

which means that the output voltage follows the input voltage. The input impedance R_i' of the voltage follower is found by applying Kirchoff's voltage law to the circuit of Fig. 6.2b. This results in $v_i = v_d + v_o = i R_i + A v_d = i R_i (1 + A) \approx i R_i A$, where we have used that $v_d = i R_i$ and $v_o = A v_d$. As the input impedance is the ratio of input voltage and input current, we obtain

$$R_i' = \frac{v_i}{i} = A R_i \qquad (6.7)$$

The approximation of R_i' by infinity is again valid since typically for op amps $R_i = 1 \text{ M}\Omega$ and $A = 10^6$, giving $R_i' = 10^{12} \Omega = 1 \text{ T}\Omega$. A million megaohms, for all practical purposes, is an open circuit.

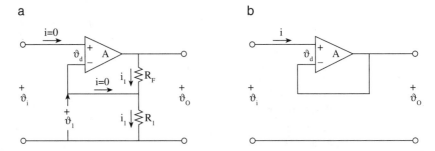

FIGURE 6.2   (a) The noninverting op amp configuration. (b) A voltage follower for which $v_o = v_i$.

---

[2]Furthermore, the output impedance of a noninverting amplifier is much less than R_o, and R_o for most op amps is about 100 Ω.

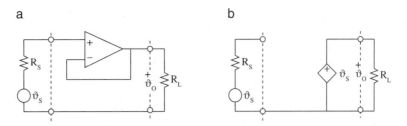

a   b

FIGURE 6.3   (a) A voltage follower is used as a buffer between a weak source and a load, isolating the source from excessive current demands. (b) The buffer is shown as an ideal buffer which draws no current from the source and has zero internal voltage drop.

Such a device, when placed between source and load, protects the source from having to deliver high load currents. It is then called a unit gain buffer. *Buffering*, or isolating the source from its load, is frequently necessary since sources, such as transducers, sensors, microphones, and tapeheads, do not produce any significant power. Connecting a low-impedance load directly to a high-impedance source would drop the available voltage to the load to negligible levels. On the other hand, a buffer with a practically infinite input impedance and an almost zero output impedance would deliver the entire source voltage to the load.[3]

EXAMPLE 6.2   A pickup device has an internal impedance of $R_s = 10$ kΩ and produces a voltage of $v_s = 2$V. If it is to drive a chart recorder which can be represented by a load impedance of 500 Ω, find the voltage and power available to the chart recorder. Repeat when a unity gain buffer is introduced between source and load.

The load voltage is $v_L = v_o = v_s 0.5/(10 + 0.5) = 0.095$ V and power delivered to the load is $P_L = v_L^2/R_L = (0.095)^2/500 = 18.14 \ \mu$W.

Placing a buffer between source and load, as shown in Fig. 6.3a, and approximating the buffer by an ideal one ($R_i = \infty$, $R_o = 0$), which means that the buffer voltage equals the source voltage (no input current flows because $R_i = \infty$) and the load voltage equals the buffer voltage (because $R_o = 0$), the load voltage is then $v_L = 2$ V and the power delivered is $P_L = v_L^2/R_L = 2^2/500 = 8$ mW. The introduction of a buffer results in a voltage gain of 2V/0.095 V = 21 and a power gain of 8 mW/18.14 μW = 441, which demonstrates the effectiveness of a buffer. It goes without saying that most pickup and transducer sources could not deliver the necessary current or power to drive low-impedance loads without the use of a buffer.   ∎

6.4   SUMMERS, SUBTRACTERS, AND DIGITAL-TO-ANALOG CONVERTERS

By applying several inputs to the inverting amplifier, as shown in Fig. 6.4a, we obtain the summer amplifier, which gives an output voltage that is the sum of input voltages.

---

[3]Of course, a noninverting amplifier provides buffering and amplification. However, there are many instances when only buffering is required.

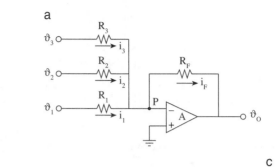

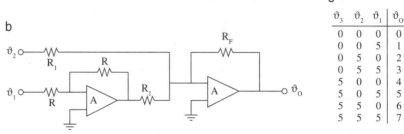

ϑ_3	ϑ_2	ϑ_1	ϑ_o
0	0	0	0
0	0	5	1
0	5	0	2
0	5	5	3
5	0	0	4
5	0	5	5
5	5	0	6
5	5	5	7

FIGURE 6.4   (a) A summing amplifier. (b) A subtracting amplifier. (c) A table giving 3-bit binary numbers (logic 0 and 1 are represented by voltages 0 V and 5 V, respectively) and their decimal equivalents.

Such a configuration can be used, for example, to mix audio signals. As pointed out in the paragraph preceding Example 6.1, point P is a current summing point as no current can flow to ground. Therefore,

$$i_1 + i_2 + i_3 \; = \; i_F \tag{6.8}$$

$$\frac{v_1}{R_1} + \frac{v_2}{R_2} + \frac{v_3}{R_3} \; = \; -\frac{v_o}{R_F} \tag{6.9}$$

which equals

$$v_o = -\frac{R_F}{R_1}(v_1 + v_2 + v_3) \tag{6.10}$$

when $R_1 = R_2 = R_3$. We refer to this as an inverting adder. A noninverting adder, that is, one giving (6.10) but without the minus sign, can be designed by connecting an inverter to output v_o. An inverter is an inverting amplifier like that of Fig. 6.1b with $R_F = R_1$, resulting in $v_o = -v_i$.

A subtracter is shown in Fig. 6.4b. Input v_1 passes through an inverter, which has a gain of -1 (that is, the output signal is phase-shifted by 180° with respect to the input signal). The resulting output voltage is thus the difference of the two input voltages, or

$$v_o = -\frac{R_F}{R_1}(v_2 - v_1) \tag{6.11}$$

By choosing $R_F = R_1$, we would obtain a simple subtracter for which $v_o = v_1 - v_2$.

A digital-to-analog converter (DAC) translates a binary number to an analog signal. For example, the table in Fig. 6.4c gives the 3-bit binary numbers 000–111 and their equivalent decimal numbers 0–7. The digital input signals are $v_1 - v_3$ and the decimal-equivalent voltage is v_o. The binary digits 0 and 1 are represented by the input voltages 0 V and 5 V, respectively.

We can select the inverting summer, Fig. 6.4a, to perform the conversion. The output voltage for such a summer, using (6.9), is

$$v_o = -R_F \left(\frac{v_1}{R_1} + \frac{v_2}{R_2} + \frac{v_3}{R_3} \right) \tag{6.12}$$

If we choose (by trial and error) $R_F = 8\,\text{k}\Omega$, $R_3 = 10\,\text{k}\Omega$, $R_2 = 20\,\text{k}\Omega$, and $R_1 = 40$ kΩ, a binary input signal 001 ($v_3 = 0$, $v_2 = 0$, $v_1 = 5$ V) applied to the summer would give us the following output voltage: using (6.12) we obtain $v_o = -8(5/40 + 0 + 0) = -1.$[4] Similarly, for input 111 ($v_3 = 5$, $v_2 = 5$, $v_1 = 5$), we would obtain $v_o = -8(5/40 + 5/20 + 5/10) = -7$. Hence, the inverting summer performs a digital to analog conversion. By adding more inputs $v_4, v_5, \ldots$, to the summer, we can handle larger binary numbers.

EXAMPLE 6.3   A noninverting summer is shown in Fig. 6.5. Analyze the circuit and show that it performs the mathematical summing operation. As the current into the noninverting terminal of the op amp is vanishingly small, the node at v_p acts as a current summing point for currents flowing in the two R resistors. Therefore,

$$\frac{v_2 - v_p}{R} + \frac{v_1 - v_p}{R} = 0 \tag{6.13}$$

which results in

$$v_2 + v_1 = 2v_p \tag{6.14}$$

At the output terminal we have that the current through R_F and R_1 is the same, that is, $(v_o - v_1')/R_F = v_1'/R_1$, because current into the inverting terminal of the op amp is

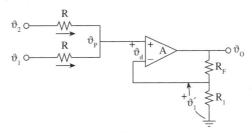

FIGURE 6.5   A noninverting summer.

---

[4]For simplicity, we are ignoring the minus sign on v_o. This can be rectified by adding an inverter to the summer, or by using a noninverting summing amplifier.

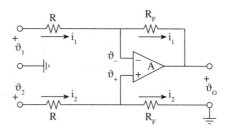

FIGURE 6.6   A differential amplifier, showing the external feedback circuitry.

vanishingly small. Noting that voltage $v_1' = v_p$, because $v_d \approx 0$, we have

$$\frac{v_o - v_p}{R_F} = \frac{v_p}{R_1} \tag{6.15}$$

Substituting into this equation v_p from (6.14) gives us the desired result

$$v_o = \frac{R_1 + R_F}{2R_1}(v_1 + v_2) \tag{6.16}$$

which is that the circuit of Fig. 6.5 performs the mathematical summing operation. A note of caution though: since the node at v_p is not at zero volts, there can be interference (cross talk) between the input signals. In that sense, the inverting summer is superior and hence more useful since point P is a virtual ground in the inverting amplifier.   ∎

6.5   THE DIFFERENTIAL AMPLIFIER

In many practical environments, such as instrumentation, biomedical applications, control systems, etc., we need to amplify the difference between two signals. In these situations, the differential amplifier, typically an op amp configuration as shown in Fig. 6.6, is used. An advantage of a differential amplifier is that any signal common to both inputs (common mode signal) is canceled and does not appear in the output voltage v_o. Undesirable signals such as noise, AC hum, DC level, drift, etc., are canceled as they are picked up equally at both inputs. On the other hand, even the smallest difference in the inputs (differential mode signal) is amplified. In situations where an interference signal is much stronger than a desired signal in need of amplification, a difference amplifier is frequently the only solution. One could conclude that an op amp by itself should serve that purpose, since after all its output is given by (6.1) as $v_o = A(v_+ - v_-) = Av_d$. Practically, this would result in a very poor, unstable differential amplifier and, as discussed previously, operation in a closed-loop configurations which introduces negative feedback is needed to give stable amplification.

Op amps, in the configuration of Fig. 6.6, are widely used differential amplifiers. To analyze this circuit, recall that current into an ideal op amp is zero, which gives for the loop equation at the inverting terminal

$$v_1 = i_1 R + v_- = \frac{v_1 - v_o}{R + R_F} R + v_- \tag{6.17}$$

At the noninverting terminal we can write either a loop equation or by voltage division obtain

$$v_+ = v_2 \frac{R_F}{R + R_F} \tag{6.18}$$

Recall that for an ideal op amp, $v_d = 0$, or equivalently $v_+ = v_-$; combining this with the above two equations gives us the desired result for the differential amplifier

$$v_o = \frac{R_F}{R}(v_2 - v_1) \tag{6.19}$$

Hence, the output voltage is proportional to the difference of the input signals amplified by the closed-loop gain R_F/R. The gain is stable as it is independent of any op amp parameters such as A, which can vary significantly with temperature, frequency, etc., and furthermore, A is specified by the manufacturer. On the other hand, the gain R_F/R depends only on the values of two resistors, and is therefore stable and can easily be changed by the circuit designer.

6.5.1 Practical versus Ideal

The above output voltage, Eq. (6.19), is due to differential-mode signals, $v_2 - v_1$, and would be the only output voltage if the op amp were perfectly balanced, that is, *ideal*. Unfortunately, practical op amps never are; they produce a small output voltage even when the differential input signal is zero. For example, let us put identical signals v_{cm} on both inputs,[5] where v_{cm} is called the *common-mode voltage* as v_{cm} contributes equally to v_1 and v_2. We now have that $v_1 = v_2$ and that $v_{cm} = v_1$ and $v_{cm} = v_2$. (By adding v_1 and v_2, we can also express the common-mode signal as $v_{cm} = (v_1 + v_2)/2$, which is just the average of the two input signals.) Any output voltage that is now obtained must be an amplified version of the common-mode signal only and therefore can be stated as $v_o = A_{cm} v_{cm}$, where A_{cm} is the common-mode gain. A_{cm} for an ideal differential amplifier is obviously zero as an ideal amplifier responds only to the difference voltage $v_2 - v_1$ and not to the common-mode voltage v_{cm}. The fact that a practical op amp responds to the average signal level can create unexpected errors. For example, if one signal is $10~\mu V$ and the second $-10~\mu V$, the output voltage will be different than if the inputs had been $510~\mu V$ and $490~\mu V$, even though the difference signal of $20~\mu V$ is the same in both cases.

---

[5]Equivalently, we could short the inputs together and apply a voltage v_{cm} to the shorted input. This would guarantee that the difference voltage $v_2 - v_1$ equals zero at the input.

To answer the question of how good a practical differential amplifier is in comparison with an ideal one, we define a common-mode rejection ratio, *CMRR*, as

$$\text{CMRR} = \frac{A_{\text{dm}}}{A_{\text{cm}}} \tag{6.20}$$

where A_{dm} is the *differential-mode gain* and from (6.19) is equal to R_F/R. Thus, CMRR for a practical op amp is a measure of quality—the larger the number, the better, as CMRR for an ideal op amp is infinite.[6] Because CMRR is normally a large number, we express it in decibels as $\text{CMRR(dB)} = 20\log\text{CMRR}$ (typical values are 50–100 dB).

Since for a practical amplifier, the output voltage (6.19) has an additional component due to the common input voltage v_{cm}, we can write the total voltage by linear superposition as

$$\begin{aligned} v_o &= (v_2 - v_1)A_{\text{dm}} + \frac{v_1 + v_2}{2}A_{\text{cm}} \\ &= (v_2 - v_1)A_{\text{dm}}\left(1 + \frac{1}{\text{CMRR}}\frac{(v_1 + v_2)/2}{v_2 - v_1}\right) \end{aligned} \tag{6.21}$$

where the voltage gain[7] for the difference-mode signal is A_{dm} and that for the common-mode signal is A_{cm}. The desired output is $(v_2 - v_1)A_{\text{dm}}$, but the presence of a common-mode input adds an error term, which is small if CMRR is large or if the common-mode input v_{cm} is small. Ideally, the two input signals should be of equal strength and 180° out of phase. Then, the common-mode signal would be zero, the differential-mode signal a maximum, and the output voltage a linear magnification of the input signal, given by the first term of (6.21), or by (6.19). Unfortunately, in most practical situations, it is just the opposite. Typically in practice, a signal of only a few millivolts rides on a common-mode signal of several volts as shown in the following example.

EXAMPLE 6.4   Find the output voltage v_o for a differential amplifier (Fig. 6.6) if the input voltages are $v_1 = 9$ V and $v_2 = 9.02$ V. Use $R = 1$ kΩ, $R_F = 120$ kΩ, and a 741 op amp for which specifications are given as a CMRR of 90 dB and an open-loop gain of $A = 2 \cdot 10^5$.

Before proceeding with the solution, we should note that only a small 20 mV signal is riding on a large common-mode voltage of 9 V. The closed-loop gain is obtained from (6.19) as $A_{\text{dm}} = R_F/R = 120$. Expressing the 90 dB common-mode rejection ratio as a numerical value gives $\text{CMRR} = 10^{90/20} = 31623$. To obtain the common-mode gain, we use (6.20), which gives $A_{\text{cm}} = A_{\text{dm}}/\text{CMRR} = 120/31623 = 0.00379$. The output

---

[6]It should be noted in passing that CMRR for an op amp alone, Fig. 6.1*a*, is the same as that for the op amp circuit with feedback, Fig. 6.6. The CMRR for the 741 op amp is given in the manufacturers spec sheet as 90 dB and thus is also the CMRR of the differential amplifier of Fig. 6.6.

[7]For example, A_{dm} can be measured by setting $v_1 = -v_2 = 0.5$ V, which results in $(v_2 - v_1) = 1$ V and $v_{\text{cm}} = (v_1 + v_2)/2 = 0$. The measured output voltage would then be $v_o = A_{\text{dm}}$, which is the gain for the difference signal. Likewise, setting $v_1 = v_2 = 1$ V results in $(v_2 - v_1) = 0$ and $v_{\text{cm}} = 1$ V; consequently $v_o = A_{\text{cm}}$ and the output voltage gives the gain for the common-mode signal.

voltage can now be given, using (6.21), as

$$v_o = (v_2 - v_1)A_{\text{dm}} + \frac{v_1 + v_2}{2}A_{\text{cm}}$$

$$= (9.02 - 9)120 + \frac{9 + 9.02}{2}0.00379$$

$$= 2.4 + 0.034 = 2.434 \text{ V}$$

The deviation from the expected value of 2.4 V caused by the common-mode signal is small in this case, only 0.034 V or 1.4%. This is due to the high CMRR value of the fine 741 op amp. A lesser value than 90 dB for CMRR would have substantially increased the percentage deviation. ∎

In some situations, the closed-loop common-mode gain A_{cm} for a differential amplifier is not known, but the open-loop (no feedback) gain A for the op amp and its CMRR is specified by the manufacturer. In these situations it is important to know that the CMRR for an op amp alone and an op amp with feedback is the same, as can be seen from an examination of (6.21). Therefore, CMRR $= A_{\text{dm}}/A_{\text{cm}} = A/A_{\text{cm,openloop}}$. This should allow calculation of A_{cm} as A_{dm} can be obtained directly from the relationship $A_{\text{dm}} = R_F/R$.

6.5.2 Interference Signals

Amplifiers with high-impedance inputs are well suited to the amplification of feeble signals since the sources for these signals in turn have high impedance. For example, a moderately high-impedance input might be 1 MΩ. If the source also has a 1 MΩ internal impedance and generates a 10 mV signal, 5 mV will then be available across the amplifier input terminals for amplification.

The difficulty with high-impedance inputs is that any stray electric or magnetic fields can induce interfering voltages, at times very large, across the input terminals. For example, electric noise fields due to power lines, lightning, sparking, switching circuits, motors, arc welders, etc., can set up IR_o voltage drops across the input impedance. Equivalently, the electric noise fields can be viewed as capacitive coupling between the noise source and the amplifier input as suggested in Fig. 6.7a. By voltage-divider action, the interfering signal at the input will be $V_{\text{noise}}R_{in}/(R_{in} + 1/j\omega C)$.[8] Even though the stray coupling capacitances are very small (less than picofarads for C), if the input impedance R_{in} or if the frequency ω of the interfering signal is high, much of the noise voltage V_n is developed across the input terminals and will be amplified with the desired signal V_s. Using shielded wires, with the outside conducting sheath connected to ground, can substantially reduce capacitive pickup as then the interfering signal will be conducted harmlessly to ground.

---

[8]To be strictly correct, R_{in} should be replaced by $R_{in} \parallel Z_s$, since R_{in} is in parallel with Z_s in Fig. 6.7a. Also, for brevity, the phasor expression for the interfering signal is given. However, since a signal must be a real number, we can obtain its magnitude by multiplying the phasor expression by its complex conjugate and taking the square root as outlined in Chapter 2 (see (2.5) or (2.11)).

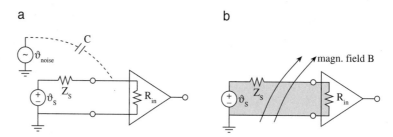

FIGURE 6.7   (a) Small, stray capacitances exist between noise source and amplifier that couple an interfering field to the input. (b) Time-varying magnetic fields due to noise sources (mainly power lines) induce a voltage in wires that close around the magnetic field.

A second source of interference are magnetic fields which are produced wherever electric current flows. Thus, power lines and power machinery are a source of stray magnetic fields. Faraday's law tells us that a time-varying magnetic field will induce a voltage, also known as induced emf, in wires which surround the magnetic field. That is, $v_{ind} = -AdB/dt$, where A is an area formed by wires that close around the magnetic field B.[9] Figure 6.7b shows such an area A, in this case a long rectangular area formed by the parallel, connecting wires between a source and the amplifier input. This area is threaded by a changing magnetic field B, most likely caused by 60 Hz power lines. Such induced voltages are unwelcome as they are amplified along with the desired signal, causing errors and distortion in the amplifier output. A remedy for this type of interference is to minimize the enclosing area by running the wires parallel and close to each other. If this does not reduce the magnetic pickup sufficiently, the wires should be twisted, which further reduces the loop area and causes cancellation of induced voltage in successive twists.

In the above two paragraphs, we considered sources of interfering signals and advanced some procedures to minimize the pickup of such signals on leads that connect a transducer, probe, or any other signal source to the amplifier. If after all these procedures, we still experience a strong noise at the source, a further decrease of such noise may be accomplished by a differential amplifier (see the following example).

EXAMPLE 6.5   With an electrocardiogram (EKG) amplifier, skin contacts are used to measure brain activity and heart activity. These high-impedance sources produce faint signals, sometimes in the microvolt range. It is not unusual to find that even after twisting and shielding the input leads, a large 60Hz voltage in addition to the desired, small signal still exists. Figure 6.8a shows pickup wires, which at one end are connected by skin contacts (electrodes) to the chest of a patient, and at the other end to an EKG amplifier (an EKG amplifier is just a differential amplifier as shown in Fig. 6.6). Since the leads are of equal length and otherwise symmetric, the noise pickup, which here

---

[9]If the magnetic field is sinusoidal with angular frequency ω, then the magnitude of $v_{ind} = -A\omega B$. The induced voltage is thus proportional to the area looped by the wires, the frequency, and the strength of the magnetic field.

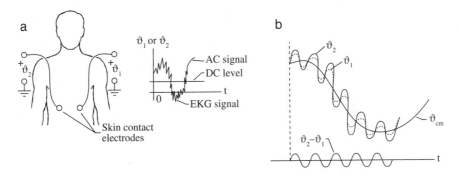

FIGURE 6.8 (a) Leads for EGK signal pickup. The ground is usually a contact point at the lower extremities of a patient, such as the right leg. (b) Signals v_1 and v_2 generated by the electrodes make up the difference signal. The common-mode signal is primarily the interfering AC voltage and the interfering DC level.

is shown as a DC level and an AC voltage, is the same for both leads. Each lead also picks up the desired EKG signal, which is shown as riding on the noise signal. The EKG signal has a complicated, almost periodic waveform (it contains frequencies in the range from 0.05 to 100 Hz). For simplicity, let us model the EKG signal by a single spectral component, say, $\sin 500t$. The total signals, picked up by leads 1 and 2, could then have the form

$$v_1 = 2 + 3 \sin 377t + 0.001 \sin 500t \cong v_{\mathrm{cm}} + 0.001 \sin 500t$$

and

$$v_2 = 2 + 3 \sin 377t + 0.002 \sin 500t \cong v_{\mathrm{cm}} + 0.002 \sin 500t$$

and are shown in Fig. 6.8b. The EKG signal, even though much smaller then the 2 V DC level or the 3V AC signal (which is due to the 60 Hz power lines: $\omega = 2\pi \cdot 60$ Hz $= 377$), is still 1 mV and 2 mV at leads 1 and 2, respectively. For an ideal differential amplifier, the output voltage would simply be

$$v_o = A_{\mathrm{dm}}(v_2 - v_1) = A_{\mathrm{dm}} \cdot 0.001 \sin 500t$$

where A_{dm} is the differential-mode gain, given by (6.19) as R_F/R. We could try to optimize the difference signal by moving the skin contact electrodes around and possibly find positions where v_1 and v_2 are almost 180° out of phase. Then, the total signal at the differential amplifier would be $v_2 - v_1 = 0.003 \sin 500t$, a threefold increase in signal strength.

If the op amp in the differential amplifier circuit of Fig. 6.6 were a 741 and its CMRR had deteriorated to 70dB, the total output voltage would then be given by (6.21) as

$$v_o = (v_2 - v_1)A_{\mathrm{dm}} + \frac{v_1 + v_2}{2}A_{\mathrm{cm}}$$

$$v_o = A_{\mathrm{dm}}\left((v_2 - v_1) + \frac{v_1 + v_2}{2\mathrm{CMRR}}\right)$$

This is the result for a practical differential amplifier. It shows that the common-mode voltage, which includes the undesirable interference signal, is reduced by the numerical value of CMRR—which, after converting from the logarithmic scale, is equal to $CMRR = 10^{70/20} = 3200$. Thus at the input, the 2 V DC level is equivalent to a $2V/3200 = 0.6$ mV and the 3 V AC signal is equivalent to only $3V/3200 = 0.9$ mV, which is smaller than the 1 and 2 mV EKG signals, but not by much. For example, using a 741 op amp with a CMRR of 80 dB would reduce the interference signals by an additional factor of 10, and a new 741 which is rated at 90 dB would reduce it by 100, which for practical purposes leaves only the desirable EKG signal to be amplified. Thus, at least for the data given in this example, an op amp with a 90 dB CMRR compares favorably with an ideal op amp for which CMRR is infinite.   ∎

There are situations where the signals are so weak that an amplifier with essentially infinite input impedance is needed. For such cases an improved version of the difference amplifier, called an *instrumentation amplifier*, is used. Basically, such an amplifier preamplifies each of the two signals before they reach the difference amplifier. The preamplifiers are noninverting op amps which are characterized by very high input impedances. As signal loading is avoided and CMRR is improved, instrumentation amplifiers also make superior EKG amplifiers.

6.6   DIFFERENTIATING, INTEGRATING, AND LOGARITHMIC AMPLIFIERS

Op amps can be used as accurate integrators and differentiators. Before proceeding, we should note that a capacitor or inductor[10] can perform integration and differentiation. For example, for a capacitor we have that $v = \frac{1}{C} \int i \, dt$. The difficulty here is that we have two variables, the current i and voltage v. What is needed is a one-variable operation: if the input is current i, the output should be the integral or derivative of i. Nevertheless, by adding one more circuit element to a capacitor or inductor, we obtain a limited integrator or differentiator in the form of a low-pass filter (Fig. 2.6 or 1.25a) or high-pass filter (Fig. 2.7). Thus, if we consider a square wave of period T as the input to a low-pass filter, the output will be the integral of the input as long as the time constant RC of the filter is much larger than the period T. Under these conditions ($RC \gg T$), we have a useful integrator for signals whose variations are much faster than the time constant of the filter. Similarly, the output will be the derivative of the input, if the time constant of a high-pass filter is $RC \ll T$. A square-wave input would then result in an output consisting of sharp up and down spikes corresponding to the leading and trailing edges of the square input pulses.

Figure 6.9a shows an op amp integrator which is an inverting amplifier with the feedback resistor replaced by a capacitor C. As for the inverting amplifier, $v_d \approx 0$, indicating that point P is at ground potential. Since P is a virtual ground, the voltage

---

[10]Inductors are avoided in integrated circuits as they are bulky, heavy, and expensive in comparison to capacitors.

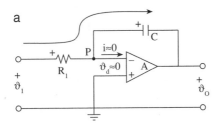 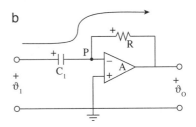

FIGURE 6.9   (a) An op amp integrator. (b) An op amp differentiator.

across R_1 is simply v_1 and the voltage across C is $v_o = -\frac{1}{C} \int i_C dt$. Also, no current flows into the op amp, $i \approx 0$, which makes point P a current summing point. Therefore, $i_{R_1} = i_C$ or

$$v_o = -\frac{1}{R_1 C} \int v_1 \, dt \tag{6.22}$$

which states that the output voltage is the integral of the input voltage. Note that for the op amp integrator, there is no restriction on the input signal as there was for the low-pass filter integrator. Only the magnitudes of the output signal of the op amp integrator cannot exceed the power supply voltage for the op amp.

EXAMPLE 6.6   In situations where several voltages have to be added and their sum integrated we would ordinarily use a *summing* amplifier followed by an *integrating* amplifier. Show that in such situations both operations can be performed by a single op amp. The circuit that can combine both operations in one is shown in Fig. 6.10. Current i is the sum of three currents flowing through resistors R_1, R_2, and R_3, that is, $i = i_1 + i_2 + i_3$. Since point P is a summing point, current i continues as current through the capacitor, or i_c. Point P is also a virtual ground which gives for the output voltage

$$
\begin{aligned}
v_o &= -\frac{1}{C} \int i_C \, dt = -\frac{1}{C} \int i \, dt = -\frac{1}{C} \int \left(\frac{v_1}{R_1} + \frac{v_2}{R_2} + \frac{v_3}{R_3} \right) dt \\
&= -\frac{1}{R_1 C} \int \left(v_1 + v_2 \frac{R_1}{R_2} + v_3 \frac{R_1}{R_3} \right) dt
\end{aligned} \tag{6.22a}
$$

which is the desired result. For $R_1 = R_2 = R_3$ we have a simple addition of the voltages in the integrand. For different resistances we have a weighted addition and the output is the integral of a weighted sum.   ■

A *differentiator* is obtained by interchanging the capacitor with the input resistor, as shown in Fig. 6.9*b*. Again equating the currents that flow through the resistor and capacitor, we obtain

$$C \frac{dv_1}{dt} = -\frac{v_o}{R}$$

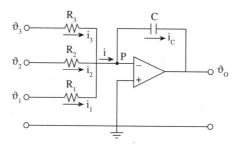

FIGURE 6.10   A summer-integrator which adds voltages v_1, v_2, v_3 and integrates the sum.

Solving for the output voltage, we obtain the desired result

$$v_o = -RC_1 \frac{dv_1}{dt} \tag{6.23}$$

which states that output voltage is proportional to the derivative of the input voltage. The op amp differentiator is not as stable as the integrator.[11] It is seldom used in practice because it has problems with noise and instabilities at high frequencies. The tendency to oscillate with some signals can be decreased by placing a low-value resistor in series with C_1, and by using the fastest op amp.

Nonlinear input–output relationships can be produced by placing a nonlinear element in the feedback path of an op amp. A *logarithmic amplifier* gives an output voltage which is proportional to the logarithm of the input voltage. Such an amplifier can handle a wide range of input voltages (a logarithm compresses the scale of a variable) which would normally saturate a linear amplifier. Essential to the operation of the logarithmic amplifier is to have a device in the feedback loop with exponential characteristics. A diode or transistor has such characteristics. Figure 6.11 shows a grounded-base transistor for which we have, using (5.11),

$$i_c = \alpha I_o \exp(-e v_o / kT) \tag{6.24}$$

where v_o is the emitter-to-base voltage. The collector current can be expressed in terms of the input voltage as $i_c = v_i / R$, recalling that P is a summing point and virtual ground. Taking the logarithm of (6.24), we obtain

$$v_o = -\frac{kT}{e} \ln \frac{v_i}{\alpha I_o R} \tag{6.25}$$

which shows that the output voltage is the logarithm of the input voltage.

If we were to place a diode or transistor in the input loop and a resistor R in the feedback loop, we would obtain an exponential or antilog amplifier. Now that we have a log and an antilog amplifier we can construct a *multiplier*, that is, a circuit that can

---

[11]The differentiator (unlike the integrator which smoothes noise) magnifies noise spikes, because of the large slopes that are present in noise voltage systems.

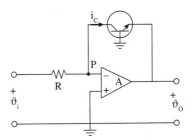

FIGURE 6.11   A logarithmic amplifier.

multiply two signals. If we sum the output of two logarithmic amplifiers and then pass the output of the combination through an antilog amplifier, the resulting signal will be proportional to the product of the two input signals.

6.7   ACTIVE RC FILTERS

Filters can be used to remove, emphasize, or deemphasize certain frequencies of a signal. This property can be used to reshape a signal or to block noise which is concentrated in a specific band of frequencies. For example, the passive, low-pass filter of Fig. 2.6 allows passage of DC and low frequencies, but attenuates the high frequencies. It is called passive, because the circuit includes only resistors, inductors, and capacitors. An active (contains sources or active elements in addition to R, L, and C), low-pass filter, on the other hand, can have the same frequency characteristics, but can also provide gain in the pass-band. Of course, we could tag on an amplifier to a passive filter to obtain gain, but it would require more circuit elements.

An *active, low-pass filter* is shown in Fig. 6.12a. The filter elements are placed directly in the feedback loop of an inverting amplifier. We can express the closed-loop gain as a function of frequency by simply replacing in (6.3) the resistances by impedances, that is,

$$A_r = -\frac{Z_F}{Z_1} = -\frac{\frac{R_F}{1+j\omega C_F R_F}}{R_1} \tag{6.26}$$

where Z_F is the parallel combination of R_F and C_F, and $Z_1 = R_1$. Normally we are interested in plotting the frequency response of the filter, which is obtained by taking the absolute value of (6.26),

$$|A_r| = \frac{R_F}{R_1} \cdot \frac{1}{\sqrt{1+(\omega C_F R_F)^2}} \tag{6.27}$$

where the cutoff, corner, or half-power frequency is $\omega_o = 1/R_F C_F$. Note that ω_o is the same as that for the previously considered, low-pass filters of (2.14) and (5.22). The above expression consists of two terms: one is simply the gain of the inverting amplifier

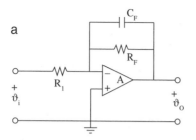

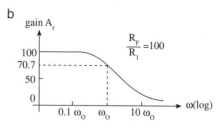

FIGURE 6.12   (a) An op amp low-pass filter. (b) Frequency response of a low-pass filter which has a gain of 100 in the pass-band.

R_F/R_1, and the other is a low-pass filter function, the same as that for a passive filter as previously considered in Fig. 2.6. The advantages of an active filter are the ease with which the gain and the bandwidth can be varied by controlling R_F/R and the corner frequency $\omega_o = 1/R_F C_F$. The frequency response of such a device for which the DC and low-frequency gain is $R_F/R = 100$ and for which the low-pass bandwidth is determined by the half-power frequency $1/R_F C_F$ is plotted in Fig. 6.12*b*.

EXAMPLE 6.7   Design a low-pass filter of the type shown in Fig. 6.12*a* with a closed-loop gain of 100 and a half-power frequency of 500 Hz. The input impedance of the device is to be 1 kΩ.

As the negative input terminal of an inverting amplifier is a virtual ground, the input impedance $Z_i = R_1 = 1$ kΩ. The low-frequency gain, which is given by $A_r = R_F/R_1$, is specified to be 100. Therefore, $R_F = A_r R_1 = 100 \cdot 1$ k$\Omega = 100$ kΩ. The remaining circuit element which must be determined is the capacitance. Since the cutoff (half-power, corner) frequency is specified and is given by $f_o = \omega_o/2\pi = 1/2\pi R_F C_F$, we can solve for $C_F = 1/2\pi f_o R_F = 1/2\pi \cdot 500 \cdot 100 = 3.2$ μF. The gain–frequency plot is given by Fig. 6.12*b* with angular half-power frequency $\omega_o = 2\pi f_o = 6.28 \cdot 500 = 3.14$ krad/s.   ∎

An *active high-pass filter* is shown in Fig. 6.13*a*. The real gain is given again by (6.3) as

$$A_r = -\frac{Z_F}{Z_1} = -\frac{R_F}{R_1 + 1/j\omega C_1} \tag{6.28}$$

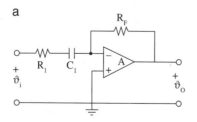

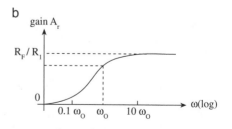

FIGURE 6.13   (a) An op amp high-pass filter. (b) Frequency response of a high-pass filter.

the absolute value of which is

$$|A_r| = \frac{R_F/R_1}{\sqrt{1 + (1/\omega R_1 C_1)^2}} \tag{6.29}$$

This expression is plotted in Fig. 6.13b and shows the typical frequency response of a high-pass filter with a half-power frequency $\omega_o = 1/R_1 C_1$ and a high-frequency gain of R_F/R_1.

Combining the responses of a low-pass and high-pass filter, we can construct a band-pass filter. An *active band-pass filter* is an op amp with a parallel combination of R_F and C_F in the feedback loop (same as in Fig. 6.12a) and a series combination of R_1 and C_1 in the input loop (same as in Fig. 6.13a). If we choose the cutoff frequency $1/R_F C_F$ of the low-pass filter to be larger than the cutoff frequency $1/R_1 C_1$ of the high-pass filter, only a band of frequencies equal to the difference of cutoff frequencies will be passed.

6.8 COMPARATORS AND ANALOG-TO-DIGITAL CONVERTERS

6.8.1 Comparator

An op amp by itself (open loop, no feedback) makes a very effective comparator, because the gain A of an op amp is so high ($\approx 10^5$). Since the output voltage v_o of an op amp cannot exceed the power supply voltage,[12] which, let us say, is ± 10 V$_{DC}$, the output voltage will swing from $+10$ V to -10 V for an input voltage swing of only $+0.1$ mV to -0.1 mV. Hence, a change of only a fraction of a millivolt at the input will give a large output voltage.

Figure 6.14a shows how an op amp is used to compare the magnitude of two signals. Applying a reference signal to the inverting input and an unknown signal to the non-inverting input, the output of the op amp is zero only when the two signals are equal. When they are not equal, the output v_o is either saturated at the supply voltage of $+10$ V when the unknown signal $v_?$ is $v_? > V_{\text{ref}}$, or $v_o = -10$ V when $v_? < V_{\text{ref}}$. Figure 6.14b shows the transfer (input–output voltage) characteristics of an op amp when the reference signal is a 5 V battery and the supply voltage is ± 10 V. It shows that even a fraction of a millivolt in the unknown signal above or below the 5 V reference signal will give a large indication of $+10$ V or -10 V in the output.

When the reference voltage is chosen as $V_{\text{ref}} = 0$ (equivalent to grounding the inverting input), we have a zero-crossing comparator. A very small input voltage (fraction of a millivolt) will swing the output to 10 V and similarly a small negative voltage will swing the output to -10 V. It is interesting to observe that the output of such a device is a square wave for practically any periodic input signal. For example, a sinusoidal

---

[12]Typically an op amp needs two power supplies, a positive voltage supply V_{BB} and a negative voltage supply $-V_{BB}$. If the op amp is to be operated from batteries, this can be a disadvantage as two separate batteries are needed as shown in Fig. 6.14a.

a   b

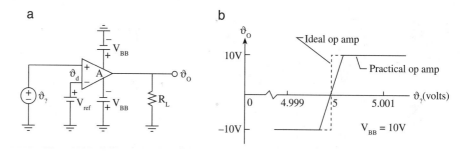

FIGURE 6.14   (a) A comparator is an op amp in the open-loop mode with $v_o = Av_d = A(v_? - V_{ref})$. Two power supplies are needed, which allows the output voltage v_o to swing between $+V_{BB}$ and $-V_{BB}$. (b) The unknown input voltage–output voltage characteristics for a reference voltage of 5 V.

input, starting at 0 V and increasing, will immediately saturate the output at 10 V. The output will stay saturated at 10 V until the input sinusoid goes negative, at which time the output saturates at -10 V and stays saturated until the sinusoid goes again positive. This repeats, generating a square-wave output voltage.

6.8.2   A/D Converter

There are many ways to convert an analog signal to a finite number of ones and zeros. Basically, the analog signal is sampled at evenly spaced, short-time intervals and the sampled values converted to binary numbers. A simple (nonsampling) *analog-to-digital converter* is shown in Fig. 6.15a which can convert a continuous signal to one of four values, that is, to a 2-bit number. The four resistors, R, constitute a voltage divider

a

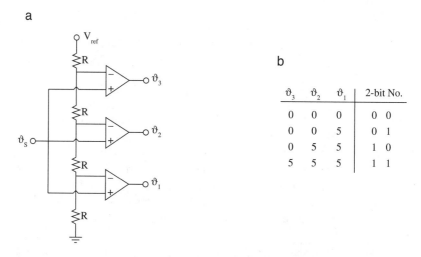

b

ϑ_3	ϑ_2	ϑ_1	2-bit No.
0	0	0	0 0
0	0	5	0 1
0	5	5	1 0
5	5	5	1 1

FIGURE 6.15   (a) A stack of three comparators can convert a continuous range of values of v_s to one of four values. (b) A table showing the coding of the outputs of $v_1 - v_3$ and a 2-bit number.

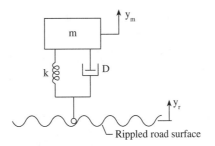

FIGURE 6.16   A simple model for the suspension of an automobile. Mass m represents the car, k represents the springs, and dashpot D the shock absorbers.

across the reference voltage V_{ref} and thus provide a fixed reference voltage for each comparator in the stack of three comparators. For example, the top comparator has a reference voltage of $(3/4)V_{ref}$ applied to the inverting input, the second $(1/2)V_{ref}$, and the bottom one $(1/4)V_{ref}$. If the analog input signal $v_s < (1/4)V_{ref}$, the output of each comparator is 0 V. As v_s increases, first the bottom comparator saturates to the op amp power supply voltage V_{BB}, then the middle one, and finally the top one. Depending on the input signal, we can have four different voltages at v_1, v_2, and v_3 which are shown in Fig. 6.15b as 0 V and 5 V (computers use 5V power supplies, i.e., $V_{BB} = 5$ V). These four values can be interpreted by a code converter or decoder (not shown in Fig. 6.15a) as $00, 01, 10$, and 11. For an 8-bit digital output, we would need a stack of 255 ($2^8 - 1$) comparators.

6.9   THE ANALOG COMPUTER

Now that we have op amp integrators, differentiators, and summers, we can use these to construct an analog computer, which can solve for us the differential equations that describe the mechanical and electrical systems which we are designing for manufacturing and production. For example, a system that all of us are familiar with is the suspension system for an automobile, which in its elementary form is a weight mounted on top of a spring and shock absorber, which in turn are connected to a wheel that moves along a usually not-too-smooth of a road. Figure 6.16 shows such a model in which a mass m (the car) moves up and down along y_m and a wheel follows a sinusoidally rippled road $y_r(t) = Y \sin \omega t$ which applies an up and down force to the spring k and shock absorber D. Summing forces on mass m

$$f_m + f_D + f_k \;=\; 0 \tag{6.30}$$

$$m\frac{d^2 y_m}{dt^2} + D\left(\frac{dy_m}{dt} - \frac{dy_r}{dt}\right) + k(y_m - y_r) \;=\; 0$$

where the vertical motion of the mass is given by $f_m = ma = md^2 y_m/dt^2$, the force on the mass transmitted by the shock absorber is modeled by a dashpot $f_D = D\,dy/dt$

(D is a velocity damping constant), and the force on the mass transmitted by the spring is $f_k = ky$, where k is a linear spring constant. Let us move all terms which represent forces exerted by the road on mass m to the right side of the equation

$$m\frac{d^2 y_m}{dt^2} + D\frac{dy_m}{dt} + ky_m = D\frac{dy_r}{dt} + ky_r \tag{6.31}$$

By modeling the road by $y_r(t) = Y \sin \omega t$, where Y is the height of the ripples in the road (road roughness), we can finally express the equation which determines the vertical motion of a car traveling along a rippled road, that is,

$$m\frac{d^2 y_m}{dt^2} + D\frac{dy_m}{dt} + ky_m = D\omega Y \cos \omega t + kY \sin \omega t \tag{6.32}$$

A solution for y_m could be used to design an optimum suspension system which would give the smoothest ride over a rippled or a washboard road.

Analog computers were widely used to solve differential equations throughout the 1960s until the digital computer largely replaced their use. However, for specialized applications such as suspension design or vibrational analysis, analog computers accurately mimic physical systems, can be easily changed for different parameters, and quickly give answers that are readily displayed on an oscilloscope. In this example, the vertical displacement y_m in the suspension system will be represented by a voltage in the electrical analog computer.

The first step is to solve for the highest-order derivative of the above differential equation, or

$$\frac{d^2 y_m}{dt^2} = -\frac{D}{m}\frac{dy_m}{dt} - \frac{k}{m}y_m + \frac{D\omega Y}{m}\cos \omega t + \frac{kY}{m}\sin \omega t \tag{6.33}$$

which is now in appropriate form for solution by double integration. We observe that the sum of the four terms on the right represents $d^2 y_m/dt^2$. The next step is to arrange an integrating amplifier which is preceded by a summing amplifier with four inputs to add the right-side terms, as shown in Fig. 6.17. In Example 6.6 it was shown that the summing and integrating operation can be combined such that only one op amp is needed (op amp 1 in Fig. 6.17). The output of the first integrator[13] is $-dy_m/dt$, which must be multiplied by D/m (op amp 2) before it can be routed to the summer (after multiplication we need to change sign by -1, which op amp 3 does). Another integration by op amp 4 gives us y_m, which after multiplying by $-k/m$ is routed as the second input to the summer. The remaining two terms in (6.33) are road condition inputs to the summer as the third input $(D\omega Y/m \cos \omega t)$ and the fourth input $(kY/m \sin \omega t)$. These are external inputs and can be changed as necessary to simulate a variety of roads. The solution to the problem is y_m; it is the desired output which can be directed to a display device such as an oscilloscope. In an analog computer such as this, the various parameters relating to the suspension system can be varied and the new performance of the suspension readily evaluated.

---

[13]In order to give the indicated integrator outputs, the RC constants must be chosen as unity according to the integrator equation (6.22).

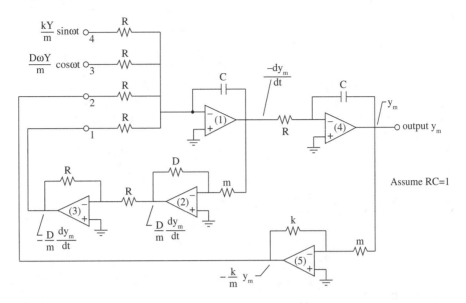

FIGURE 6.17 Analog computer solution to Eq. (6.30), which describes a suspension system for an automobile.

EXAMPLE 6.8 Design an efficient analog computer program for the solution of the familiar mass–spring–damper problem in mechanics as shown in Fig. 6.18a. The vibrating mass m_o is acted upon by an external, sinusoidal force $f = F \sin \omega t$ N (newtons). The solution is obtained when the displacement and velocity of mass m_o is known. Assume $m_o = 500$ kg, the damping coefficient $D = 2.5 \cdot 10^4$ N · s/m, and the spring constant $k = 10^6$ N/m. Specify all resistor and capacitor values in the op amp circuitry of the analog computer.

The equation describing the motion of the vibrating mass m_o is the same as that for the automobile suspension given by (6.31) except for the right side of (6.31), which is the external driving force and which in this example is given by f. Therefore, summing forces on mass m_o, we obtain in this case

$$m_o \frac{d^2 y}{dt^2} + D \frac{dy}{dt} + ky = -F \sin \omega t \tag{6.34}$$

Rearranging in a form suitable for analog simulation, we have

$$\frac{d^2 y}{dt^2} = -50 \frac{dy}{dt} - 2000y - \frac{F}{m_o} \sin \omega t$$

where $D/m_o = 50$ and $k/m_o = 2000$. Integrating once gives

$$\frac{dy}{dt} = -\int \left(\frac{F}{500} \sin \omega t + 2000y + 50 \frac{dy}{dt} \right) dt \tag{6.35}$$

a

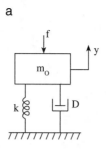

b

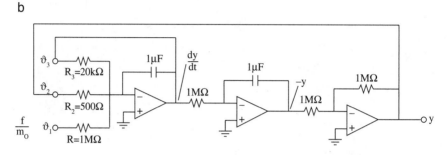

FIGURE 6.18 (a) Vibrating mass supported by spring k and damper D and acted upon by an external force f. (b) Analog computer simulation of vibrating mass. The external force input is f.

In order to find the resistance values in the summer-integrator op amp, let us refer to Fig. 6.10 and compare (6.35) to (6.22a). If we identify $R_1 C$ with unity ($R_1 = 10^6 \, \Omega = 1$ MΩ, $C = 10^{-6}$ F $= 1 \, \mu$F) and identify v_1 with $F \sin \omega t / 500$, v_2 with y, and v_3 with dy/dt, then $R_1/R_2 = 2000$ and $R_1/R_3 = 50$. The values of the remaining resistors are therefore $R_2 = 500 \, \Omega$ and $R_3 = 20$ kΩ. A circuit diagram for the analog computer simulation is given in Fig. 6.18b. Note that this is a more efficient implementation than that outlined in Fig. 6.17, where for pedagogical[14] reasons we have shown the summer-integrator with four identical input resistors R. Using the same R's required two additional op amps (op amp 2 to change the scale and op amp 3 to invert). As (6.22a) shows, we can use the input resistors of a summer-integrator to change the scale of the inputs, which we have done in this example with the result that two op amps were eliminated.   ■

6.10   SUMMARY

An op amp is a high-gain, direct-coupled integrated circuit amplifier of about 20 transistors on a tiny silicon chip. By itself, it is not particularly stable with a gain that varies

---

[14]It might be even better to first show the summer-integrator as two separate op amps, one for summing and one for integrating, and only then proceed to the combined summer-integrator of Fig. 6.17 and finally to the most efficient implementation shown in Fig. 6.18b.

with frequency, dropping off to a negligible level at about a megahertz. However, applying negative feedback[15] to the op amp, we obtain a moderate-gain amplifier that is very stable and responds to frequencies much higher than a megahertz. Combining that with the characteristics of high input impedance and low output impedance, we almost have an ideal amplifier that is used by itself or as a building block in numerous applications in many diverse fields. In analog design, they have surpassed transistors as the basic building block. The versatility of this amazing chip allows it to be cascaded without loading problems, to be used as an oscillator, to be used as a differential amplifier that amplifies a difference signal but attenuates a common-mode (interfering) signal such as hum, and to be used widely in "operation" applications which we started to explore in this chapter (comparators, D/A converters, integrators, etc.). The op amp fundamentals learned in this chapter should be readily applicable to new situations encountered by designers in many engineering fields. Because designing with them is so simple, even casual users should be able to construct simple amplifiers and filters.

- There are two rules that govern the design of op amp circuits. The first is that the two input terminals of an op amp in any circuit are at the same voltage, and the second is that no current flows into either of the two input terminals. These were elaborated on in Section 6.2, where the application of the two rules to the inverting amplifier showed that point P, for practical purposes, is at ground level—a virtual ground—and also that point P is a summing point—current flowing into the input resistor continues into the feedback resistor. A thorough understanding of these rules can expedite the design of op amp circuitry.

- There are some physical limitations of op amps, some of which are obvious and others which are more subtle. An obvious one is the limit on the output voltage of an op amp which cannot be larger than the power supply voltage.

The finite bandwidth is another limitation that can be overcome to some extent by applying negative feedback—the more feedback, the larger the bandwidth, but the gain is proportionately reduced. To quantify it, we introduce the *gain-bandwidth product*, which is constant for any given op amp circuit. Hence reducing gain increases bandwidth and vice versa. Initially op amps were limited to a gain-bandwidth product of 1 MHz, which restricted their use to low-frequency applications, primarily audio. At present, since the gain-bandwidth product has been pushed to 500MHz, op amps are suited for video applications as well.

Another limitation of op amps is the presence of a small voltage, called the *input offset voltage*, which due to imbalances in the internal op amp circuitry is always present. Maximum values are quoted in manufacturers' data sheets. For the same reason, a small *input bias current* is present at the inverting and noninverting terminals of an op amp that is connected in a circuit.

---

[15] Routing a portion of the output signal back to the input, such that it is $180°$ out of phase with the input signal, is called negative feedback. It results in decreased gain but improved frequency characteristics, and reduces waveform distortion. In positive feedback, a portion of the output signal that is fed back is in phase with the input signal, usually leading to an unstable situation in which the circuit oscillates.

A further limitation is the slowed response of an op amp to a quickly changing input. If the input changes too quickly, there will be a delay in the output, given by the *slew rate limit*. Typically, the vertical part in an ideal step input will be reproduced slanted.

A final limitation is the common-mode rejection ratio (CMRR) in differential amplifiers, which are amplifiers that provide a high degree of discrimination against commonmode (interfering) signals while amplifying a differential signal. CMRR is a ratio of differential gain to common-mode gain and was fully explored in this chapter. A popular, general-purpose and cheap op amp is the 741. Its specifications are $R_i = 2$ MΩ, $R_o = 75\ \Omega$, $A = 2 \cdot 10^5$, CMRR = 90 dB, and supply voltage = ± 15 V.

Problems

1. Explain why a high input resistance and a low output resistance are desirable characteristics of an amplifier.

2. Calculate the gain of the inverting op amp given in Example 6.1 without initially assuming that $v_d = 0$. Use the resistance values specified in the example and compare the gain to the value of -100 obtained by using the gain expression $-R_F/R_1$.
   Ans: Error $= -0.02\%$.

3. In the inverting amplifier of Fig. 6.1*b*, find the input impedance if $R_1 = 10$ kΩ and $R_F = 200$ kΩ. Assume the op amp is ideal.

4. A particular microphone which produces an open-circuit voltage of 50 mV can be modeled by 50 mV source voltage (v_s) in series with a 10 kΩ source resistance (R_s). If the microphone voltage needs to be amplified to a level of 5 V, design an inverting amplifier to accomplish it.

5. For the microphone case described in Problem 4, design a noninverting amplifier.
   Ans: If $R_1 = 1$ kΩ, then $R_F = 99$ kΩ in Fig. 6.2*a*.

6. A mediocre op amp (see Fig. 6.1*a*) with $A = 10^4$, $R_i = 100$ kΩ, and $R_o = 0.5$ kΩ is to be used as a unit gain buffer (Fig. 6.2*a*). By writing the circuit equations for Fig. 6.2*a*, show that $v_o \approx v_i$ and show that the input impedance $Z_i = (1 + A)R_i$. Give numerical values.

7. Repeat the calculations in Example 6.2 when the input device is a microphone with $v_s = 10$ mV (rms) and source impedance $R_s = 50$ kΩ. Draw a conclusion regarding the effectiveness of a buffer.

8. An op amp such as the 741 is used in an inverting amplifier (Fig. 6.1*b*). If the input impedance to the inverting amplifier is to be 2 kΩ, design the amplifier for a gain of -50, that is, find R_1 and R_F.
   Ans: $R_F = 100$ kΩ, $R_1 = 2$ kΩ.

9. An inverting amplifier of the type shown in Fig. 6.1*b* uses a 741 op amp that is powered by a ± 15 V supply. If the input current i_1 is not to exceed $\pm 20\ \mu$A, design the circuit for a gain of -100, that is, find R_1 and R_F. Assume that $v_{o,\max}$ is limited by the ± 15 V power supply voltage.

10. Two voltages v_1 and v_2 are to be added by a summing amplifier to give an output that is $v_o = -v_1 - 5v_2$. Design a summer of the type shown in Fig. 6.4*a*. Use $R_F = 10$kΩ.

Ans: $R_1 = 10$ kΩ, $R_2 = 2$ kΩ.

11.

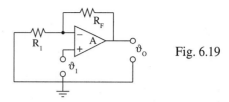

Fig. 6.19

Find the gain $A_r = v_o/v_1$ of the op amp amplifier circuit shown. What operation does it perform?

12. Repeat Problem 10 but change the desired output voltage to $v_o = -v_1 + v_2$.

13. A voltage-to-current converter converts a voltage signal to a proportional output current. Show that for the circuit in Fig. 6.20 we have $i_L = v_s/R_1$. As the circuit is basically a noninverting amplifier (Fig. 6.2a), the load current is independent of the source impedance R_s and load impedance R_L. Hence, the amplifier requires very little current from the signal source due to the very large input resistance of a noninverting amplifier.

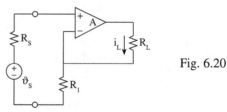

Fig. 6.20

14. A current-to-voltage converter converts an input current to a proportional output voltage. Show that for the circuit in Fig. 6.21 we have $v_o = -i_s R_F$. Note that due to the virtual ground at the inverting amplifier input, the current in R_s is zero and it flows through the feedback resistor R_F.

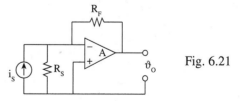

Fig. 6.21

15.

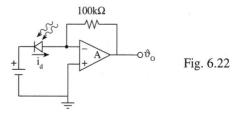

Fig. 6.22

A reverse-biased photodiode in the circuit of Fig. 6.22 generates 0.4 μA of current (i_d) per 1 μW of light power falling on it. Find the voltage v_o that a voltmeter would read when the diode is illuminated with radiant power of 50 μW.

16. Two differential amplifiers, differing only in their CMRRs, are available. If the inputs are $v_1 = 1$ mV and $v_2 = 0.9$ mV, calculate the output voltages for the two amplifiers and show the effect that a higher CMRR has.

    (a) Differential-mode gain 2000, common-mode gain 100
    (b) Differential-mode gain 2000, common-mode gain 10^4
        Ans: (a) 219 mV with 9.5% error; (b) 200.2 mV with 0.1% error.

17. A 1 μV signal is to be amplified in a differential amplifier for which the CMRR is specified as 100 dB. Find how large the magnitude of an interfering signal can be before its output magnitude is equal to that of the desired signal.

18. Find the magnitude of the common-mode output signal of an op amp for which the CMRR = 80 dB, $A_d = 10^4$, and an interfering common-mode signal of 2 V exists.
    Ans: $v_{o,\mathrm{cm}} = 2$ V.

19. In the op amp circuit of Fig. 6.9, if capacitor C is replaced by an inductor L, what operation would the circuit perform?

20. Show that the output voltage v_o of a circuit in which R and the transistor in Fig. 6.11 are interchanged is that of an exponential or antilog (inverse log) amplifier.
    Ans: $v_o = \alpha R I_o \exp(-ev_i/kT)$; the orientation of the *npn* transistor is such that the input is at the emitter of the transistor.

21. An op amp low-pass filter is to have a cutoff frequency of 100 Hz and a gain of magnitude 50. Determine the remaining parameters if the capacitor is specified as $C = 1$ μF.

22. An op amp low-pass filter (Fig. 6.12*a*) has $R_1 = 2$ kΩ, $R_F = 22$ kΩ, and $C_F = 0.1$ μF. Determine the corner frequency and the DC gain.
    Ans: 72.3 Hz, -11.

23. Find the unity-gain bandwidth for the low-pass filter of Problem 22; that is, find the frequency at which the gain has dropped to unity.

24. Design a high-pass op amp filter with a high-frequency gain of -100 and a cutoff frequency of 1 kHz. Resistors with values of 1 kΩ, 10 kΩ, and 100 kΩ are available. Determine the capacitance and resistance values of the filter.

25. A 741 op amp for which $A = 2 \cdot 10^5$ has a supply voltage of ±15 V. If it is used as a comparator (Fig. 6.14*a* with $V_{\mathrm{ref}} = 5$ V), determine the change in $v_?$ which will drive the output voltage from negative saturation (-15 V) to positive saturation.
    Ans: 4.99985 V to 5.00015 V.

26. Repeat Problem 25, except for a zero reference voltage.

27. Design a comparator circuit to set off an alarm when the temperature in a boiler reaches 160°C. You have available a temperature-to-voltage transducer which generates a voltage of 5 V at 160°C and an alarm which activates at -15 V and is off for larger voltages. Sketch the transfer characteristics.

28. A zero-crossing op amp comparator is powered by two batteries, one with $V_{\mathrm{BB}} =$

15 V and the other with $V_{BB} = -15$ V, as shown in Fig. 6.14a. Such an arrangement has an output voltage v_o which is $+15$ V when $v_? > 0$ and -15 V when $v_? < 0$. If the input voltage is given by $v_? = 0.01 \sin \omega t$ V, sketch the output voltage.

29. Design an analog computer to determine the current $i(t)$ for $t > 0$ in the series R_L circuit shown in Fig. 1.26a. The switch is turned on at $t = 0$ and connects a battery V to RL. The resulting current is determined by $di/dt = V/L - Ri/L$.

30. A typical solution for Problem 6-29 involves more than one op amp. It is possible, by combining addition and integration in one op amp (see Fig. 6.10), to design an efficient, single-op-amp solution to Problem 29. Show the circuit.

31. Determine the differential equation whose analog is shown in Fig. 6.23.

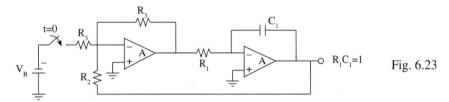

Fig. 6.23

32. Design an analog computer to solve $\frac{d^2x}{dt^2} + 3x = \cos \omega t$.

Ans:

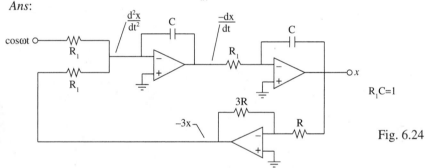

Fig. 6.24

33. Design an analog computer to solve for the current $i(t)$ in the series RLC circuit of Fig. 2.2a. The current is the solution to (2.3), which can be restated here as

$$\frac{dv(t)}{dt} = R\frac{di(t)}{dt} + L\frac{d^2i(t)}{dt^2} + \frac{i(t)}{C}$$

CHAPTER 7

Digital Electronics

7.1 INTRODUCTION

7.1.1 Why Digital?

The best answers to "why digital?" are the existence of digital computers and immunity to noise in digital transmission of information. Computers continue to be important as digital processors but computers are also increasingly important as new sources of digital information, such as in publishing and entertainment media.

Transmission of information in analog form (AM and FM, for example), which is still widely used, degrades the signal irreversibly. This is easily understood since an analog signal such as speech includes many small variations in amplitude which are easily corrupted by noise. An example of this would be the degradation of your favorite radio station signal when driving away from the station in an automobile. As you continue driving, a time will come when the signal becomes so noisy that a new station must be selected. If, on the other hand, this signal were coded in digital form (PCM, or pulse code modulation), your digital receiver would receive a sequence of 0 and 1 signals. The two levels of the signal are separated by a sufficiently large voltage that the small noise voltages that are received by the antenna and are added to the 0 and 1 signals do not degrade reception—in other words, your receiver is able to identify the 0 and 1's in the presence of noise. You might correctly point out that eventually even the 0 and 1 signals will be corrupted by noise when sufficiently far from the source. A peculiar thing happens with digital equipment at that point—the reception, even though perfect a second ago, will suddenly cease. Digital equipment will suddenly stop working when the digital signal is on the order of the noise signal, whereas analog equipment will continue amplifying the combined noise and signal even though the result is mostly noise. In that sense both transmissions become useless. However, a digital transmission can be perfectly restored to its original form at some distance from the station at which the 0 and 1's are not so contaminated by noise that they cannot be recognized as 0 and 1's. Therefore, it should be possible, using repeated restoration, to send a digital signal

a   b

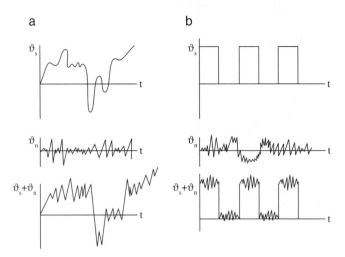

FIGURE 7.1   (a) An analog signal is permanently altered when noise is added. In a digital signal to which noise is added, the separation in voltage between the 0 and 1's is sufficiently large that the original 0 and 1's are still recognizable. (b) A digital receiver simply makes a decision to recognize the corrupted pulses as the original 0 and 1's.

over long distances without distortion. An analogous situation occurs when copying information in digital form, such as digitally encoded music. It is possible to make repeated copies without loss of fidelity, a feat not possible with analog recordings such as tapes or vinyl records for which the fidelity is reduced in the copy.

Summarizing, digital signals can be transmitted perfectly, copied perfectly, and stored perfectly because only a 0 or 1 decision must be made correctly for digital signals, as opposed to analog where it is necessary to exactly reproduce a complex waveshape in which the small variations in the signal are easily distorted by the presence of noise. Figure 7.1 shows the differences between analog and digital signals when noise is added.

7.1.2   Digital Signals in an Analog World

Even though we live in an analog world in which physical systems are characterized by continuously varying signals, the advantages of digital signals which are immune to noise and can be processed by the digital computer and by digital hardware that typically is small, low cost, and high speed are so great that the trend in laboratories and industry is irreversibly toward complete digitization in instruments and consumer products. For the same reasons telecommunications, including long-distance telephone, HDTV, and cable, throughout the world are moving to a common digital network. Because an analog signal becomes contaminated by noise that cannot be removed, changing the analog signal to a digital one will allow transmission of the original analog signal without degradation by noise.

Digital electronics provides for us the hardware that takes us into the digital world and back. We start with digitization of analog signals which are produced by microphones, cameras, sensors, etc., usually referred to as analog–digital (A/D) converters, which includes sampling of the analog signal at periodic intervals and storage of the sampled data. The digitized data can now be processed, routed, compressed, and multiplexed. Time-division multiplexing, which interleaves the digital signal with 0 and 1's from other messages, leads to efficient and economical use of a transmission medium such as cable, wireless, and optical fibers. Digital processing is powerful because it uses the digital computer, and it is flexible because a computer can be reprogrammed. None of this is available in analog processing which uses expensive and inflexible equipment that is dedicated to perform only the function for which it was built (cannot be reprogrammed). A digital signal, after the desired processing and transmission, can be converted to analog form. This is accomplished by digital–analog (D/A) converters.

7.2   DIGITAL SIGNAL REPRESENTATION

We have described a digital signal in terms of 0 and 1's, suggesting voltage pulses of 0 V and 1 V propagate down a cable or in free space as wireless signals. Such an interpretation is not wrong but is very restrictive since any difference in voltage can represent two states, for example, 10 V and 0 V or 3 V and 2 V, as long as the voltage difference is clearly larger than any noise voltage that is present—in other words, the noise should not be able to change the high voltage to a low voltage and vice versa. Figure 7.2a shows a common range of voltages in a digital computer that are identified with 0 and 1's. Figure 7.2b shows a simple switching arrangement to generate 0 and 1's: with the switch open the output v_o is high at 5 V, which is identified with a 1, whereas closing the switch results in a zero.

In general, we associate a **HIGH** voltage with **1** and a **LOW** voltage with **0**. We can now speak of a *binary signal* which is just like a *binary number* and is used as such

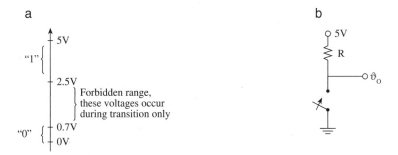

FIGURE 7.2   (a) Two distinct voltage ranges, separated by a forbidden range, can be identified with the two states of digital logic which is operated from a 5 V power supply. (b) Closing and opening the switch will generate a digital signal like that shown at the top of Fig. 7.1*b*.

when performing binary arithmetic computation. For convenience, we associate binary words with binary numbers. The shortest binary word consists of a single symbol, or *bit*,[1] which can express two words: the word **LOW** or **0** and the word **HIGH** or **1**. If we have two signal lines, each transmitting 0 and 1's, we can create a longer word (a 2-bit word) which makes four words possible: 00, 01, 10, and 11. Similarly, eight binary lines can be used to express $2^8 = 256$ different words; each such word is called a *byte*. In general, the number of binary words that can be created with k lines is given by 2^k. Thus, whenever distances between components are short, such as inside a computer, *parallel* transmission is used: an n-bit word is transferred on n wires (called a *bus*), one wire for each bit. On the other hand, when distances are long, as, for example, in the use of telephone lines to connect modems to servers, *serial* transmission of data is used: successive bits of a word are transferred one after the other over a single wire and are collected and recombined into words at the receiving end. Parallel transmission is faster but it is too expensive to run multiwire lines over long distances.

It is also possible to use digital signals for purposes other than computation. We can use the **0** and **1**'s to represent the **FALSE** (also called **LOW** or **OFF**) and **TRUE** (also called **HIGH** or **ON**) of Boolean logic. The processing of digital signals in computers or in digital hardware is done by logic circuits which are described by the simple but powerful rules of Boolean algebra. The simplest logic circuits are called gates and perform the elementary operations of **AND**, **OR**, and **NOT**. Logic gates can have two or more inputs but only one output. Output signals in logic circuits switch between high and low as the information being represented changes. We will refer to these signals as logic variables and represent them by capital letters.

7.2.1   Combinatorial and Sequential Logic

In digital electronics digital outputs are generated from digital inputs. If the output of the logic circuit depends only on the present input values, we refer to the system as not having memory. Systems without memory are also known as *combinatorial logic circuits* because they combine inputs to produce the output. Combinatorial circuits can be constructed with gates alone. If, on the other hand, the output of the logic circuit depends on present as well as past input values, we then refer to such a circuit as having memory, because such circuits remember past input values. Systems with memory are also known as *sequential logic circuits*. Such circuits are more complicated and require some form of memory (flip-flops) and the presence of a clock signal to regulate the response of the circuit to new inputs, ensuring that the necessary operations occur

---

[1] One should not assume that only analog signals represent real-life situations. There are many circumstances when a digital signal is appropriate. For example, situations such as whether a switch is on or off, whether a signal such as a radio station carrier is present or absent, park or drive, win or loose, to go or not to go, etc., can be represented by a 1-bit signal, i.e., a 1-bit word. More complex situations can be represented by more complex words: a description of the outcomes of tossing two coins requires a 2-bit word (00, 01, 10, 11).

in proper sequence—hence the name sequential logic circuit. We will first consider combinatorial circuits and then proceed to sequential ones.

7.3 COMBINATORIAL LOGIC

Basic to digital electronics are *switching gates*. They control the flow of information which is in the form of pulses (0 and 1's). If we list all possible combinations of inputs to a gate and the corresponding outputs we obtain what is called a *truth table*. We will now use truth tables to show how memoryless combinatorial tasks are performed by **AND**, **OR**, and **NOT** gates.

7.3.1 The AND Gate

The logic of an **AND** gate can be demonstrated by a battery, two switches (denoted by **A** and **B**) in series, and a light bulb, as shown in Fig. 7.3*a*. The bulb lights only if both switches are on; for any other state of switches the bulb is off. Using **0** and **1**'s to represent off and on of the switches, we construct a truth table in Fig. 7.3*b*. The "**A AND B**" operation is denoted by **A · B** or simply by **AB**.

A simple **AND** gate consisting of a 5 V source and two diodes is shown in Fig. 7.3*c* (the resistor is only there to limit current). The input signals at *A* or *B* are either 5 V or 0 V signals; we say the input is **HIGH** or **1** when it is at 5 V (with respect to ground) and the input is **LOW** or **0** when it is at 0 V (such a **NO** input can be simply mimicked by

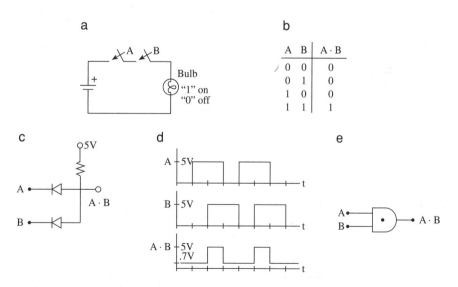

FIGURE 7.3 (a) A two-switch **AND** gate. (b) Truth table for an **AND** gate. (c) A two-input diode **AND** gate. (d) Typical voltage pulses at the input and output of an **AND** gate. (e) Logic symbol for an **AND** gate.

shorting the input terminal to ground). Unless both inputs are **HIGH**, one of the diodes will conduct, strapping the output to **LOW** (the **LOW** will not be 0 V but 0.7 V, which is the forward-bias voltage of the conducting diode). Examples of input and output voltages are given in Fig. 7.3*d*, showing that the output is **HIGH** only when both inputs are **HIGH**. The logic symbol for an **AND** gate is shown in Fig. 7.3*e*. **AND** gates can have more than two inputs, which all must be **HIGH** for the output to be **HIGH**.[2]

7.3.2   The OR Gate

The logic of an **OR** gate can be demonstrated by a battery, two switches in parallel, and a bulb, as shown in Fig. 7.4*a*. The bulb lights if either switch is on. Using **0** and **1**'s to represent off and on of the switches, we construct a truth table in Fig. 7.4*b*. The "**A** OR **B**" operation is denoted by **A** + **B**; that is, the Boolean symbol for **OR** is +.

A simple **OR** gate is shown in Fig. 7.4*c*. The input signals at **A** or **B** are either 5 V or 0 V signals. Unless both inputs are **LOW**, one of the diodes will conduct, transferring the 5 V of the input to the output, thereby making it **HIGH**. Examples of input pulses are given in Fig. 7.4*d*, which show that the output will be **HIGH** if any one input is **HIGH**.

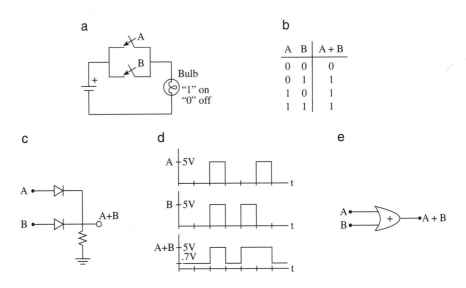

FIGURE 7.4   (a) A two-switch **OR** gate. (b) Truth table for an **OR** gate. (c) A two-input diode **OR** gate. (d) Typical **OR** gate voltage pulses. (e) Logic symbol for an **OR** gate.

---

[2]An example of the use of the **AND** operation is when a word processing or a drawing program must find an exact match, like selecting all red in an image. The computer compares the binary digits for each color in an image against the digits that denote red. If the digits match, the result of the **AND** operation is a 1 (or **TRUE**), which means the color matches red and is selected. If there is the slightest mismatch, the result is **0** (or **FALSE**), and the color is not selected, since it is not red.

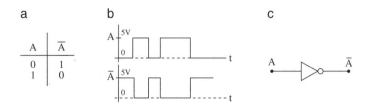

FIGURE 7.5   (a) Truth table for a NOT gate. (b) Input–output pulses for a NOT gate. (c) Symbol for a NOT gate (in general, a bubble placed on any terminal of a gate symbol denotes negation).

The logic symbol for an **OR** gate is given in Fig. 7.4*e*. Again, more than two inputs are possible, and if any one is **HIGH** the gate will give a **HIGH** output.[3]

7.3.3   The NOT Gate

A **NOT** gate performs an inversion: the output is the complement of the input, denoted by a bar over the symbol for the logic variable. Thus, if **A** is the input, the output is $\bar{A}$ (read as "**NOT A**"). If **A** is **0**, $\bar{A}$ is **1**, and vice versa. If **A** means the switch is open, $\bar{A}$ means the switch is closed.[4] The truth table, response to input pulses, and the symbol for a **NOT** gate, which is a one-input device, are shown in Fig. 7.5*a*, *b*, and *c*.

Digital electronics is at its most powerful with sequential logic circuits, but even combinations of the simple **AND**, **OR**, and **NOT** gates (combinatorial logic) will already yield useful devices as the next example demonstrates.

EXAMPLE 7.1   A popular switching circuit found in many two-story houses is one that controls a light by two switches located on separate floors. Such a circuit, shown in Fig. 7.6*a*, implements the logic **F** is **HIGH** when **A AND B OR NOT A AND NOT B** (the light is on when the switches are in positions *A* and *B* or in positions $\bar{A}$ and $\bar{B}$). These switches are known as SPDT (single pole, double throw) switches and are somewhat more complicated than ordinary on–off switches which are SPST (single pole, single throw) switches.

Figure 7.6*b* shows a logic gate implementation of the function $\mathbf{F} = \mathbf{AB} + \mathbf{\bar{A}\bar{B}}$. Two **AND** gates, one of which is preceded by **NOT** gates, feed into an **OR** gate to give the output **F**. It is basically an equality comparator: the output is **HIGH** when the inputs are equal.   ■

---

[3]Like the AND operation, the OR operation also combines two or more numbers bit by bit. But it outputs 1 if any bit is 1 (or both are). It outputs 0 only if all bits are 0. The OR operation is used, for example, when searching for text. By using OR operations, the (word) processor can be told that whether the letters are uppercase or lowercase, the text still matches the search string.

[4]An example of the NOT operation is what the (word) processor does when you choose a menu item and the normally black-on-white text becomes white-on-black.

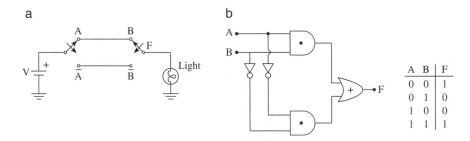

FIGURE 7.6   (a) The circuit for controlling a light by two switches. (b) A logic gate circuit which mimics the two-switch system. (c) The truth table.

The **NOT** operation is elegantly implemented by the simple transistor switch shown in Fig. 7.7a. For a zero-voltage input (0), the input junction of the transistor is not forward-biased, the transistor does not conduct, and the output floats to 5 V, i.e., to **HIGH** (1). We say the switch is open. If the input receives a 5 V pulse (a **HIGH** or **1**), the transistor becomes forward-biased, conducts, and straps the output voltage to almost zero volts (a **LOW** or **0**). We say the switch is closed.

Let us point out a subtle difference in use of a transistor in digital and in analog electronics. The digital nature of a transistor switch is highlighted by the output characteristics with the load line drawn in Fig. 7.7b. For a pulsating input (like the square wave in Fig. 7.1b), the output voltage flips between two states on the load line. The output voltage is **HIGH** when the transistor is near cutoff, the switch is open, and we label it as a 1. The other state has the transistor in saturation, the output voltage is approximately 0.3 V, and we have a **0**. Hence, in digital electronics a transistor has only two stable points on the load line, whereas in analog electronics, where the input signals

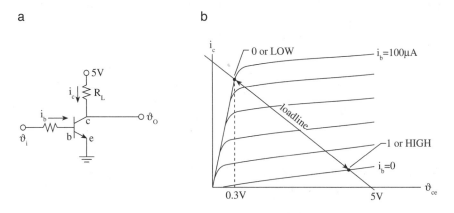

FIGURE 7.7   (a) A transistor switch. (b) The two points on the load line corresponding to the on–off states of a transistor switch are shown.

Prove DeMorgan's theorems by constructing truth tables for the expressions in (7.4).

The truth table for the first expression in (7.4) is obtained as follows: Entering first all possible combinations for **A** and **B** in the first two columns (equivalent to counting 0 to 3 in binary), we proceed to construct the entries in the remaining columns.

A	**B**	$\overline{\mathbf{A} \cdot \mathbf{B}}$	$\bar{\mathbf{A}} + \bar{\mathbf{B}}$
0	0	1	1
0	1	1	1
1	0	1	1
1	1	0	0

$$(7.5)$$

We see that the entries in the last two columns are the same, and hence we conclude that (7.4) is correct. In terms of logic gates, this theorem states that a two-input **AND** gate followed by a **NOT** gate (i.e., a two-input **NAND** gate) is equivalent to a two-input **OR** gate, provided the two inputs first pass through **NOT** gates.

The truth table for the second expression in (7.4) is as follows

A	**B**	$\overline{\mathbf{A} + \mathbf{B}}$	$\bar{\mathbf{A}} \cdot \bar{\mathbf{B}}$
0	0	1	1
0	1	0	0
1	0	0	0
1	1	0	0

$$(7.6)$$

Again the entries in the last two columns are the same, implying the correctness of (7.4). In terms of logic gates, this theorem states that a two-input **OR** gate followed by a **NOT** gate (i.e.. a two-input **NOR** gate) is equivalent to a two-input **AND** gate, provided the two inputs first pass through **NOT** gates.   ∎

We have postulated that a logic variable can have only two values: true (T) or false (F), or 1 or 0. As a relation between two or more logic variables can be represented by a logic equation or a truth table, we will now give more examples designed to make the student quickly familiar with the interrelationship of logic equations, truth tables, and logic circuits. We will show that given a truth table, the corresponding logic function can be constructed and vice versa. We will show that given a logic circuit, the corresponding logic function can be determined and vice versa.

EXAMPLE 7.4   Simplify $\mathbf{F} = \mathbf{A}(\mathbf{B} + \mathbf{C}) + \mathbf{A}\overline{\mathbf{BC}}$ using Boolean rules.

Factoring out **A** and simplifying, we obtain

$$
\begin{aligned}
\mathbf{F} &= \mathbf{A}(\mathbf{B} + \mathbf{C} + \overline{\mathbf{BC}}) \\
&= \mathbf{A}(\mathbf{B} + \mathbf{C} + \bar{\mathbf{B}} + \bar{\mathbf{C}}) \\
&= \mathbf{A}\big((\mathbf{B} + \bar{\mathbf{B}}) + (\mathbf{C} + \bar{\mathbf{C}})\big) \\
&= \mathbf{A}(1 + 1) = \mathbf{A}
\end{aligned}
$$

∎

EXAMPLE 7.5   Find the logic function **F** which the given truth table represents.

A	B	C	F
0	0	0	1
0	0	1	1
0	1	0	0
0	1	1	1
1	0	0	0
1	0	1	1
1	1	0	1
1	1	1	1

Using the six "1" entries in the column for **F**, we can state **F** as

$$\mathbf{F = \bar{A}\bar{B}\bar{C} + \bar{A}\bar{B}C + \bar{A}BC + A\bar{B}C + AB\bar{C} + ABC}$$

The next step would be to try to simplify this expression. However, because there are only two entries for $\mathbf{\bar{F}}$ (i.e., two "0" entries for **F**), it is simpler in this case to write the complement of **F** as follows:

$$\mathbf{\bar{F} = \bar{A}B\bar{C} + A\bar{B}\bar{C}}$$

This expression can be simplified by applying DeMorgan's theorems. Thus, taking the complement of both sides and applying (7.4), we obtain

$$\mathbf{F = AB + \bar{A}\bar{B} + C}$$ ∎

EXAMPLE 7.6   Determine a truth table for the logic relation $\mathbf{F = \left(\bar{A}B + A\bar{B}\right)AB}$.

Entering first all possible combinations for **A** and **B** in the first two columns, we proceed to construct entries for the remaining columns.

A	B	$\mathbf{\bar{A}B + A\bar{B}}$	AB	F
0	0	0	0	0
0	1	1	0	0
1	0	1	0	0
1	1	0	1	0

This is an interesting logic function as the **F** column indicates. All entries in that column are zeros, implying that no matter what values **A** and **B** assume, the output is always **LOW**. Hence, in an actual logic circuit, the output denoting **F** would be simply grounded. As an additional exercise, the student should verify this result directly from the logic function **F**. *Hint*: use Boolean rules to reduce the logic function to $\mathbf{F = 0}$.   ∎

EXAMPLE 7.7   For the logic circuit shown in Fig. 7.10, determine **F** in terms of **A** and **B**. Simplify the resulting expression so it has the fewest terms. Then check the simplified expression with the original by constructing a truth table.

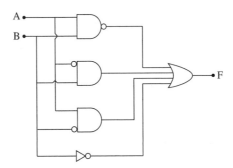

FIGURE 7.10   A logic circuit which is equivalent to the simpler NAND gate.

The output **F** is obtained by first stating the outputs of the gates that feed into the **OR** gate, noting that the bubbles on the gates always denote negation. The output of the **OR** gate is then

$$\mathbf{F} = \overline{\mathbf{A} \cdot \mathbf{B}} + \bar{\mathbf{A}}\mathbf{B} + \mathbf{A}\bar{\mathbf{B}} + \bar{\mathbf{B}}$$

Simplifying, using Boolean rules,

$$
\begin{aligned}
\mathbf{F} &= \bar{\mathbf{A}} + \bar{\mathbf{B}} + \bar{\mathbf{A}}\mathbf{B} + \mathbf{A}\bar{\mathbf{B}} + \bar{\mathbf{B}} \\
&= \bar{\mathbf{A}}(1 + \mathbf{B}) + \bar{\mathbf{B}}(1 + \mathbf{A}) \\
&= \bar{\mathbf{A}} + \bar{\mathbf{B}} \\
&= \overline{\mathbf{A} \cdot \mathbf{B}}
\end{aligned}
$$

Next we construct a truth table

A	B	$\overline{AB}$	$\bar{A}\,B$	$A\bar{B}$	$\bar{B}$	F	$\overline{AB}$
0	0	1	0	0	1	1	1
0	1	1	1	0	0	1	1
1	0	1	0	1	1	1	1
1	1	0	0	0	0	0	0

The entries in the last two columns check and we conclude that the simplification of **F** is correct.   ∎

EXAMPLE 7.8   Design a logic circuit that will implement the function $\mathbf{F} = \mathbf{A}\bar{\mathbf{B}}\mathbf{C} + \mathbf{ABC} + \overline{\mathbf{A} \cdot \mathbf{B}}$.

First let us see if we can simplify **F**. After applying DeMorgan's theorems to the last term of **F** we can factor out $\bar{\mathbf{B}}$ and obtain $\mathbf{F} = \bar{\mathbf{B}} + \mathbf{ABC} + \bar{\mathbf{A}}$. The circuit which implements this function must include a triple-input **OR** gate, a triple-input **AND** gate, and two **NOT** gates. The schematic is shown in Fig. 7.11.   ∎

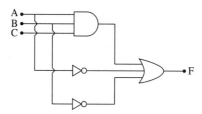

FIGURE 7.11   A logic circuit for **F**.

7.4   COMBINATORIAL LOGIC CIRCUITS

7.4.1   Adder Circuits

An important, practical example of combinatorial logic, that is, logic that does not have memory, is adder logic. The addition of binary numbers is basic to numerical computation. Multiplication and division is accomplished simply by repeated addition and subtraction. Of course, the power of computers is that these rather mundane operations can be performed at lightning speeds. Binary addition is performed by logic gates and is similar to conventional addition with decimal numbers.[5] It is a two-step process in which the digits in each column are added first and any carry digits are added to the column immediately to the left (the column representing the next higher power of two).

7.4.2   The Half-Adder

Figure 7.12*a* shows the addition of two binary numbers, resulting in the sum 10101. The addition is straightforward, except for the column in which **A** = **1** and **B** = **1**, which produces a sum of **0** and a carry of **1** in the column to the left. A circuit that can produce a sum and a carry has the truth table shown in Fig. 7.12*b* and is called a *half-adder* (in comparison, a full adder can also handle an input carry). The truth table shows that the sum **S** of two bits is **1** if **A** is **0** AND **B** is **1**, OR if **A** is **1** AND **B** is **0**,

---

[5]In principle, there is no difference between the binary and the decimal number systems. The decimal system is based on the number of fingers we possess—not surprisingly, it is not suited for digital devices such as computers. The binary system, on the other hand, is a natural for digital devices, which are based on switching transistors which have two states, off and on, which can be used to represent two numbers, 0 and 1. Similar to the decimal system which has 10 numbers, 0 to 9, the binary system has two, 0 and 1. The binary number 11011, for example, is a shorthand notation for the increasing powers of two. That is,

$$
\begin{aligned}
11011 &= (1 \cdot 2^4) + (1 \cdot 2^3) + (0 \cdot 2^2) + (1 \cdot 2^1) + (1 \cdot 2^0) \\
&= 16 + 8 + 0 + 2 + 1 \\
&= 27
\end{aligned}
$$

The binary number 11011 represents the same quantity as the decimal number $27 = 2 \cdot 10^1 + 7 \cdot 10^0 = 20 + 7$. However, the binary number can be operated on by a digital device such as a computer. Binary numbers in a computer are represented by voltage signals which have waveforms with regularly spaced pulses of uniform amplitude.

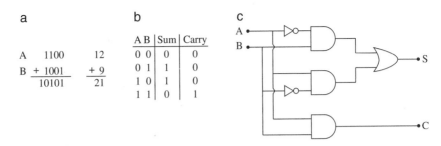

FIGURE 7.12 (a) The addition of two binary numbers (decimal equivalents are also shown). (b) Truth table for binary addition. (c) A half-adder logic circuit.

which can be stated in logic as

$$S = \bar{A}B + A\bar{B} \tag{7.7}$$

The carry **C** is just simple AND logic

$$C = AB \tag{7.8}$$

Several realizations of half-adder logic are possible. The simplest is obtained by a straightforward application of the above two logic statements and is shown as Fig. 7.12c. We can obtain a circuit with less logic gates though by applying DeMorgan's theorems to (7.7). Using that $\bar{\bar{S}} = S$ (see (7.2)), we obtain

$$
\begin{aligned}
\bar{A}B + A\bar{B} &= \overline{(A + \bar{B})(\bar{A} + B)} \\
&= \overline{\overline{AB} + \bar{A}\bar{B}} \\
&= (A + B)(\bar{A} + \bar{B}) \\
&= (A + B)\overline{AB} \\
&= A \oplus B
\end{aligned}
\tag{7.9}
$$

where we have used that $A\bar{A} = 0$ and the symbol $\oplus$, referred to as *exclusive or*, is used to denote the sum operation of (7.9).

Two implementations of a half-adder using (7.9) are shown in Fig. 7.13. The *exclusive or* operation of (7.7) or (7.9) has applications other than in a half-adder, because the $\oplus$ operation is an inequality comparator: it gives an output of **1** only if **A** and **B** are not equal ($0 \oplus 0 = 0, 1 \oplus 0 = 1, 0 \oplus 1 = 1, 1 \oplus 1 = 0$). Figure 7.13c shows the logic gate symbol for *exclusive or*.

7.4.3 The Full Adder

For general addition an adder is needed that can also handle the carry input. Such an adder is called a full adder and consists of two half-adders and an OR gate in the arrangement shown in Fig. 7.14a. If, for example, two binary numbers **A** = 111 and **B** = 111

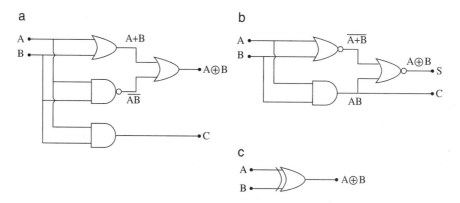

FIGURE 7.13   (a, b) Two different logic circuits of a half-adder. (c) The symbol for the *exclusive or* gate.

are to be added, we would need three adder circuits in parallel, as shown in Fig. 7.14b, to add the 3-bit numbers. As a carry input is not needed in the least significant column (A_o, B_o), a half-adder is sufficient for this position. All other positions require a full adder. The top row in Fig. 7.14b shows the resultant sum 1110 of the addition of the two numbers **A** and **B**. Note that for the numbers chosen the addition of each column produces a carry of 1. The input to the half-adder is digits from the first column, $A_o = 1$ and $B_o = 1$; the input to the adjacent full adder is a carry $C_o = 1$ from the half-adder

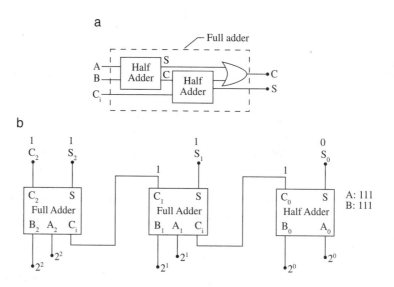

FIGURE 7.14   (a) The circuit of a full adder designed to handle an input carry C_i. (b) An adder which can add two 3-bit numbers A and B. The sum is given as $C_2 S_2 S_1 S_0$.

and digits $A_1 = 1$ and $B_1 = 1$ from the second column, which gives $C_1 = 1$ and S_1 as the output of the first full adder. Ultimately the sum $C_2 S_2 S_1 S_o = 1110$ is produced.

For the addition of large numbers such as two 32-bit numbers, 32 adders are needed; if each adder requires some 7 logic gates, about 224 logic gates are required to add 32-bit numbers. Clearly such a complexity would be unwieldy were it not for integrated circuits, which are quite capable of implementing complex circuitry in small-scale form.

7.4.4 Encoders and Decoders

Other examples of combinatorial (memoryless) logic are *decoders* (devices that select one or more output channels according to a combination of input signals), *encoders* (convert data into a form suitable for digital equipment), and *translators* (change data from one form of representation to another).

Representing numbers in binary form is natural when the processing is to be done by computers. People, on the other hand, prefer to use decimal numbers. An encoder could be designed to register a binary number whenever a decimal number is entered on a computer keyboard. For example, the decimal number 105 would register as the binary 1101001. We could also use *binary-coded decimals* (BCD) as an intervening step between binary machine code and the decimal input. When BCD is used, the bits are arranged in groups of four bits with each group representing one of the decimal digits 0 to 9. For example, when a computer operator enters the decimal number 105, he first strikes the "1" key, and the computer encodes it as 0001. Then he strikes the "0" key, and the computer enters it as 0000, and finally he strikes the "5" key, and the computer enters it as 0101. When the entry sequence is finished, the computer has the binary-coded decimal 0001,0000,0101 for decimal 105. As is readily seen, such a representation is not as efficient as the binary representation 1101001, but it allows entry of the number in three steps, and as will be shown in the next section, BCD representation is particularly suited for digital displays.

EXAMPLE 7.9 Design a two-line to four-line decoder that will translate a 2-bit address and specify 1 of 2^2 bits.

Such a device will have two incoming lines which can be set four different ways (00,01,10,11)—we call this an *address*. The output of this device has four lines. The objective is to pick one line corresponding to one of the input combinations. This is done by taking the chosen line HIGH while all other lines are LOW. The truth table for the 2-to-4 decoder is shown in Fig. 7.15a.

A circuit that will implement this truth table requires four AND gates and two NOT gates connected as shown in Fig. 7.15b. In this example, the address or input word 10 yields a 1 at O_2 ($A\bar{B}$) and specifies that output word O_2 be read or, depending on the application, the line corresponding to output word O_2 be selected or connected to some other device. Such a decoder is also known as a 1-of-4 decoder for obvious reasons.

If the input address consists of 3 bits, we can have a 1-of-8 decoder, as 3 bits can specify eight numbers. Such a 3-to-8 decoder will be used in the next section to activate

a   b

A	B	O_0	O_1	O_2	O_3
0	0	1	0	0	0
0	1	0	1	0	0
1	0	0	0	1	0
1	1	0	0	0	1

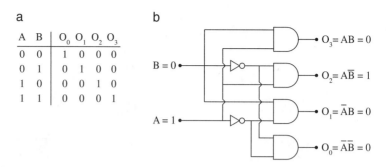

FIGURE 7.15   (a) Truth table for a 2-to-4 decoder. For illustration, the address 10 is chosen, which in (b) selects the third line from the bottom by sending O_2 HIGH. (b) A logic circuit of the decoder.

the segments of a digital display. Similarly, a 4-bit address line can be used in a 1-of-16 decoder, a 5-bit address line in a 5-to-32 decoder, and so forth.   ■

We have seen that decoders can convert binary numbers to BCD and BCD to binary, and can perform other arithmetic operations on numbers such as comparing numbers, making decisions and selecting lines.

EXAMPLE 7.10   Another practical use of combinatorial gates is *triple redundancy sensing*. For example, three temperature sensors *A*, *B*, and *C* are mounted on a jet engine at critical locations. An emergency situation is defined if the engine overheats at least at two locations, at which time action by the pilot is called for. The truth table in Fig. 7.16*b* represents this situation: **F** is **HIGH** when the majority of inputs are **HIGH**.

Design a logic circuit that will give a **HIGH** output (**F** = 1) when at least two sensors are **HIGH**, that is,

$$\mathbf{F = AB\bar{C} + A\bar{B}C + \bar{A}BC + ABC}$$

This expression is obtained by entering the product term for each row of the truth table in which **F** is equal to **1**. Such a circuit is also known as a vote taker, because the output is 1 (**HIGH**, **TRUE**, or **YES**) when the majority of inputs are **1**. A straightforward implementation of the above expression would be four triple-input **AND** gates feeding into a quadruple-input **OR** gate whose output would then be **F**.

A simpler circuit results if the rules of Boolean algebra are applied to the above expression. Using that $\mathbf{A + \bar{A} = 1}$,

$$\mathbf{F = AB\bar{C} + A\bar{B}C + BC}$$

and after applying distribution rules, we finally obtain

$$\mathbf{F = A(B + C) + BC}$$

This is a simpler expression which results in a simpler circuit. A triple redundancy sensor consisting of two **AND** gates and two **OR** gates is shown in Fig. 7.16*c*.   ■

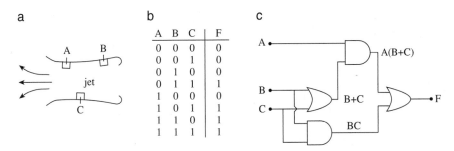

FIGURE 7.16 (a) A jet engine with three temperature sensors. (b) Truth table for triple redundancy sensing. (c) A logic circuit implementation.

7.4.5 Seven-Segment Display

Calculator and digital watch displays typically consist of light-emitting diodes (LEDs) or liquid-crystal displays (LCDs) with seven segments, as shown in Fig. 7.17a. Numerals 0 to 9 are displayed by activating appropriate segments a–g. Figure 7.17b shows the digits 0–9 and the corresponding segments that need to be activated. Noting that 8 combinations (2^3) are possible with a 3-bit word and 16 combinations (2^4) with a 4-bit word, we must choose a 4-bit word even though we need only ten combinations to represent the digits 0 to 9. Thus, 0 will be represented by 0000, 1 by 0001, 2 by 0010, 3 by 0011, 4 by 0100, and so forth, while 10 (1010) to 15 (1111) will not be used. Representing decimal numbers this way makes use of binary-coded decimals.

What is now needed is a decoder having four inputs and seven outputs that will translate the 4-bit BCD words into 7-bit words. Each 7-bit word will activate one combination of the a–g segments, thus lighting up one of the 10 decimal digits. For example, the digit 4 needs to have segments a,d,e **LOW** (**0**) and segments b,c,f,g **HIGH** (**1**). Figure 7.18a illustrates how a single-digit display[6] is driven by a four-line to seven-line decoder to display the digit 4.

FIGURE 7.17 (a) A seven-segment decimal display. A segment, if taken **HIGH** (logic **1**, which in a typical circuit means 5 V), will become visible. (b) Taking the appropriate segments **HIGH** will display the digits 0 through 9.

---

[6]It is assumed that a segment, if taken **HIGH**, will be activated. For example, a **1** to segment b will make it visible, whereas a **0** would make it dark (common-cathode display). LED displays can also be connected so that a luminous segment is state **0** and a dark segment is state **1** (common-anode display).

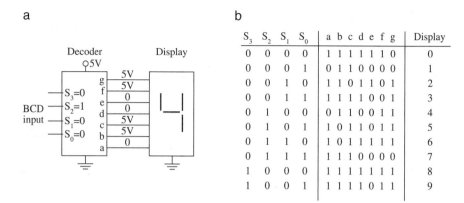

a   b

S_3	S_2	S_1	S_0	a	b	c	d	e	f	g	Display
0	0	0	0	1	1	1	1	1	1	0	0
0	0	0	1	0	1	1	0	0	0	0	1
0	0	1	0	1	1	0	1	1	0	1	2
0	0	1	1	1	1	1	1	0	0	1	3
0	1	0	0	0	1	1	0	0	1	1	4
0	1	0	1	1	0	1	1	0	1	1	5
0	1	1	0	1	0	1	1	1	1	1	6
0	1	1	1	1	1	1	0	0	0	0	7
1	0	0	0	1	1	1	1	1	1	1	8
1	0	0	1	1	1	1	1	0	1	1	9

FIGURE 7.18   (a) A four-to-seven decoder driving a digital display. (b) A truth table specifying the lit segments for each numeral in a digital display.

A truth table specifying the lit segments for each numeral is given in Fig. 7.18*b*. Hence, the function of the decoder is to accept the four input signals corresponding to a **BCD** word and provide the proper output signal at each of its seven output terminals.

7.5   SEQUENTIAL LOGIC CIRCUITS

The potential to do all sorts of exciting things greatly increases when we add "memory" to logic gates. Whereas the output of combinatorial logic circuits at a given instant of time depends only on the input values at that instant, sequential circuits have outputs that also depend on past as well as present values. Flip-flops are the principal memory circuits that will store past values and make them available when called for.

The sequential circuit environment is more complicated as present and past inputs have to be made available in an orderly fashion. Operation of sequential circuits is therefore regulated by a clock signal which permeates the entire circuit and consists of periodic pulses of the type shown in the top Fig. 7.1*b* (a clock signal is basically a square wave). For example, when the clock signal is **HIGH** (logic 1), a new input could become available to a gate, and when it is **LOW** (logic **0**), it would not be available. A clock signal guarantees that operations occur in proper sequence.[7]

---

[7]This is also why such circuits are referred to as *sequential logic circuits*. For example, in a 100 MHz computer, most logic operations are synchronized by a clock signal that changes hundred million types per second.

7.5.1 Flip-Flop: A Memory Device

The simplest flip-flop is constructed with two **NOR** gates, as in Fig. 7.19a. Recall that if either input to a **NOR** gate is **1** the output is **0** (see Fig. 7.8).

Combining two **NOR** gates as shown results in the flip-flop having two outputs, **Q** and $\bar{\text{Q}}$, which are complements of each other. The cross-coupling of the outputs gives the combination two stable states ($\text{Q}\bar{\text{Q}} = \textbf{10}$ or $\textbf{01}$), and it must remain in one of the states until instructed to change. Assume, for the time being, that the outputs are **Q = 0** and $\bar{\text{Q}} = \textbf{1}$. Now suppose that the **S** input (which stands for **SET**) is set to **1** and the **R** input (for **RESET**) is set to **0**. Let us check if input **SR = 10** is consistent with the assumed output $\text{Q}\bar{\text{Q}} = \textbf{01}$. From the circuit of the flip-flop, we see that the output should be $\text{Q} = \overline{\text{R} + \bar{\text{Q}}} = \overline{0 + 1} = \textbf{0}$, which checks with the above assumption. The other output is $\bar{\text{Q}} = \overline{\text{S} + \text{Q}} = \overline{1 + 0} = \textbf{0}$, which does not check with the above assumption and is contradictory as **Q** and $\bar{\text{Q}}$ cannot both be **0**; hence the system becomes unstable and the flip-flop must flip, so the output becomes $\text{Q}\bar{\text{Q}} = \textbf{10}$, which is now consistent with the **SR** input (that is, $\text{Q} = \overline{\text{R} + \bar{\text{Q}}} = \overline{0 + 0} = \textbf{1}$, and $\bar{\text{Q}} = \overline{\text{S} + \text{Q}} = \overline{1 + 1} = \textbf{0}$). We conclude that **S = 1** sets the output to **Q = 1**, which is now a stable state. Similar reasoning would show that changing **RESET** to **R = 1** resets the output to **Q = 0**.

If we started with one of the two input combinations (**SR = 10** or **01**) and then changed to **SR = 00**, we would find that the output remains unchanged.[8] We conclude that we have a memory device as the flip-flop remembers which of the inputs (**R** or **S**) was **1**. This is illustrated in Fig. 7.19b, which shows the truth table for the RS flip-flop. The logic or circuit symbol for the RS flip-flop is shown in Fig. 7.19c.

The combination **SR = 11**, on the other hand must be avoided as it creates an unacceptable state, an ambiguity with **Q** and $\bar{\text{Q}}$ both equal to **0** (practical circuits are designed to avoid such a possibility. Furthermore, a change from this state to **SR = 00** could give either **0** or **1** for an output, depending which gate has the faster response, therefore making the output unpredictable).

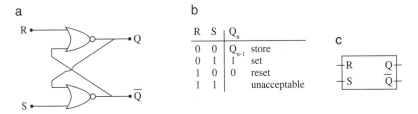

a

b

R	S	Q_n	
0	0	Q_{n-1}	store
0	1	1	set
1	0	0	reset
1	1		unacceptable

c

FIGURE 7.19 (a) A RS (reset-set) flip-flop, a basic memory unit, formed by cross-connecting two **NOR** gates. (b) Truth table for the flip-flop. Q_n is the present state of the output; Q_{n-1} is the previous state. (c) Symbol for the flip-flop.

---

[8]Assuming that **S = 1** sets the output to **Q = 1** (**SR = 10** and $\text{Q}\bar{\text{Q}} = \textbf{10}$), we see that changing **S** to **0** will leave the output unaffected, that is, $\text{Q} = \overline{\text{R} + \bar{\text{Q}}} = \overline{0 + 0} = \textbf{1}$ and $\bar{\text{Q}} = \overline{\text{S} + \text{Q}} = \overline{0 + 1} = \textbf{0}$.

7.5.2   Clocked Flip-Flops

In the previous flip-flop, a $S = 1$ sets the output Q to 1 and a 1 to R resets the output to
0. As there is no limitation as to when the inputs must occur, we can state that the flip-
flop responds to *asynchronous* inputs. In circuits in which there are many flip-flops, it
might be necessary to activate all flip-flops at the same time. This can be accomplished
with a clock signal, which is a square-wave voltage used to synchronize digital circuits.
In addition to the clock signal C, we need to add two **AND** gates at the inputs to the flip-
flop, as shown in Fig. 7.20a. The flip-flop is now enabled only when the clock signal C
is **HIGH**; that is, data can be read into the flip-flop only when the clock signal is **HIGH**.
When the clock signal is **LOW**, the outputs of the **AND** gates are also **LOW**, irrespective
of the signal at R and S, effectively disabling the inputs to the flip-flop. The flip-flop is
then in the **HOLD** or **STORE** mode. Once again, when the clock signal goes **HIGH**, the
output of the **AND** gates is immediately equal to the signal on R and S and the flip-flop
is either **SET**, **RESET**, or remains unchanged (**RS** $=$ **01, 10,** or **00**). R and S inputs
that are recognized or enabled only when a clock signal is **HIGH** are called *synchronous
inputs*. Figure 7.20b shows an example of the response of the clocked flip-flop to some
waveforms. Figure 7.20c shows the truth table, and Fig. 7.20d the symbol for a clocked
flip-flop.

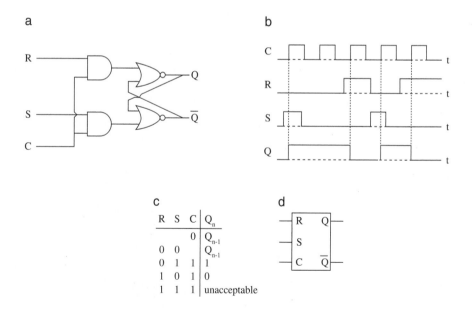

FIGURE 7.20   (a) A RS flip-flop that is enabled by the clock signal. (b) A sample of timing waveforms.
(c) The truth table. If Q_n is the nth state of the output, $Q_n = Q_{n-1}$ means that the previous state is preserved.
(d) Symbol for RS flip-flop with enable.

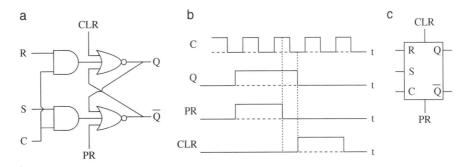

FIGURE 7.21 (a) A flip-flop with **CLEAR** and **PRESET** capability. (b) Timing waveforms, showing that **CLR** and **PR** override RS inputs and the clock signal. (c) Symbol for the flip-flop.

7.5.3 Clocked RS Flip-Flop with Clear and Preset

Adding an extra input to each **NOR** gate of the clocked **RS** flip-flop shown in Fig. 7.20a allows us to preset **Q** to **1** or clear **Q** to **0** independent of the clock signal or any other inputs. Most practical flip-flops have this feature. The connections of such a flip-flop are shown in Fig. 7.21a. Recall that a **1** to any input of a multiple-input **NOR** gate results in a **0** output. Thus a **1** to CLR (**CLEAR**) gives a **0** at **Q**, and a **1** to PR (**PRESET**) gives a **1** at the output **Q**. The PR and CLR inputs are called *asynchronous* because their effects are not synchronized by the clock signal, in contrast to inputs to **R** and **S**, which are *synchronous*. Figure 7.21b shows some timing waveforms and Fig. 7.21c the symbol for such a flip-flop.

7.5.4 Edge-Triggered Flip-Flops

In the flip-flops considered thus far, the input signals were enabled or disabled by the clock signal. We showed that during the entire time interval that the clock signal was **HIGH** (in Fig. 7.20b, for example), the RS inputs were transferred to the flip-flop, and conversely when the clock was **LOW**, the inputs were effectively removed from the flip-flop. *Edge-triggered flip-flops*, on the other hand, are enabled or respond to their inputs only at a transition of the clock signal. The output can therefore change only at a downward transition (trailing-edge triggered) or at the upward transition (leading-edge triggered) of the clock signal. During the time that the clock signal is steady, either **HIGH** or **LOW**, the inputs are disabled.

7.5.5 D Flip-Flop

A common type of memory element is the *D flip-flop*, short for *delay* flip-flop. It is an arrangement of two flip-flops in series, commonly referred to as a master–slave combination. In practice flip-flops are invariably used in some form of master–slave arrangement and are available as inexpensively packaged integrated circuits (ICs). Figure 7.22a

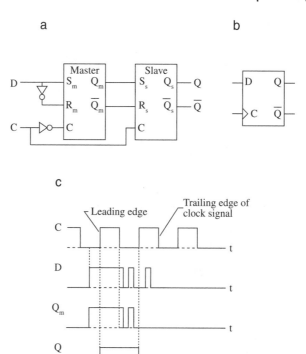

FIGURE 7.22   (a) The diagram for the D flip-flop. It is used whenever we need to maintain the present state at **Q** while we read in a new state at **D**. (b) Symbol for a D flip-flop. (c) Sample timing waveforms: clock signal **C**, input signal **D**, and output signal **Q**.

shows the arrangement for the D type. A **NOT** gate is connected between the RS inputs of the first flip-flop which eliminates the unacceptable input **RS = 11**. Furthermore, a **NOT** gate is also connected between the incoming clock signal and the clock input (also known as the enable input) of the master flip-flop. Hence, the master is enabled when the clock signal goes **LOW**, but the following slave is enabled when the clock signal goes **HIGH**. The effect is that Q_m follows the input **D** whenever the clock is **LOW**, but any change in the output **Q** is delayed until the slave becomes enabled, which is at the next upward transition of the clock signal. Figure 7.22b shows the symbol for a D flip-flop. The wedge stands for "*positive-edge-triggered*" flip-flop, which means that **D** is transferred to **Q** on the leading edge of the clock signal. Figure 7.22c shows some timing waveforms, and how the signal at **D** is first transferred to Q_m and then to **Q**. It also shows that two sudden but random inputs do not effect the output.

7.5.6   JK Flip-Flop

Even more popular in memory circuits than the D flip-flop is the *JK flip-flop*. In comparison to the D type, it has two data inputs, and unlike the D type it is triggered on the

a

b

J	K	Q_n	
0	0	Q_{n-1}	Store
0	1	0	Reset
1	0	1	Set
1	1	$\overline{Q}_{n-1}$	Toggle

FIGURE 7.23 (a) Device symbol of a JK flip-flop. The angle or wedge denotes "edge-triggered" and the bubble denotes "negative edge." (b) Truth table showing that **00** input corresponds to memory and **11** to toggle.

trailing edge of the clock signal (negative-edge triggering is depicted in Fig. 7.23a by the small circle or bubble at the clock input of the JK flip-flop symbol). Similar to the D flip-flop, the present state at **Q** is maintained while a new state at **J** is read in which is then transferred to the output **Q** after the clock pulse.

We will forego showing the internal connections of the JK master-slave and go directly to the truth table which is shown in Fig. 7.23b. The first entry in the table is for the case when **J** and **K** are both **LOW**. For this case, the output does not change, that is, the previous state Q_{n-1} is equal to the present state Q_n. The next two entries are when **J** and **K** are complements. In this situation, **Q** will take on the value of the **J** input at the next downward clock edge. The final entry shows the most interesting aspect of a JK flop-flop: if **J** and **K** are both held **HIGH**, the flip-flop will toggle, that is, the output **Q** will reverse its state after each clock pulse (recall that previously a **11** input was a disallowed state). For example, we can connect the **J** and **K** inputs together[9] and hold the input **HIGH** while applying a clock signal to the clock terminal. The output **Q** will then toggle after each clock cycle, which gives us an output that looks like a clock signal but of half-frequency. The following example demonstrates this.

EXAMPLE 7.11 A 10 kHz square-wave signal is applied to the clock input of the first flip-flop of a triple JK flip-flop connected as shown in Fig. 7.24a. Show the outputs of all three flip-flops.

Since the JK inputs of the flip-flops are connected together and held **HIGH** at the power supply voltage of 5 V, the output of each flip-flop will toggle whenever the clock input goes **LOW**. Therefore at Q_1 we have a 5 kHz square wave, a 2.5 kHz square wave at Q_2, and a 1.25 kHz square wave at Q_3 as shown in Fig. 7.24b. ■

---

[9]A JK flip-flop that has its inputs tied together and therefore has only one input, called the T input, is known as a T flip-flop. If **T** is held high, it becomes a divide-by-two device as the output **Q** is simply the clock signal but of half-frequency. Similarly, if the clock input responds to a sequence of events, the T flip-flop divides by two. If **T** is held **LOW**, output **Q** is stored at whatever value it had when the clock went **LOW**. Summarizing, the JK flip-flop is a very useful device: it can toggle or hold its last state. It can mimic D and T flip-flops.

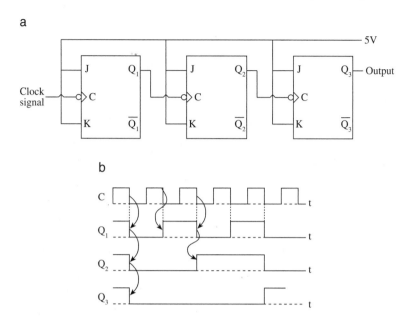

FIGURE 7.24   (a) Three JK flip-flops connected in series as shown will produce (b) half-frequency square waves at their respective outputs. Hence, Q_1 is a divide-by-two output, Q_2 a divide-by-four output, and Q_3 a divide-by-eight output.

7.5.7   Shift Registers

We have shown that a flip-flop can store or remember 1 bit of information (one digit of a binary number). Many flip-flops connected together can store many bits of data. For example, an array of 16 flip-flops connected in parallel is called a 16-bit shift register and can store and transfer a 16-digit binary number on command of a clock pulse.

A shift register is a fundamental unit in all computers and digital equipment. It is used as a temporary memory, typically by the CPU (*central processing unit*), while processing data. It is also used to convert data from one form to another. Data can be loaded into a computer serially or in parallel. Serial loading is naturally slower since a 32-bit word, for example, has to be loaded in 1 bit at a time, which normally would take 32 clock cycles. The advantage of serial loading is that only a single-wire line is required, which is an advantage when transmission distances are long. Serially loaded messages are used in modems which connect a remote computer over telephone lines to a server, in faxing, etc. Parallel loading, on the other hand, is faster since, for example, all 32 bits of a 32-bit word are loaded in at the same time, which normally would take 1 clock cycle. The disadvantage is that it requires a bus of 32 wires, clearly prohibitive for long-distance transmission but exclusively used inside computers where flat, multiwire strip busses are common, for example, between the floppy drive and CPU or between a hard disk and CPU. Next, we will show how a shift register converts a serial stream of bits into parallel form and vice versa. We will show that shift registers generate an orderly and predictable delay, i.e., a shifting of a signal waveform in time.

We can distinguish four types of shift registers: *serial in-serial out; serial in-parallel out; parallel in-serial out*; and *parallel in-parallel out.*

7.5.8 "Serial In–Parallel Out" Shift Register

A common situation is for data to be loaded serially (1 bit at a time) and read out in parallel, as, for example, when a serially coded message from a modem, normally transmitted over telephone lines, has to be loaded into the CPU of a server, which accepts data in parallel form only. Or when a serial stream of bits coming from a single line has to be converted to drive simultaneously all segments of a display. Figure 7.25*a* shows a 4-bit serial shift register made up of four negative-edge triggered, JK flip-flops. The **NOT** gate between the JK inputs allows loading either by a **1** ($\mathbf{JK} = \mathbf{Q\bar{Q}} = \mathbf{10}$) or by a **0** ($JK = Q\bar{Q} = \mathbf{01}$); in other words, **J** = **K** is not allowed. Also because the clock inputs are all connected together, the circuit operates synchronously.

Let us now show how the number **0101** is placed in the register. We start with the most significant bit[10] by placing the **0** at the input. At the trailing edge of the first clock pulse that **0** is transferred to the output $\mathbf{Q}_0$ of the first flip-flop. At the second clock pulse it is shifted to $\mathbf{Q}_1$ of the next flip-flop and so on until the **0** appears at the last flip-flop on the fourth clock pulse. At the next (fifth) clock pulse the **0** disappears from $\mathbf{Q}_3$ and the following **1** is loaded in. Position $\mathbf{Q}_3$ is identified as the most significant bit position in the parallel read-out. The waveforms of Fig. 7.25*b* show how the number shifts through the register and that after four clock pulses, 4 bits of input data have been transferred into the shift register such that the number is available in parallel form as $\mathbf{Q}_3\mathbf{Q}_2\mathbf{Q}_1\mathbf{Q}_0 = \mathbf{0101}$. An alternative way of viewing the shifting of bits through a register is given in Fig. 7.25*c*. Again, at the trailing edge of the first clock pulse, a **0** is shifted to $\mathbf{Q}_0$, then to $\mathbf{Q}_1$, and so forth.

From the above it is clear that this is a right-shifting register: at each clock pulse the bits shift one position to the right. If we ignore the first three outputs, we have a **SERIAL IN–SERIAL OUT** shift register. In this mode, the data pattern is read out at $\mathbf{Q}_3$ and is delayed by four clock pulses. A delay by itself is important in many applications. When so used the **SERIAL IN–SERIAL OUT** shift register is referred to as a *digital delay line*.

7.5.9 A "Parallel In–Serial Out" Shift Register

In the operation of a computer, parallel data (*n* bits present simultaneously on *n* individual lines) need to be transmitted between computer and monitor and between computer and keyboard. The link is a serial data path (1 bit after another on a single wire), whereas the computer, monitor, and keyboard operate with parallel words. Striking a key, for example, creates a parallel binary word which is at once loaded in parallel into

---

[10]Depending on the application, we could start with the least significant bit or the most significant bit.

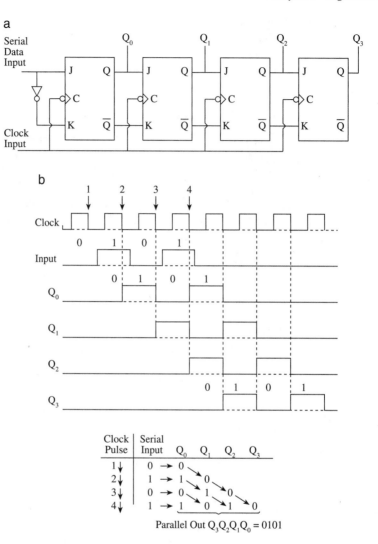

FIGURE 7.25 (a) A 4-bit **serial in–parallel out** shift register. (b) Timing waveforms at input and output. (c) Viewing how **0101** at the serial input terminal is placed in the register.

a shift register, converted, and transmitted serially to the computer, where the process is reversed, that is, the bits are loaded serially into a shift register and the binary word is read out in parallel.

Such a register can be constructed with positive-edge-triggered D flip-flops with asynchronous **CLEAR** and **PRESET** inputs, as in Fig. 7.26, which shows a 4-bit register. At the positive edge of a pulse to the **CLEAR** input, the register is cleared. This *asynchronous clearing* operation can occur anytime and is independent of the clock sig-

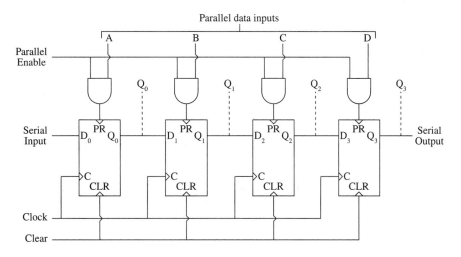

FIGURE 7.26 A universal 4-bit shift register.

nal. Data in parallel form are applied to the **A**, **B**, **C**, and **D** inputs. Now, at the rising edge of a short pulse applied to **PARALLEL ENABLE**, the data at the **ABCD** inputs appears at the respective **Q**'s of each flip-flop, thus loading 4 parallel bits into the register. For example, **ABCD = 0101** will set the second and last flip-flop HIGH, while the first and third remain cleared, so that $Q_0Q_1Q_2Q_3 = 0101$. At the next upward transition of the clock pulse, the data shifts to the right by one flip-flop. The serial output now reads $Q_3 = 0$ (the previous $Q_3 = 1$ is now lost; also note that the data at the serial output read backward). When the serial output has received the entire word, an asynchronous **CLEAR** can again be applied, a new word read into the parallel data inputs, and the process repeated.

The shift register of Fig. 7.26 can practically serve as a universal shift register. By placing parallel data on the **ABCD** inputs, the data will remain at $Q_0Q_1Q_2Q_3$ until the next rising edge of the clock signal. Hence it acts as a **PARALLEL IN–PARALLEL OUT** storage register. By ignoring $Q_0Q_1Q_2$ it can also act as a **SERIAL IN–SERIAL OUT** register. The data pattern at the **serial input** now appears at Q_3 but is delayed by four clock periods. This suggests the use of a **SERIAL IN–SERIAL OUT** register as a digital delay line. If we choose $Q_0Q_1Q_2Q_3$ as outputs, we have a **SERIAL IN–PARALLEL OUT** register. We should also mention in passing that division by two is equivalent to shifting a binary number to the right by one position. Hence shift registers are common in the arithmetic units of computers.

Digital Counters

A counter is a device used to represent the number of occurrences (random or periodic) of an event. Counters like registers are made possible by connecting flip-flops together. Typically, each flip-flop is the JK type operating in the toggle mode (T flip-flop). In this

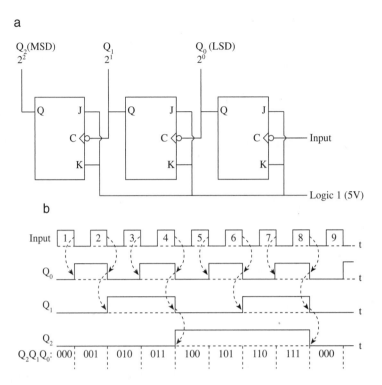

FIGURE 7.27 (a) A three-stage ripple counter (also referred to as a 3-bit binary counter). (b) The arrows which originate at the trailing edge of each pulse are used to show the toggling of the following stage. After seven pulses, the counter is in the 111 state, which means that after the next pulse (the eighth), the counter returns to the 000 state.

operation the **J** and **K** terminals are tied to a **HIGH** voltage (5 V) to remain at logic **1**. The input pulses are applied to the clock (C) terminal. The number of flip-flops equals the number of bits required in the final binary count. For example, a 2-bit counter with two flip-flops counts up to binary **11**, or decimal 3. A 5-bit counter with five flip-flops counts up to binary **11111**, or decimal 31.

By cascading flip-flops, so the **Q** output of each flip-flop is connected to the clock input of the next flip-flop, as shown in Fig. 7.27a, we will find that the first flip-flop toggles on every pulse (giving the 2^0 position of a binary number that will represent the total pulse count), the second flip-flop toggles on every second pulse (2^1), the third flip-flop on every fourth pulse (2^2), and so forth. Hence, three flip-flops can count eight distinct states, that is, zero to seven, which in binary is (000) to (111). Such a counter would be called a modulo-8 counter, because after counting eight pulses it returns to zero and begins to count another set of eight pulses. Figure 7.27b demonstrates this. Assuming all flip-flops were initially set to zero, after eight pulses the counter is again at zero.

As shown in Fig. 7.27a, three JK flip-flops are cascaded. The JK inputs are tied together and held **HIGH** at 5 V (logic **1**) so each flip-flop toggles at the trailing edge of an input pulse placed at the clock terminal. We have chosen to have the input terminal on the right side so that a digital number stored in the register is presented with its most significant digit (MSD) on the left as it is normally read.

Such a counter is called an asynchronous ripple counter, because not all flip-flops change at the same time (because of time delay, each flip-flop will trigger a tad later than its predecessor)—the changes ripple through the stages of the counter.

We have shown that a modulo-8 counter counts from 0 to 7 and then resets. Such a counter could also be called a divide-by-eight circuit, as was already shown in Example 7.11. It is obvious that one flip-flop can count from 0 to 1 (modulo-2) and is a divide-by-two circuit, two flip-flops count from 0 to 3 (modulo-4) and is a divide-by-four circuit, and so on.

7.5.10   A Decade Counter

Counters are most frequently used to count to 10. What we need then is a modulo-10 counter that counts from 0 to 9 and then resets. As three flip-flops, whose natural count is eight, are insufficient, four flip-flops are needed whose natural count is 16 (modulo-16). By appropriate connections, counters of any modulo are possible. Figure 7.28a shows a binary-coded decimal, or BCD, counter which counts 10 pulses and then resets. In other words, it counts 0, 1, 2, 3, 4, 5, 6, 7, 8, 9, 0, and after 9 it resets to zero. The resetting is accomplished by a **NAND** gate connected to the Q_1 and Q_3 outputs, because at the count of 10 the binary pattern is $Q_3 Q_2 Q_1 Q_0 = \mathbf{1010}$ and, as shown in Fig. 7.28b, the combination $Q_1 Q_3 = \mathbf{11}$ occurs for the first time in the count to 10. At 10 the **NAND** gate goes **LOW** and resets the counter to zero (the bubble at the CLR terminal implies that the flip-flops are cleared to 0 if **CLR** goes **LOW**).

If it is desired to store each digit of a decimal number separately in binary form, we can use a four-flip-flop register, which counts from 0 to 9, to represent each digit of the decimal number. To represent larger numbers additional registers are used. For example, three registers can count and store decimal numbers from 0 to 999, five registers from 0 to 99,999, and so forth.

EXAMPLE 7.12   Design a modulo-7 counter.

We will use JK flip-flops with **CLEAR** capabilities. By connecting the **J** and **K** inputs together and holding the connection **HIGH**, the flip-flops will be in the toggle mode. The counter should count from 0 to 6, and on the seventh input pulse clear all flip-flops to 0. If we examine the counting table in Fig. 7.28b, we will find that on the seventh count, the outputs of all flip-flops are 1 for the first time in the count. Sending these outputs to a **NAND** gate will generate a **LOW**, which when applied to the **CLEAR** inputs will reset the flip-flops to **0**. Such a counter using a triple-input **NAND** gate is shown in Fig. 7.29.

Similarly, a modulo-5 counter can be designed by observing that on the fifth count $Q_0 Q_2 = \mathbf{11}$ for the first time. Sending this output to a **NAND** gate, a **CLEAR** signal can be generated on the fifth pulse.   ∎

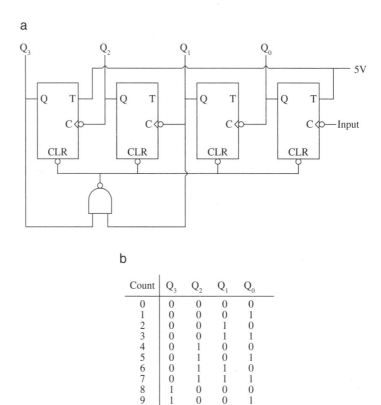

FIGURE 7.28   (a) Four T flip-flops connected as a BCD decade counter (recall that a T flip-flop is a JK flip-flop with the **J** and **K** terminals tied together). Holding T **HIGH** puts the flip-flops in the toggle mode. (b) The output states of a decade counter.

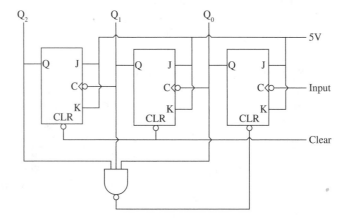

FIGURE 7.29   A modulo-7 counter which uses a **NAND** gate to clear the register at the count of 7.

7.5.11 Synchronous Counters

In the asynchronous ripple counter each flip-flop is clocked by the output of the previous flip-flop. As with any physical device or system, it takes a flip-flop time to respond to an input. Due to such time delays, each flip-flop in the ripple counter will trigger at a small time interval later than its predecessor. If ripple counters are used in complex circuits, even small time delays can introduce serious timing errors. Furthermore, since changes ripple through the counter, the accumulated time delays (usually refered to as *propagation delays*) can result in slow operation. These disadvantages do not exist in synchronous counters, in which the clock signal is applied to all flip-flops simultaneously with the effect that all flip-flops trigger at the same moment.

Figure 7.30a shows a synchronous, modulo-8 counter using three JK flip-flops. Since the input flip-flop has the **J** and **K** inputs tied together and held **HIGH**, it will toggle on every positive edge of the clock signal[11] and its output gives the 2^0 position of the count. The second flip-flop will toggle on the positive edge of a clock pulse only if Q_0 is **HIGH** (logic **1** and **HIGH** have the same meaning). Hence, the output of the second flip-flop, which is Q_1, gives the 2^1 position. The third flip-flop, because of the **AND** gate, will toggle only if both previous flip-flops are **HIGH** (Q_0 **AND** $Q_1 = 1$) and its output thus gives the 2^2 position of the count. The truth table of Fig. 7.30b gives the state of the flip-flops for each count. On the eighth count, the register will reset to zero and the counting sequence repeats.

EXAMPLE 7.13 Design a 4-bit, modulo-16, synchronous counter.

We can begin with the 3-bit synchronous counter shown in Fig. 7.29a and add a flip-flop for the next position (2^3) in the binary count. Because this flip-flop must toggle on the 7th and 15th clock pulse, the input to that flip-flop must be preceded by a triple-input **AND** gate which will give a 1 when $Q_0 Q_1 Q_2 = \mathbf{111}$. The connections of the four flip-flops are shown in Fig. 7.31a, and one counting cycle, giving the state of each flip-flop, is shown in Fig. 7.31b. Analyzing the circuit, we find that the first flip-flop acts as a toggle for output Q_0. The second flip-flop with output Q_1 toggles when $Q_0 = \mathbf{1}$. The third flip-flop with output Q_2 must toggle when the previous two flip-flops are **FULL** (**FULL** is sometimes used in technical jargon to mean **HIGH** or logic **1**), that is, $J_2 = K_2 = Q_0$ **AND** Q_1. Finally, the last flip-flop must toggle when $Q_2 Q_1 Q_0 = \mathbf{111}$ or when $J_3 = K_3 = Q_0$ **AND** Q_1 **AND** Q_2. ■

If we compare ripple counters (Fig. 7.27) to synchronous counters (Fig. 7.30), which are faster and less prone to error, we will find that in general, synchronous counters require a few extra gates.

---

[11] Given the connections of the JK flip-flops in Fig. 7.30, we could have replaced all JK flip-flops by T flip-flops. Furthermore, we are showing a positive-edge-triggered JK flip-flop. Since JK flip-flops are commonly negative-edge triggered, a **NOT** gate at the clock input would give the same behavior as a positive-edge-triggered JK flip-flop.

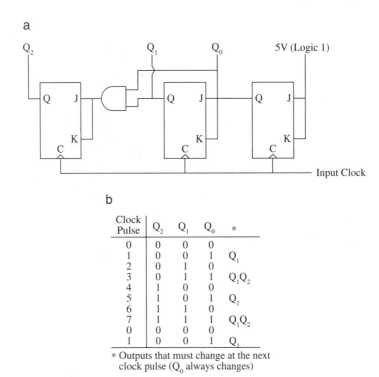

FIGURE 7.30   (a) A synchronous, 3-bit, modulo-8 counter. (b) One counting sequence of the counter.

7.6   MEMORY

In Section 7.5, we stated that the simplest memory device is an RS flip-flop. It can hold 1 bit of information. To make such a device practical, we have to endow it with **READ** and **WRITE** capabilities and a means to connect it to an information line or information bus. Such a device will then be called a 1-bit memory cell. A byte (8 bits) would require eight cells. A thousand such cells could store one kilobit of information (1 kbit) and would be called a *register*. Typically, registers hold many words, where each word is either 8 bits, 16 bits, 32 bits, or some other fixed length. The location of each word in the register is identified by an *address*.

All instructions that a computer uses to function properly are stored in memory. All data bases, such as typed pages and payroll data, are stored in memory. All look-up tables for special functions such as trigonometric ones are stored in memory. Basically, any datum a digital device or a computer provides, such as time, date, etc., is stored in memory. Flip-flop memory is *volatile*, which means that the register contents are lost when the power is turned off. To make registers *nonvolatile*, either the computer is programmed to automatically refresh the memory or batteries can be used to refresh the registers (lithium batteries are common as they can last up to 10 years).

a

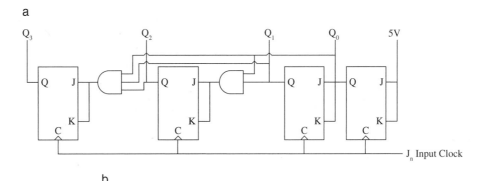

b

Clock Pulse	Q_3	Q_2	Q_1	Q_0	*
0	0	0	0	0	
1	0	0	0	1	Q_1
2	0	0	1	0	
3	0	0	1	1	$Q_1 Q_2$
4	0	1	0	0	
5	0	1	0	1	Q_1
6	0	1	1	0	
7	0	1	1	1	$Q_1 Q_2 Q_3$
8	1	0	0	0	
9	1	0	0	1	Q_1
10	1	0	1	0	
11	1	0	1	1	$Q_1 Q_2$
12	1	1	0	0	
13	1	1	0	1	Q_1
14	1	1	1	0	
15	1	1	1	1	$Q_1 Q_2 Q_3$
0	0	0	0	0	
1	0	0	0	1	Q_1

* Outputs that must change at the next clock pulse

FIGURE 7.31   (a) A 4-bit, synchronous counter. (b) One counting sequence of the modulo-16 counter.

A computer uses many types of memory, each suitable for a particular task. Hard and floppy disks can permanently store huge amounts of data, which are nonvolatile because the data are laid down as a sequence of tiny permanent magnets along circular tracks. Access time is slow (about 10ms), but it is the primary medium for storing the operating system and programs such as word processors and spreadsheets.[12] While processing, the main processor of the computer uses *random access memory* (RAM), now referred to as read-and-write memory, which is able to store and retrieve information very fast, which in turn makes fast computation and processing possible. Such memory is volatile as it is lost when the computer is turned off. In addition, most computers have a small amount of cache memory, which is located on the microprocessor chip itself and is very fast (but very expensive) as it stores data that were just accessed by the processor and are therefore likely to be needed again.

[12]A newer permanent memory called flash memory (also known as flash RAM), which is a nonvolatile solid-state memory, is also available for applications such as digital cellular phones and PC cards for notebook computers in which memory requirements are not in the gigabyte range. For example, in digital cameras,flash memory cards, which are the size of credit cards, can provide convenient storage capacity of 64 or 128 MB.

Finally, we have *read-only memory* (ROM), which as the name suggests can only be accessed and cannot be changed. This nonvolatile memory is permanently programmed at time of manufacture and is used by computers to provide the information needed for start-up when the computer is first turned on. It also provides look-up tables (trigonometric, logarithmic, video displays, character representation, etc.) and anything else that is needed on a permanent basis. Even though the information in ROMs cannot be changed, programmable ROMs (PROMs) which can be programmed by the user are available. Again, once programmed, they cannot be erased or reprogrammed.

7.6.1   RAM Cell

By adding a few logic gates (two inverters and three **AND** gates) to a RS flip-flop, as shown in Fig. 7.32, we can construct a basic 1-bit memory cell which can store 1 bit of information. It has a **SELECT** input, which when **HIGH** enables the cell for reading and writing. It has a **READ/WRITE** (R/W) input, which when **HIGH** activates the **READ** operation and when **LOW** activates the **WRITE** operation. It has an Output which during the **READ** operation transfers the contents of the cell to whatever line is connected to this terminal. It has an Input for changing the contents of the cell during the **WRITE** operation.

The **READ** operation is selected when Select is **1** and **READ/WRITE** is **1**. The input to the SR terminals of the flip-flop is then **00**, the cell remains unresponsive to changes in input, the contents of the flip-flop remain unchanged, and **Q** is connected to the output for reading. On the other hand, when R/W goes to **0** and Select remains at **1**, the cell is activated for writing and can respond to a write signal to store new information in the cell. For example, a **1** at the input appears as **SR = 10** at the flip-flop, implying[13] that a **1** is loaded in the flip-flop and is available for reading at output **Q**. Thus, the input bit has been successfully stored in the cell. A **0** at the input appears as **SR = 01**; hence a **0** is stored and is available for reading, and once again the new input bit has been successfully stored.

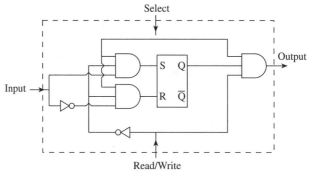

FIGURE 7.32   An elementary 1-bit memory cell with one output and three input terminals (one for selecting the cell, one for choosing between read and write and one for inputting new information).

---

[13]See Section 7.5.1, "Flip-Flop: A Memory Device."

7.6.2   RAM

A piece of memory consists of an array of many cells. Typically a $m \times n$ array can store m words with each word n bits long. The Select signal for a particular word is obtained from an address which is decoded to produce a **1** on the desired line. For example, the decoder considered in Fig. 7.15 decodes the information on two address lines and selects one of four lines by making that particular line **HIGH**. Such a decoder could select one word out of four; that is, given a $4 \times n$ array, it would pick one word that is n bits long. Should the word size be 8 bits, then 32 elementary cells would be needed for the 4×8 memory. Should the word size be 4 bits, 16 cells would be needed. Such a 16-bit RAM memory module is shown in Fig. 7.33.

The CPU of a computer uses RAM to store and retrieve data. As pointed out before, RAM is volatile. Any data stored in RAM are lost when power is turned off. Therefore, data that need to be saved must be transferred to a magnetic disk, magnetic tape, or flash memory.

7.6.3   Decoding

Decoding was already considered in the previous section. Since memories can have a capacity of thousands and even millions of words and each word can be as large as 64 or 128 bits, accessing precisely each word in the memory is done by decoding an address code. Given k address lines, we can access 2^k words. Thus a decoder with k inputs will have 2^k outputs, commonly referred to as a $k \times 2^k$ decoder. Each of the decoder outputs

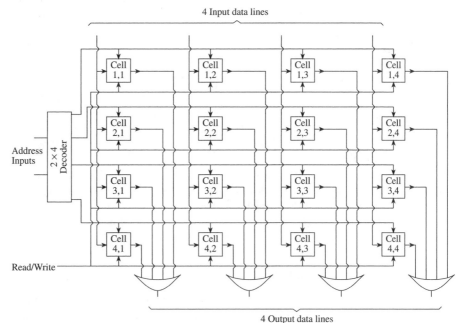

FIGURE 7.33   Connection diagram of 16 memory cells in a 4×4 RAM module.

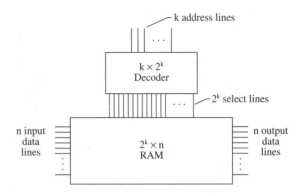

FIGURE 7.34   Representation of a RAM module showing explicitly the cell array and the decoder.

selects in the memory one word which is n bits long for reading or writing. The size of the memory is therefore an array of $2^k \times n$ cells. A schematic representation of such a RAM module is shown in Fig. 7.34.

Word size is normally used to classify memory. Microprocessors that use 4-bit words have 4-bit registers. Small computers use byte-size words that require 8-bit registers, and powerful microcomputers use 64-bit registers. To access words in memory an address is used. As pointed out before, two address lines, after decoding the address, can access four words. In general, to access 2^k words requires a bus of k address lines. Ten address lines can access 1024 words of a program or data memory ($2^{10} = 1024$, commonly referred to in computerese as 1K). A 32-bit address bus can locate four billion words ($2^{32} = 4,294,967,296$), and so on.

7.6.4   Coincident Decoding

In large RAM modules, the memory cells are arranged in huge rectangular arrays. Linear addressing, described above, activates a single word-select line and can become unwieldy in huge arrays, necessitating very long addresses. In linear addressing, a decoder with k inputs and 2^k outputs requires 2^k **AND** gates with k inputs per gate. The total number of **AND** gates can be reduced by employing a two-part addressing scheme in which the X address and the Y address of the rectangular array are given separately. Two decoders are used, one performing the X selection and the other the Y selection in the two-dimensional array. The intersection (coincidence) of the X and Y lines identifies and selects one cell in the array. The only difference is that another Select line is needed in the cell structure of Fig. 7.32, which is easily implemented by changing the three **AND** gates to quadruple-input **AND** gates.

7.6.5   ROM

A ROM module has provisions for **READ** only, and not for **WRITE**, which makes it simpler than RAM. ROM is also nonvolatile, which means that data are permanently stored and appear on the output lines when the cell is selected. In other words, once a pattern in memory is established, it stays even when power is turned off.

Typically the manufacturer programs ROMs that cannot be altered thereafter. PROMs are field programmable, which allows the user to program each cell in the memory. A PROM comes to the user with all cells set to zero but allows the user to change any zeros to ones. Again, once the user programs the PROM, it is irreversible in the sense that it cannot be reprogrammed or erased. Changing one's mind about the contents of the memory implies that the memory chip must be discarded. There is, however, a type of erasable–programmable read-only memory called EPROMS which can be erased and programmed repeatedly. The process can be tedious but in the development stage of a product it is invaluable.

7.7 SUMMARY

A basic knowledge of digital electronics is required just to stay current in the rapidly changing field of engineering, which is increasingly dominated not just by the ubiquitous computer but by digital circuitry of all kinds. We began our study with binary arithmetic, Boolean algebra, and Boolean theorems, including DeMorgan's theorems, which form the basis of logic circuits. We introduced the truth table which states all possible outcomes for a given logical system. This provided sufficient background to proceed to logic gates, which are the building blocks of any digital system such as the computer. The fundamental logic gates **AND**, **OR**, and **NOT** were presented, followed by the only slightly more complicated **NAND** and **NOR** gates. An observation important in integrated circuit (IC) design was that any logic system can be constructed using only a single gate type such as the **NAND** gate or the **NOR** gate. For example, **NAND** gates can be connected in such a way as to mimic any gate type, be it a **OR** gate or a **NOT** gate. Therefore, complex chips that contain thousands of gates can be more reliably manufactured if only one gate type is used.

Even though logic gates are elementary building blocks, in complex digital systems such as the microprocessor and the computer, larger building blocks are common. These include flip-flops, memories, registers, adders, etc., which are basic arrangements of logic gates and to which the remainder of the chapter was devoted. Complex digital systems are beyond the scope of this book but many books and courses exist which treat the interconnection of such building blocks to form practical systems. The larger building blocks were divided into two groups, depending on the logic involved. Combinatorial logic, which is memoryless, has outputs that depend on present input values only, with adders and decoders being prime examples of such building blocks. Sequential logic building blocks incorporate memory; hence, outputs depend on present as well as on past input values. Sequential logic circuits open the digital world to us with devices such as flip-flop memories, shift registers, and counters. Out of all of the flip-flops considered, it is the JK edge-triggered flip-flop which is the workhorse of sequential systems. Because it is edge-triggered, that is, it changes state precisely at the time the periodic clock signal makes its **HIGH**-to-**LOW** or **LOW**-to-**HIGH** transition, clocked flip-flops bring a high degree of order in complicated digital systems that are prone to chaos.

Problems

1. Perform a decimal-to-binary conversion (converting a numeral written in base 10 to the equivalent numeral written in base 2) of the following decimals: 2, 5, 12, 19, 37, and 101. To reduce confusion, the base can be denoted by a subscript. *Hint*: See the footnote on p. 248 in Section 7.4.
   Ans: $2_{10} = 10_2, 37_{10} = 100101_2$.

2. Convert the following binary numbers to decimal numbers: 11, 00111, 01101, 101010.
   Ans: $01101_2 = 13_{10}$.

3. The octal system uses the digits 0 to 7 for counting. Convert the following octal numbers, denoted by subscript 8, to decimals: $5_8, 12_8, 502_8, 6745_8$.
   Ans: $5_8 = 5_{10}, 12_8 = 10_{10}, 502_8 = 322_{10}$.

4. Express the decimal numbers 4, 9, 15, and 99 in the octal system (see Problem 3).
   Ans: $4_{10} = 4_8, 9_{10} = 11_8$.

5. In the hexadecimal system (base 16), commonly used in microprocessor work, the digits 0 to 15 are used for counting. To avoid using double digits, the 10 decimal digits $0, \ldots, 9$ are supplemented with the letters A, B, C, D, E, and F, that is, 0, 1, 2, 3, 4, 5, 6, 7, 8, 9, A, B, C, D, E, F. Express the decimal numbers 3, 14, 55, 62, and 255 as hex numbers.
   Ans: $3_{10} = 3_{16}, 62_{10} = 3C_{16}, 255_{10} = FF_{16}$.

6. Convert the binaries 0011, 1111, 11000011, and 11111111 to hex.
   Ans: $0011_2 = 31_{16}, 11111111_2 = FF_{16}$.

7. Construct a truth table for a triple-input **AND**, **OR**, **NAND**, and **NOR** gate.
   Ans: (partial)

A	B	C	A · B · C
0	0	0	0
0	0	1	0
0	1	0	0
0	1	1	0
1	0	0	0
1	0	1	0
1	1	0	0
1	1	1	1

8. Three binary waveforms are shown in Fig. 7.35. If these are used as inputs to an **AND** and a **NOR** gate, show the outputs of the gates.

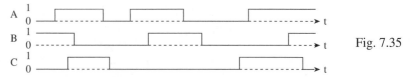

Fig. 7.35

9. Similar to Figs. 7.3*a* and 7.4*a*, which illustrate an **AND** and **OR** gate, connect two switches, *A* and *B*, a battery, and a bulb to illustrate a **NAND** gate and a **NOR** gate.

Verify the truth tables in Fig. 7.8 for the **NAND** and **NOR** gate.

10. A bus consisting of 16 wires can carry how many binary words?
    Ans: $2^{16} = 65,536$.

11. (a) Show that both voltages **A** and **B** must be **HIGH** for the **AND** gate circuit of
    Fig. 7.3*c* to give a **HIGH** output.
    (b) Show that when one of the voltages is zero, the output is 0.7 V, which is logic
    LOW.

12. For each of the logic circuits shown in Fig. 7.36 write down the logic expression
    for the output F.
    Ans: $F_a = (A + B) \cdot \bar{C}$, $F_c = (A + B)(\bar{A} + \bar{B})$.

a

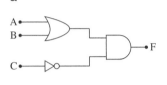

b

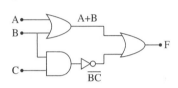

c   Fig. 7.36

13. Prove the identity of each of the following Boolean equations:

    (a) $\bar{A}B + A\bar{B} + AB + \bar{A}\bar{B} = 1$
    (b) $A\bar{B} + AB + \bar{A}B = A + B$
    (c) $\bar{A} + AB + A\bar{C} + AB\bar{C} = \bar{A} + B + \bar{C}$
    (d) $\bar{A}BC + \bar{A}B\bar{C} + AC = \bar{A}B + AC$
    (e) $A + A \cdot B = A$
    (f) $A \cdot (A + B) = A$

14. Draw a logic circuit to simulate each of the Boolean expressions given.

    (a) $F = A + \bar{B}C$
    (b) $F = (A + B) \cdot \bar{C}$
    (c) $F = \overline{A + B} \cdot A\bar{B} + \overline{AB}$
    Ans: for (b) see Fig. 7.37.

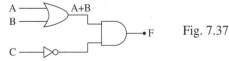

Fig. 7.37

15. Use only **NAND** gates to construct a logic circuit for each of the Boolean functions
    given.

    (a) $F = A + B$

(b) $\mathbf{F} = \mathbf{A} \cdot (\mathbf{B} + \mathbf{C})$

(c) **exclusive or** $\mathbf{A} \oplus \mathbf{B}$

16. Repeat Problem 15 except use only **NOR** gates.

17. A logic circuit and three input waveforms **A**, **B**, and **C** are shown in Fig. 7.38. Draw the output waveform F. *Hint*: find the simplest form for **F** and make a truth table for **F** from which the waveform can be drawn.

a   b

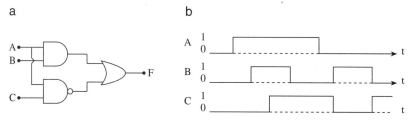

Fig. 7.38

18. Construct an adder that can parallel-add two 2-bit binary numbers **A** and **B**, that is, $\mathbf{A_1 A_0} + \mathbf{B_1 B_0} = \mathbf{C_1 S_1 S_0}$, where **S** stands for *sum* and **C** for *carry*.

19. Design a three-line to eight-line (3-to-8) decoder which will decode three input variables into eight outputs. An application of this decoder is a binary-to-octal conversion, where the input variables represent a binary number and the outputs represent the eight digits in the octal number system.

20. Design a 2-to-4 decoder with **Enable** input. Use only **NAND** and **NOT** gates. The circuit should operate with complemented **Enable** input and with complemented outputs. That is, the decoder is enabled when **E** is equal to **0** (when **E** is equal to **1**, the decoder is disabled regardless of the values of the two other inputs; when disabled, all outputs are **HIGH**). The selected output is equal to **0** (while all other outputs are equal to **1**).

Ans: See Fig. 7.39.

a   b

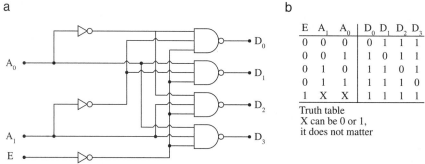

E	A_1	A_0	D_0	D_1	D_2	D_3
0	0	0	0	1	1	1
0	0	1	1	0	1	1
0	1	0	1	1	0	1
0	1	1	1	1	1	0
1	X	X	1	1	1	1

Truth table
X can be 0 or 1,
it does not matter

Fig. 7.39

21. An encoder performs the inverse operation of a decoder. Whereas a decoder selects one of $2k$ outputs when k address lines are specified, an encoder has $2k$ input lines and k output lines. The output lines specify the binary code corresponding to the input values. Design an octal-to-binary encoder that utilizes three, multiple-input **OR** gates. Use the truth table in the answer to Problem 20, but reverse it, i.e., use eight inputs, one for each of the octal digits, and use three outputs that generate the corresponding binary number. Assume that only one input has a value of **1** at any given time. *Hint*: $\mathbf{A} = \mathbf{O}_1 + \mathbf{O}_3 + \mathbf{O}_5 + \mathbf{O}_7$, $\mathbf{B} = \mathbf{O}_2 + \mathbf{O}_3 + \mathbf{O}_6 + ?$, $\mathbf{C} =$ you are on your own.

22. How many states does an SR flip-flop have?

    Ans: Three states—set (**SR = 10**, output **1**); reset (**SR = 01**, output **0**); hold (**SR = 00**, output stays unchanged).

23. How many states does a JK flip-flop have?

    Ans: Four states—the three states of an SR flip-flop plus a toggle mode (**JK = 11**).

24. Sketch the output waveform of an RS flip-flop if the input waveforms are as shown in Fig. 7.40.

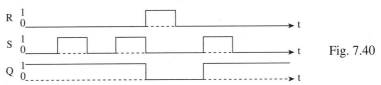

Fig. 7.40

25. If only JK flip-flops are available, show how they can be converted to D and T flip-flops. Sketch the circuit diagrams.

    Ans: For the D flip-flop connect a **NOT** gate between the **J** and **K** terminals and a **NOT** gate between the clock signal and the clock input of the JK flip-flop (a **NOT** gate before the clock input of a JK flip-flop will give a positive-edge trigger; however, the **NOT** gate must be a fast-acting one so as to maintain synchronous operation). For the *T* flip-flop connect the input terminals of JK flip-flop together.

26. Connect a JK flip-flop to act as a divide-by-two circuit.

27. A T flip-flop can act as a divide-by-two circuit. Compare its operation to that of the JK flip-flop of Problem 7.26.

28. Two JK flip-flops are connected in series in a manner similar to Fig. 7.24*a*. If a 10 kHz square wave is applied to the input, determine and sketch the output.

29. Using additional gates (a **NOT** gate, two **AND** gates, and a **OR** gate), convert a D flip-flop into a positive-edge-triggered JK flip-flop. Note that the **D** input to the D flip-flop is defined by the Boolean expression $\mathbf{D} = \mathbf{J}\bar{\mathbf{Q}} + \bar{\mathbf{K}}\mathbf{Q}$.

30. A 4-bit shift register is shifted six times to the right. If the initial content of the register is **1101**, find the content of the register after each shift if the serial input is **101101**.

    Ans: Initial value of register **1101**; input **1**, register after first shift **1110**; **0, 0111**; **1, 1011**; **1, 1101**; **0, 0110**; **1, 1011**.

31. Design a 2-bit shift register, using D flip-flops, that moves incoming data from left to right.

32. The serial input to the shift register of Fig. 7.25*a* is the binary number **1111**. Assume the register is initially cleared. What are the states after the first and third clock pulse?

Ans: $Q_0Q_1Q_2Q_3 = 1000$, $Q_0Q_1Q_2Q_3 = 1110$.

33. Design a 2-bit binary counter using two negative-edge-triggered JK flip-flops and describe its output for the first four clock pulses. Assume the counter is initially cleared ($Q_0 = Q_1 = 0$).

34. Design a modulo-5 counter that will count up to 4 and on the fifth pulse clear all three flip-flops to 0. Use either T or JK flip-flops.
    Ans: Since the count is larger than 4, we need three flip-flops with CLEAR capability. The register needs to be cleared when binary 101 is reached, which can be accomplished by adding a NAND gate and connecting its inputs to Q_0 and Q_2 and the output to all three CLEAR terminals.

35. The counter in Problem 7.33 is a modulo-4 ripple counter (also referred to as a divide-by-four ripple counter). The difficulty with such counters is that the output of the preceeding flip-flop is used as the clock input for the following flip-flop. Since the output of a flip-flop cannot respond instantaneously to an input, there is a small time delay in each successive clock signal which is known as a ripple delay. If many flip-flops are involved, the cumulative time delay can cause the counter to malfunction. For speedy clock signals, ripple delay cannot be tolerated. Convert the 2-bit ripple counter of Problem 33 to a synchronous counter in which the clock signal is applied simultaneously to all flip-flops.

36. A counter uses flip-flops which have a 5 ns delay between the time the clock input goes from **1** to **0** and the time the output is complemented.

    (a) Find the maximum delay in a 1-bit binary ripple counter that uses these flip-flops.

    (b) Find the maximum frequency the counter can operate reliably.
        Ans: 50 ns, 20 MHz.

37. How would you modify the RAM cell in Fig. 7.32 to act as a ROM cell?

38. What is the length of each word in a 256×8-bit RAM?
    Ans: 8 bits long.

39. How many bits can be stored in a 256×8-bit ROM?

40. How many words of data can be stored in a 256×8-bit ROM?
    Ans: 256.

41. In the elementary RAM cell of Fig. 7.32 find the Output if $S = 1$, **Read/Write** $= 0$, and Input $= 1$.

42. If each word in RAM has a unique address, how many words can be specified by an address of 16 bits? If each word is byte long (8 bits long), how many memory cells must the RAM have?
    Ans: 65536, 524288.

43. Arrange a 6-bit RAM as a 2×3-bit RAM and show its connection diagram. Label the input lines, the output lines, the decoder, and the address lines.

CHAPTER 8

The Digital Computer

8.1 INTRODUCTION

This chapter and the following one deal with applications of the material developed in the preceding chapters. Needless to say, there are numerous other examples for which analog and digital electronics have served as the underlying base. But perhaps for society today, the most pronounced developments in technology have been the digital computer and more recently digital communication networks. Because these two disciplines share the same binary language of 0's and 1's, computing and communications have merged and advanced with revolutionary results as exemplified by the *Internet* and its *World Wide Web*. The information revolution, also known as the "third" revolution, which the digital computer heralded 50 years ago has profoundly changed the way society in general works and interacts. This revolution followed the 18th century industrial revolution, which in turn followed the agricultural revolution of about 10,000 years ago. All have been technology-based, but the "third" revolution is by far the most sophisticated. This chapter will try to provide you with a working knowledge of the computer—specifically the personal computer.

The first digital computers were built on college campuses by John Atanasoff in 1937–1942 and by John Mauchly and Presper Eckert, Jr., in 1946 with government sponsorship. These were soon followed by the commercial machines UNIVAC 1 and IBM 701. These computers had no operating system (OS), but had an assembler program which made it possible to program the machine. Soon primitive operating systems and the first successful higher-order language FORTRAN (formula translation) followed, which allowed engineers and scientists to program their problems with relative ease. The birth of the personal computer (PC) was more humble. In a garage with meager personal funding, Steve Wozniak and Steve Jobs breadboarded the first personal computer in 1976 and started the PC revolution. Their first series of Apple computers was followed by the Macintosh, which had a new and novel operating system (the Mac OS) based on a graphical user interface (GUI) with the now familiar click-on icons that many, even today, claim is a superior operating system for PCs. However, Apple's refusal to license the Mac OS allowed the IBM-compatible PC to dominate the market. It was under Andrew Grove that the Intel Corporation pioneered the famous 8086 family of chips (286, 386, 486, Pentium, Pentium II, Pentium III, etc.) that became the basis for the IBM-compatible PCs. This, in turn, inspired another computer whiz, Bill Gates, and the then-fledgling Microsoft Corporation to create the DOS and Windows operating system software that now dominates the computing scene.

8.2   THE POWER OF COMPUTERS—THE STORED PROGRAM CONCEPT

Before the computer, machines were dedicated, which means that they were single-purpose machines. They pretty much were designed and built to do a single job. An automobile transports people on land, a ship transports in the water, and an airplane transports in the air. It is true that an airplane can do the job of a fighter or a bomber, but to do it well it must be designed and built for that purpose. In other words the differences between the Queen Mary, an Americas Cup yacht, and a battleship can be more profound than their similarities. Imagine a machine that could do all the above. A computer comes very close to that. Its design is a general machine, which is referred to as the hardware and which operates from instructions which we refer to as software. As the applications can be very broad and can come from many disciplines, the tasks that a computer can perform are also very broad. It is like a fictitious mechanical machine which, according to the type of program that is inserted in it, acts as an airplane or as a ship, automobile, tank, and so on. A computer is that kind of a machine. It can act as a word processor; it can do spreadsheets; it can do mathematics according to instructions from *Mathematica, Maple, Matlab*, etc.; it can play games; it can analyze data; it facilitates the Internet; it executes money transfers; and on and on. This then is the stored program concept: instructions and data are loaded or read into an electrically alterable memory (random access memory or RAM) which the computer accesses, executes, and modifies according to intermediate computational results. It has the great advantage that the stored programs can be easily interchanged, allowing the same hardware to perform a variety of tasks. Had computers not been given this flexibility, that is, had they been hardwired for specific tasks only, it is certain that they would not have met with such widespread use. For completeness it should be stated that two inventions sparked the computer revolution: the stored program concept and the invention of the transistor in the late 1940s which gave us tiny silicon switches, much smaller than vacuum tubes, and which in turn gave birth in 1959 to integrated circuits with many transistors on a single, small chip. This set in motion the phenomenal growth of the microelectronics industry, which at present produces high-density chips containing millions of transistors at lower and lower cost, thereby making possible the information revolution.

8.2.1   Computational Science

Until the age of powerful computers, new theories and concepts introduced by engineers and scientists usually needed extensive experiments for confirmation. Typically models and breadboards (a board on which experimental electronic circuits can be laid out) needed to be built, often at prohibitive costs and sometimes not at all due to the complexity and large scale of the experiment. Today, experiments often can be replaced by computational models which can be just as accurate, especially in complex situations such as weather prediction, automobile crash simulations, and study of airflow over airplane wings. Thus to theory and experimentation, the backbone of scientific inquiry, we have added a third discipline, that of *computational science*,[1] which provides solutions to very complex problems by modeling, simulation, and numerical approximation.

8.2.2   Microcontrollers, Microprocessors, and Microcomputers

In a sense all three of the above can be referred to as computers. Besides the processor, all three types have some memory in the form of RAM and ROM (read-only memory) and some control circuitry. Microcontrollers are self-contained ICs (integrated circuits)—frequently referred to as single-chip computers. Generally they are programmed at the factory to perform a specific function by themselves. They are designed typically for applications which do not involve significant numerical computation but may require modest amounts of communication with other devices and nonnumerical capabilities such as serial input/output, precise measuring of time, and analog-to-digital and digital-to-analog conversion. Such applications include traffic light controllers (controlling relays to turn up to 20 or so different sets of lights on and off) and MIDI systems (serial communication—MIDI stands for *musical instrument digital interface*). In the literature they are also characterized as microprocessors with simpler system structures and lower performance. Controllers are embedded in cordless telephones, electronic cash registers, scanners, security systems, automobile engines, disk drives, and a variety of home and industrial applications such as refrigerators and air conditioners. In fact, electric lights are almost the only electrically powered devices that do not use microcontrollers. Even though microcontrollers and microprocessors share many architectural features, they differ in important respects. A microcontroller is generally a one-chip integrated system meant to be embedded in a single application. The chip, therefore, is likely to include a program, data memory, and related subsystems which are needed for the computer aspect of the particular application. By contrast, a microprocessor drives a general-purpose computer whose ultimate application is not known to the system designers.

The single-chip computer is evolving rapidly. Low-cost consumer items such as microwave ovens, shavers, toasters, tape players, and toys are served by 4-bit controllers which process information in groups of 4 bits at a time. Four-bit chips are ill-suited to running programs written in high-level languages,[2] which typically are byte or word

---

[1] We have to distinguish this term from *computer science*, which is the study of computers and computation. Computational science is that aspect of any science that advances knowledge in the science through the computational analysis of models. Like theory and experimentation, it is now one of the three legs of that science.

[2] See the next section on programming languages.

oriented (a byte is 8 bits; a word is usually 2 bytes). Four-bit chips would therefore have to execute multiple operations to do the simplest things. As 4-bit controllers have little compiler support, programming must be done in assembly language, which is slow and tedious. In part for these reasons 8-bit controllers dominate the market, especially in embedded-control applications such as television sets, disk drives, and car radios, as well as in personal computer peripherals such as printers, joysticks, mice, and modems. In more advanced applications, 16-bit controllers are deployed in disk drives, automobile engine control, and generally in industrial control. In still more advanced applications 32-bit chips are employed in communication boards, laser printers, and some video games. In highly competitive fields such as video games, 64-bit embedded controllers are now common, with 128-bits for high-end game stations. The ability to process 128-bit words at a time makes these controllers very powerful indeed. It should be understood that since all conversations with the control unit of a computer are in binary language, the power and speed with which a computer operates is proportional to the ease with which it can handle large binary words. The ability to accumulate, store, operate upon, and output these very large binary words is achieved by assembling on a single chip massive arrays of the basic circuits and devices discussed in the previous chapter.

Summarizing, we can state that the microcontroller is a single-chip computer dedicated to a single task, mainly control applications. A microprocessor is an integrated circuit that contains all the arithmetic, logic, and control circuitry to perform as the central processing unit of a computer, i.e., a complete CPU on a single IC chip. Adding memory to the chip and some input/output (I/O) ports, it becomes a computer-on-a-chip. Adding external memory in the form of high-speed semiconductor chips (RAM and ROM) and peripherals such as hard and floppy disks and CD-ROM drives for storage of software programs, as well as various I/O devices such as monitors, keyboards, and mice, it becomes a microcomputer, of which the personal computer is the best example. PCs are low-cost machines that can perform most of the functions of larger computers but use software oriented toward easy, single-user applications. It is not worthwhile to make precise distinctions between types of computers in a rapidly changing environment as, for example, PCs and workstations can now perform tasks that seemingly only yesterday minicomputers and only mainframes could do.

8.2.3   Communicating with a Computer: Programming Languages

Programming languages provide the link between human thought processes and the binary words of machine language that control computer actions, in other words, instructions written by a programmer that the computer can execute. A computer chip understands machine language only, that is, the language of 0 and 1's. Programming in machine language is incredibly slow and easily leads to errors. Assembly languages were developed that express elementary computer operations as mnemonics instead of numeric instructions. For example, to add two numbers, the instruction in assembly language is ADD. Even though programming in assembly language is time consuming, assembly language programs can be very efficient and should be used especially

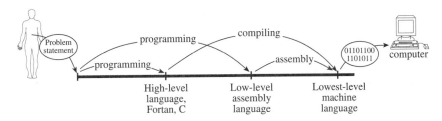

FIGURE 8.1   Programming languages provide the link between human thought statements and the 0 and 1's of machine code which the computer can execute.

in applications where speed, access to all functions on board, and size of executable code are important. A program called an assembler is used to convert the application program written in assembly language to machine language. Although assembly language is much easier to use since the mnemonics make it immediately clear what is meant by a certain instruction, it must be pointed out that assembly language is coupled to the specific microprocessor. This is not the case for higher-level languages. Higher languages such as C and Pascal were developed to reduce programming time, which usually is the largest block of time consumed in developing new software. Even though such programs are not as efficient as programs written in assembly language, the savings in product development time when using a language such as C has reduced the use of assembly language programming to special situations where speed and access to all a computer's features is important. A compiler is used to convert a C program into the machine language of a particular type of microprocessor. A high-level language such as C is frequently used even in software for 8-bit controllers and is almost exclusively used in the design of software for 16-, 32-, and 64-bit microcontrollers.

Figure 8.1 illustrates the translation of human thought to machine language by use of programming languages.

Single statements in a higher-level language, which is close to human thought expressions, can produce hundreds of machine instructions, whereas a single statement in the lower-level assembly language, whose symbolic code more closely resembles machine code, generally produces only one instruction.

Figure 8.2 shows how a 16-bit processor would execute a simple 16-bit program to add the numbers in memory locations X, Y, and Z and store the sum in memory location D. The first column shows the binary instructions in machine language. Symbolic instructions in assembly language, which have a nearly one-to-one correspondence with the machine language instructions, are shown in the next column. They are quite mnemonic and should be read as "Load the number at location X into memory location A; add the number at location Y to the number in memory location A; add the number at location Z to the number in memory location A; store the number in memory location A at location D." This series of assembly language statements, therefore, accomplishes the desired result. This sequence of assembly statements would be input to the assembler program that would translate them into the corresponding machine language (first column) needed by the computer. After assembly, the machine language program would

Machine language instructions	Assembly language instructions	FORTRAN language instructions
0110 0011 0010 0001	LDA X	$D = X + Y + Z$
0100 0011 0010 0010	ADA Y	
0100 0011 0010 0011	ADA Z	
0111 0011 0010 0100	STA D	

FIGURE 8.2   Three types of program instructions. Machine language gives instructions as 0 and 1's and is the only language that the computer understands. Assembly language is more concise but still very cumbersome when programming. A high-level language such as FORTRAN or C facilitates easy programming.

be loaded into the machine and the program executed. Because programming in assembly language involves many more details and low-level details relating to the structure of the microcomputer, higher-level languages have been developed. FORTRAN (FORmula TRANslator) was one of the earlier and most widely used programming languages and employs algebraic symbols and formulas as program statements. Thus the familiar algebraic expression for adding numbers becomes a FORTRAN instruction; for example, the last column in Fig. 8.2 is the FORTRAN statement for adding the three numbers and is compiled into the set of corresponding machine language instructions of the first column.

8.3   ELEMENTS OF A COMPUTER

Fundamental building blocks for digital systems were considered in the previous chapter. We can treat logic gates as elementary blocks which in turn can be arranged into larger building blocks such as registers, counters, and adders, which are the fundamental components of the digital computer. A personal computer is basically the addition of memory, control circuitry, I/O devices, and a power supply to a microprocessor that acts as the CPU.[3] The basic computer architecture of a PC is given in Fig. 8.3. A discussion of each subsystem will now be given. For the subsystems that are more complex such as the CPU, a more detailed study will be given in subsequent sections.

In the construction of a modern computer, Fig. 8.3, the CPU, memory, clock circuits, and I/O interface circuits are placed on a single printed-circuit board, typically called the motherboard, main logic board, or system board. This board, which normally has extra sockets and expansion slots for additional memory and custom logic boards, is then placed in a housing which also contains a power supply, the hard drive, floppy drives, a CD-ROM drive, speakers, etc., and is then collectively known as the computer. In a computer such as a PC there is no need for an external bus as shown in Fig. 8.3, since the system is self-contained. An external bus, however, would be needed when

---

[3]For purposes of this chapter we can assume CPU and microprocessor to mean the same thing.

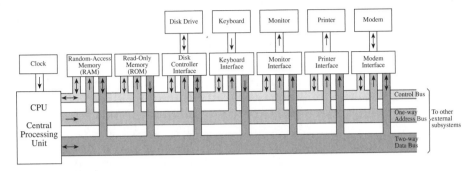

FIGURE 8.3   The architecture of a digital computer. The keyboard, monitor, and so on that are connected by buses to the CPU are referred to as input/output devices or peripherals. The connection for any given peripheral is referred to as an I/O port.

the computer is used in a laboratory, for example, controlling or receiving data from instruments.

8.3.1   The Central Processing Unit

The CPU can be likened to a conductor in an orchestra; with the help of a clock it synchronizes the efforts of all the individual members of an orchestra. The CPU is the most complicated and mysterious of all the subsystems in a computer and frequently is referred to as its brains. Contained within the CPU is a set of registers, an arithmetic and logic unit (ALU), and a control unit as shown in Fig. 8.8*b*. These three sections of the CPU are connected by an internal bus within the CPU which may not neccessarily be the same width as the bus external to the CPU. Before running a program, the software for the program must first be fetched from storage (such as a hard disk) and installed in RAM. To execute a program, the microprocessor (CPU) successively transfers instruction codes from the external program memory (RAM) to an internal memory circuit (register) and executes these instructions. Registers are made up of high-speed CPU memory (small internal memory located on the CPU chip) and hold and store bytes between execution cycles of the computer. The calculating functions are accomplished by the ALU, which contains, at a minimum, an adder for two binary words. The addition of bytes is the most important task of the ALU since the functions of subtraction, multiplication, etc., can be carried out in terms of it. By means of the clock, the control unit synchronizes all the digital circuits in the CPU.

Instructions and data to and from a CPU are carried on a bus which is a set of electrical lines that interconnect components. These lines can be traces on a printed circuit board or parallel wires imbedded in a flat plastic ribbon. The number of wires is proportional to the word size in bits that the processor can handle. For example, if the processor can execute 8 bits at a time (an 8-bit word), the bus will have 8 wires and the processor is referred to as an 8-bit CPU. More powerful processors run at faster clock speeds (execute instructions faster) and can handle larger words at a time, which means that such things as saving files and displaying graphics will be faster. The early Zilog Z80, an 8-bit CPU, had a 16-bit address and an 8-bit data bus; a 16-bit processor like the MC68000 has a 24-bit address and 16-bit data; a 32-bit processor like the 80386 has a 32-bit address and 32-bit data; and the Pentium (which is not called a 80586 because numbers cannot be trademarked) and the Power PC have 64-bit data and a 32-bit address. It can be expected that data size in more powerful processors will continue to increase beyond 64 bits, as, for example, 128 bits in PlayStation 2, but address buses larger than 32 bits, which can already locate $2^{32} = 4,294,967,296$ addresses, might not be essential. On the other hand, the Intel Epic architecture and the alpha processor family are pure 64-bit architectures, operating on 64-bit data and addresses.

8.3.2   Clock

In Section 7.5 we showed that in order for sequential logic circuits to operate in an orderly fashion they were regulated by a clock signal. Similarly for a computer, the timing control is provided by an external clock (a crystal oscillator circuit)[4] that produces a clock signal which looks like a square wave shown in Fig. 7.20b. This regular and steady signal can be considered the heartbeat of the system. All computer operations are synchronized by it. The square-wave clock signal provides two states (top and bottom of the pulse) and two edges (one rising, one falling) per period that are used in switching and timing various operations. Edge-triggering is preferred as this leads to devices with more accurate synchronization since edges are present only a short time in comparison with the tops or bottoms of pulses.

Figure 8.4 shows a basic CPU operating cycle which is an instruction-fetch interval followed by an instruction-execute time, which together occupy two periods of the clock signal. (Note that instructions requiring data from external memory and complex instructions like multiply and divide can take several cycles.) It should be clear now that the speed of a computer is intimately related to the instruction-fetch–execute sequence.

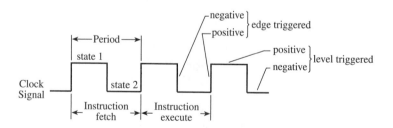

FIGURE 8.4   A CPU instruction cycle.

---

[4]A crystal oscillator is a thin, encapsulated vibrating piezoelectric disk which generates a precise sinusoidal signal. By some additional circuitry, this signal is shaped into the desired square wave and is made available to all circuits of the computer. The crystal is typically external to the CPU and is located somewhere on the motherboard.

Slower or older computers required many clock cycles to perform elementary operations whereas modern computers can do complex ones within a single clock period.[5] Processor speed is usually stated in millions of clock cycles per second, or megahertz (MHz). No instruction can take less than one clock cycle—if the processor completes the instruction before the cycle is over, the processor must wait. Common processors operate at speeds from 8 MHz to 1000 MHz (1 GHz). At 8 MHz, each clock cycle lasts 0.125 millionths of a second (0.125μs); at 100 MHz, 0.01μs; and at 300 MHz, 0.0033μs $= 3.3$ns (nanoseconds).

In addition to raw megahertz speed, smart design augments the newest processors' capabilities. These processors contain a group of circuits that can work on several instructions at the same time—similar to a factory that has several assembly lines running at the same time. The more instructions a processor can work on at once, the faster it runs. The Motorola 68040 microprocessor, for example, can work on six instructions at once. The older 68030 is limited to four. That is why a computer with a 25 MHz 68040 is faster than one with a 25 MHz 68030.

8.3.3   RAM

Random access memory is an array of memory registers in which data can be stored and retrieved; it is short-term memory and is sometimes called read–write memory. It is memory that is external to the microprocessor, usually in the form of a bank of semiconductor chips on the motherboard (logic board) to which the user can add extra memory by purchasing additional chips. RAM is volatile, meaning that it is a storage medium in which information is a set of easily changed electrical patterns which are lost if power is turned off because the electricity to maintain the patterns is then lost.[6] For this reason magnetic disks or tapes which have the advantage of retaining the information stored on them even when the computer is off are used for permanent storage. They can do this because they store information magnetically, not electrically, using audio and video tape technology which lays down the information as a sequence of tiny permanent magnets on magnetic tape. The downside of disk storage is that it is about 200,000 times slower in transfer of information than RAM is (typically 60 ns for RAM and 10 ms for hard disks). Hence, if disk storage has to be used when working with an application program in which information and data are fetched from memory, processed, and then temporar-

---

[5]Efficiency is directly related to computer speed. The fewer clock cycles per instruction, the better. For example, the Motorola 68020 microprocessor takes 6 clock cycles to move data from a register to the logic board's main memory, while the newer 68040 takes just 1 clock cycle. The 68040, one of the fastest of the older processors, executes instructions at an average rate of 1.25 clock cycles each. Today's processors achieve an average of close to one instruction per cycle. Clock speed, expressed in MHz (megahertz), is how fast the processor runs. For ordinary tasks such as word processing one might not notice how fast a computer is. However, with processor-intensive tasks such as working with and displaying graphical images or calculating π to 10,000 digits, processor speeds less than 100 MHz become painfully slow.

[6]RAM chips store data in rows and columns in an array of transistors and capacitors and use a memory-controller circuit to retrieve information located at specific addresses. The chips must be constantly refreshed with electrical pulses to keep the charges current.

ily stored, and this cycle is repeated over and over during execution of a program, one can see that the program would run terribly slow. It is precisely for this reason that high-speed RAM is used during execution of a program and is therefore referred to as the main memory. The slower disk storage is referred to as secondary memory.

Virtual memory is a clever technique of using secondary memory such as disks to extend the apparent size of main memory (RAM). It is a technique for managing a limited amount of main memory and a generally much larger amount of lower-speed, secondary memory in such a way that the distinction is largely transparent to a computer user. Virtual memory is implemented by employing a *memory management unit* (MMU) which identifies what data are to be sent from disk to RAM and the means of swapping segments of the program and data from disk to RAM. Practically all modern operating systems use virtual memory, which does not appreciably slow the computer but allows it to run much larger programs with a limited amount of RAM.

A typical use of a computer is as follows: suppose a report is to be typed. Word-processing software which is permanently stored on the hard disk of a computer is located and invoked by clicking on its icon, which loads the program from hard disk into RAM. The word-processing program is executed from RAM, allowing the user to type and correct the report (while periodically saving the unfinished report to hard disk). When the computer is turned off, the RAM is wiped clean—so if the report was not saved to permanent memory, it is lost forever. Since software resides in RAM during execution, the more memory, the more things one is able to do. Also—equivalently—since RAM is the temporary storage area where the computer "thinks," it usually is advantageous to have as much RAM memory as possible. Too little RAM can cause the software to run frustratingly slow and the computer to freeze if not enough memory is available for temporary storage as the software program executes. A minimum amount for modern PCs begins with 16 megabytes (MB) for ordinary applications such as word processing, increases to 32 MB for data bases and spread sheets, and then to 128 MB (or as much as possible in excess of that) for graphics work, which is very RAM-intensive. In other words, the more RAM you have, the more efficient your computer will be. Typical access times for memory chips is 50–100 ns. If a CPU specifies 80 ns memory, it can usually work with faster chips. If a slower memory chip is used without additional circuitry to make the processor wait, the processor will not receive proper instruction and data bytes, and will therefore not work properly.

In the 1980s capacities of RAMs and ROMs were 1 M × 1 bit (1-megabit chip) and 16 K × 8 bit, respectively, and in the mid-1990s 64 M × 1-bit chips became available. Memory arrays are constructed out of such chips and are used to develop different word-width memories; for example, 64 MB of memory would use eight 64 M × 1-bit chips on a single plug-in board. A popular memory size is 16 MB, consisting of eight 16-megabit chips. (Composite RAM, which has too many chips on a memory board, tends to be less reliable. For example, a 16 MB of composite RAM might consist of thirty-two, 4-megabit chips, while an arrangement with eight, 16-megabit chips would be preferable.) The size of memory word width has increased over the years from 8 to 16, 32, and now 64 bits in order to work with advanced CPUs which can process larger words at a time. The more bits a processor can handle at one time, the faster it can work; in other words, the inherent inefficiencies of the binary system can be overcome by raw processing power. That is why newer computers use at least 32-bit processors, not 16-bit processors. And by processing 32 bits at a time, the computer can handle more complex tasks than it can when processing 16 bits at a time. A 32-bit number can have a value between 0 and 4,294,967,295. Compare that to a 16-bit number's range of 0 to 65,535, and one sees why calculations that involve lots of data—everything from tabulating a national census count to modeling flow over an airplane wing or displaying the millions of color pixels (points of light) in a realistic image on a large screen—need 32-bit processors and are even more efficient with 64-bit processors. A simple 16×8-bit memory array is shown in Fig. 8.5.

8.3.4   ROM

Read-only memory is permanent and is indelibly etched into the ROM chip during manufacture. ROM contains information (usually programs) that can be repeatedly read by the computer but cannot be modified. For example, when a computer is first turned on, there are no operating instructions because the computers main memory (RAM) is "empty."[7] For a computer to start up it first executes a bootstrap program residing in ROM. The bootstrap program, which is permanently stored on ROM chips, is an initialization program (located at a special place in the addresss space) which actually is a small part of the operating system.[8] The instructions on the ROM cause the computer to look for the remaining and major part of the operating system, which can be on a floppy disk or CD-ROM, but usually is stored on the hard disk, and install the operating system into the computer's working memory, or RAM. It is only now that the desired application software can be loaded from floppies, CD-ROM, or the hard disk into RAM, after which the computer is ready to do useful work.

Besides containing the bootstrap program, ROMs can give a particular type of computer some of its character. Some makes are better suited for number crunching, while, for example, Macintosh computers are especially suitable for a graphical user interface because Macintosh ROM contains elements of the software that implements that interface, even though, generally speaking, it is the operating system that gives a computer its character.

---

[7] *Empty* implies that the RAM does not contain information that can be understood or interpreted by the computer.

[8] There are two primary categories of software: operating systems and applications. Operating system software is designed to perform system and computer management utility and/or "housekeeping" functions such as directing the central processor in the loading, storage, and execution of programs and in accessing files, controlling monitors, controlling memory storage devices, and interpreting keyboard commands. Examples of operating systems include Windows 3.1/95/98/2000/NT, Mac OS, MS-DOS, Xenix, Unix, Novell, etc. This special software serves as the interface between the computer, its peripheral devices—such as disk drives, printers, and displays—and the application programs.

8.3.5   Interfaces

A glance at the basic architecture of a digital computer, Fig. 8.3, shows that after considering the clock, ROM, and RAM, which are the nearest and most essential components to the CPU, the remaining items are the internal and external peripherals. Common to these peripherals are interfaces. These are positioned between the peripherals and the CPU. For example, the disk drive, typically an internal component, requires a disk controller interface to properly communicate with the CPU, as is the case for external components such as the keyboard, monitor, printer, and modem. Why is an interface needed at the point of meeting between the computer and a peripheral device? An analogous situation exists when a German-speaking tourist crosses into France and finds that communication stops, and can only resume if an interface in the form of translator or interpreter is available. The need for special communication links between the CPU and peripherals can be summarized as follows:

(a) Since many peripherals are electromechanical devices, a conversion of signals is needed for the CPU, which is an electronic device.

(b) Electromechanical devices are typically analog, whereas CPUs are digital.

(c) To represent the 0 and 1's, the CPU data bus uses fixed voltages (0 and 5 V, or 0 and 3.3 V). Peripherals most likely use other voltage levels or even nonelectrical signals such as optical ones.

(d) The data transfer rates of peripherals are typically slower than the transfer rate of the CPU. For proper data communication, synchronization and matching of transmission speeds is required, and is accomplished by an interface.

(e) A given microprocessor has a fixed number of bus lines which can differ from that used in peripherals.

(f) Peripherals and CPUs can use different data codes because there are many ways to encode a signal: serial, parallel, different bit size for the words, etc.

The matching or interfacing between the CPU and peripherals is usually in terms of hardware circuits (interface cards). For example, a disk drive is matched to the CPU by a disk controller card, which is a circuit board with the necessary control and memory chips to perform the interfacing; one of the interfaces for connecting a computer to the Internet is an Ethernet card which is placed in one of the free expansion slots of the computer. In addition to hardware, the interface between the CPU and peripherals can also include software (emulation programs). There are trade-offs between hardware and software. If hardware is used for the interface, the advantage is speed, whereas the disadvantages are cost and inflexibility. The advantage of software is versatility, whereas its main disadvantage is its slow speed.

In Fig. 7.32 a 1-bit memory cell is shown, and Fig. 7.33 shows how such single memory cells are used to form a 4×4-bit RAM. Combining this with the decoder of Fig. 7.34, it is now straightforward to show a CPU–RAM interface. For simplicity a 16×8-bit memory array is shown interfacing with a CPU in Fig. 8.5. The 16×8-bit array is formed from eight 16×1-bit memory chips. The various memory locations are accessed via the address bus, and the contents of each memory location are transferred

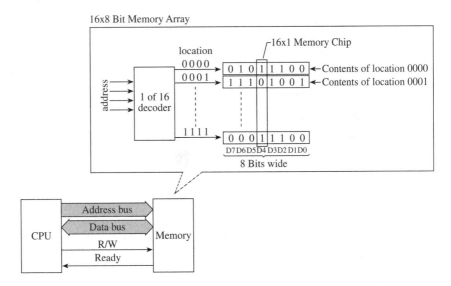

FIGURE 8.5   The interface between the CPU and RAM.

over the data bus. Ready is a handshake signal from the memory to the CPU which indicates that the desired memory location has been accessed, its contents are stable, and the next operation can proceed. It is an asynchronous interface signal not directly related to the system clock. The read and write control signals on the R/W bus control the memory operation in progress at any time. Even though it appears that the CPU–RAM interface consists of connecting wires only, and therefore is not shown as a separate block in Fig. 8.3, it can be more complicated, including buffers and gates.

8.3.6   Interrupts

When the computer is communicating with peripherals that have a low data speed such as the keyboard, the much faster computer has to wait for the data to come from these devices. To avoid unnecessary waiting, the CPU is designed with special interrupt inputs which allow the processor to carry out its normal functions, only responding to I/O devices when there are data to respond to. On receipt of an interrupt, the CPU suspends its current operation and responds to the I/O request; that is, it identifies the interrupting device, jumps ("vectors") to a program to handle it, and performs the communication with it, which may be acquisition of new data or execution of a special program. After the interrupt routine is completed, the CPU returns to the execution of the program that was running at the time of the interrupt. This is an efficient procedure in the sense that the processor communicates with devices only when they are ready to do so rather than have the processor continually asking I/O devices whether they have any data available.

An interrupt interface is not explicitly shown in Fig. 8.3 but is assumed to be part of the interface block between each I/O device and the CPU. There are other features of interrupts, as, for example, the priority interrupt: a typical computer has a number of I/O devices attached to it with each device being able to originate an interrupt request. The first task of the interrupt system is to identify the source of the interrupt and, if several sources request service simultaneously, decide which is to be serviced first.

8.3.7   The Three Buses

The remaining subsystem of a computer that we need to consider are the three buses, shown in Fig. 8.3, that run from the CPU and connect together all subsystems, both internal and external (keyboard, monitor, etc.). These are parallel buses, consisting of parallel wires, allowing for parallel transfer of data (see bus discussion, p. 285). For example, if a word were a byte long (8 bits long) and needed to be transferred from memory to the CPU, then the data bus would consist of eight parallel wires with each wire carrying a specific bit. A parallel-wire bus makes it possible for all 8 bits stored in a specific memory location to be read simultaneously in one instant of time. This is the fastest way to transfer data between the CPU and peripherals, and since speed in a computer is extremely important, parallel transfer of data inside a computer is invariably used. Serial transfer of data, which takes a byte and transmits each bit one at a time over a single wire, is much slower and is used in links where stringing eight wires would be prohibitively expensive, for example, when interconnecting remote computers by telephone modems. In a typical computer, both serial and parallel ports are provided for communication interface with the outside world.

The *address bus* carries the location in memory where data are to be found or placed (read or written). Addresses are binary coded, meaning that with a 16-bit address bus (16 parallel wires), $2^{16} = 65536$ locations can be selected. The address bus is a unidirectional line in the sense that information flows from the CPU to memory or to any of the I/O peripherals.

The *control bus* carries instructions when the CPU requires interactions with a subsystem, turning on the subsystem of interest and turning off other subsystems. Clock pulses, binary signals to initiate input or output, binary signals to set memory for reading and writing, binary signals to request services by means of interrupts, bus arbitration to determine who gets control of the bus in case of conflict, etc., are carried on the control bus. Individual lines may be bidirectional, that is, data flow may be going in both directions to the CPU and the subsystem, or unidirectional, going only to the CPU or the subsystem.

The *data bus*,[9] usually the widest bus in modern computers (64 bits or larger in

---

[9]Frequently, the data-carrying bus internal to the microprocessor—the processor path—is referred to as the data bus, while external to the CPU it is called the I/O bus. The bit width (word size) of the processor path is the number of bits that the processor is designed to process at one time. Processor bit width typically is equal to or less than the external data path. For example, a Pentium computer is usually characterized as a 64-bit computer because the external I/O bus is 64 bits wide. However, the internal processor path for a Pentium is 32 bits. Hence, the larger amount of data carried by the I/O bus must que up to the CPU for processing. Thus, the larger the data path, the more data can be queued up to be immediately processed. On processors with internal cache memory, the data path between the cache and processing unit is the same as the processor path.

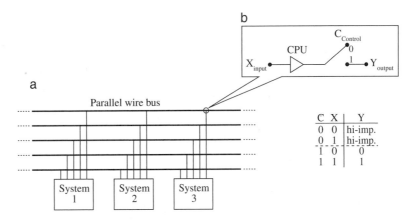

FIGURE 8.6   (a) All subsystems or units are connected in parallel to the bus, thus reducing the number of interconnections. (b) A tristate logic circuit showing high impedance when control is low (0 or off) and low impedance when C is high (1 or on).

workstations), carries the program instructions and transfers data between memory, I/O devices, and the CPU. The data bus is bidirectional as it is used for sending and receiving of data.

The bus system solves a difficult computer problem. The number of internal connections in a complicated electronic circuit such as a computer would be tremendous—it would be unacceptably high if made in the conventional way. The bus is a completely different connection structure. All circuits or systems intended to communicate with each other are connected in parallel to the bus as shown in Fig. 8.6a or Fig. 8.3 with the information flow time-multiplexed to allow different units to use the same bus at different times. With such an arrangement it is easy to add or remove units without changing the structure of the overall system. For example, it is straightforward to add additional memory, a modem, or a CD-ROM drive to a computer. Data reception can take place by all the connected systems at the same time because each has a high input impedance. That is, each system on the bus will see all of the data placed on the data bus lines. However, the data on the data bus are usually intended for a specific unit with a particular address. The address bus then determines if the available information is to be processed or ignored. If the address of a unit matches that on the address bus, the data are provided to that unit with the remaining units not enabled. Transmission of data, however, is restricted to only one transmitting system or unit at a time. The transmitting unit is enabled when it assumes a low output impedance. Figure 8.6b shows how the CPU and the control bus determine a high and low impedance—each line is driven with a tristate logic device (true, false, and hi-Z) with all drivers remaining in the disconnected state (hi-Z, meaning high impedance) until one is specifically enabled by a properly addressed control signal.

Even though the bus structure offers a high degree of flexibility, it has a disadvantage in the sense that only one word at a time can be transmitted along the bus. However, another further advantage is that a computer system can be readily extended by expanding its bus. This is done by designing and building computers with several onboard expansion slots which are connected to the bus. A user can now expand the basic system beyond its original capabilities simply by adding plug-in boards into its expansion slots. There are numerous boards available for the personal computer—in fact many plug-in boards far exceed the host computer in terms of power and sophistication.

A critical issue in the PC market is the type of expansion bus architecture. As stated before, the "wider" the bus, the faster one is able to do things like save files and display graphics. Hence, along with processor speed, the bus type helps determine the computer's overall speed. The most common PC bus architectures are the 16-bit ISA (Industry Standard Architecture), the 32-bit VLB (VESA local bus), and the most recent, 64-bit PCI (Peripheral Component Interconnect). PCI is a high-performance expansion bus architecture that was originally developed by Intel to replace the traditional ISA and EISA (Enhanced Industry Standard Architecture) buses found in many 8086-based PCs. The fast PCI bus is used for peripherals that need fast access to the CPU, memory, and other peripherals. For fastest performance, the choice would be a PC with a PCI bus, although a versatile PC should include some ISA expansion slots as well as PCI expansion slots. A PC with the older ISA bus architecture is acceptable as PCI is backward-compatible (PCI cards fit in ISA slots). Newer Macintosh computers use the PCI bus running at 33 MHz with a maximum throughput of 132 MB per second, whereas older Macintoshes used the NuBus, which is a 32-bit data bus, 32-bit address bus running synchronously at 10 MHz with a maximum throughput of 40 MBps.

Bus speeds must constantly improve to keep pace with ever-increasing microprocessor speeds. High-end computers have CPUs that run at gigahertz rates with bus speeds that run at 200 MHz. Because the speed of a bus affects the transfer of bits, a faster bus means faster transfer of information in and from the CPU. A bus relies on its own internal clock, which may be slower than the CPU clock. Slow bus clocks affect the rate at which instructions are executed. Because buses cannot bring instructions to the CPU as quickly as the CPU can execute them, a bottleneck is created on the way to and from the CPU, causing CPU delays in instruction execution. Graphics-intensive programs require high bandwidth. Recall that bandwidth is the amount of data that can be transmitted in a certain period of time. Thus graphics can create bottlenecks and would benefit the most from a fast bus. The *accelerated graphics port* (AGP) is a new fast bus which runs between the graphics controller card and main RAM memory and was developed and is dedicated to graphics. The function of this additional bus is to relieve the main bus from carrying high-bandwidth graphics, with the result that data traffic on the main bus is reduced, thus increasing its throughput.

Ports are physical connections on a computer through which input or output devices (peripherals) communicate with the computer (PC). Peripherals can be serial or parallel in nature, requiring serial or parallel ports for connection with the computer. For example, keyboards, mice, and modems connect to serial ports, while printers generally connect to parallel ports. In *serial communication* the bits are sent one by one over a single wire, which is slow, cheap (single wire), and reliable over long distances. In *parallel communication* all bits of a character are sent simultaneously, which is fast, expensive, (every bit has its own wire—16 wires for a 2-byte character, for example), and not as reliable over long distances. However, inside a computer where distances are short, parallel communication with multiwire buses (which resemble belts) is very reliable. Typically an interface card (video, disk drive, modem, etc.) plugs into the motherboard. The card contains a port into which an external device plugs into and thereby allows the external device to communicate with the motherboard. For example, SCSI (Small Computer System Interface) is an interface that uses parallel communication to connect numerous devices (by daisy-chaining) to a single port. SCSI needs a fast bus and is usually plugged into the PCI bus. Another parallel interface is IDE; it is used to connect hard disks and CD-ROM and DVD drives. As already mentioned, a PC uses ISA and PCI bus architecture, with PCI the newer and faster bus. The capability of SCSI to link many peripherals was combined with the speed of the PCI bus to develop two new buses. USB (Universal Serial Bus) is used for medium-speed devices with a transfer rate of 12 Mbps. It can link 127 devices to a single USB port. The second bus is *Firewire* (IEEE 1394). It is a very fast serial bus that can transfer large amounts of data with a transfer rate of 400 Mbps and can connect 63 devices. It is typically used for video and other high-speed peripherals. Both newer buses support *plug-and-play*, which is a user-friendly technology that requires little set up effort by the user: the PC identifies a new device when plugged in and configures it, and the new device can be added while the PC is running.

8.3.8   The Peripherals: Hard Disk, Keyboard, Monitor, Printer, and Modem

The combination of the peripherals interconnected by the three buses with the CPU and memory is what constitutes a computer.

The medium most commonly used for storing information generated using a computer is called a disk. The two principal types of storage devices[10] are floppy diskettes and hard drives. *Floppy diskettes* are usually used for installing newly acquired software programs and storing user-generated data files. The *hard disk* (also known as the *hard drive* or *disk drive*) is the permanent storage space for the operating system, programs, configuration files, data files, etc. Obviously, the larger the drive, the more files it can save. New machines come with drives larger than 1 GB (gigabyte). The large space is needed when large graphics or sound files are stored. As a minimum one needs 400 MB for system software, 32 MB of RAM and at least as much as for virtual memory/swap space as you do RAM, 100 MB for network drivers/communication software and Internet applications and 300 MB for user applications such as Microsoft Office, the popular integrated set of tools for word processing, data base and spreadsheet creation. Graphics programs and graphics data quickly exceed such minimum specifications.

---

[10]For extensive use of graphics or sound, CD-ROM and DVD drives, which use optical, massive storage media, are preferred. Because graphic and sound files are so large, many libraries of these files and applications that use them are sold on CD-ROM.

Hard Drives

Hard drives are usually not removable from the computer. The disks are made of metal with a magnetic coating rather than plastic with a magnetic coating (as for floppies) and their rigidity allows them to be spun faster than a floppy disk can be, resulting in faster information storage and retrieval times. To store information on disk, the logic board converts digital signals from its memory into analog signals—varying electrical currents—and sends them to the disk drive. Inside the disk drive, this current flows through a tiny electromagnet called a read–write head, which floats above the magnetic coating of a rotating disk. Most disk drives have two heads, one for each side of the disk. The variations in the current going through the head set the orientation of magnetic particles in the coating. The pattern of information—which is laid down as a sequence of tiny permanent magnets in the coating—encodes the data.

When the drive retrieves information, the differently oriented magnetic particles induce a changing electrical current in the read–write head. The drive's circuitry converts the changing current into digital signals that the logic board can store in its memory.

The drive arranges recorded information on the disk in concentric bands called tracks; each track is divided into consecutive sectors.[11] A 1.44 MB floppy disk, for example, has 80 tracks with a total of 1440 sectors on each side of the disk (18 sectors per track, and 512 bytes per sector). To position the read–write heads over a specific track, a *stepper motor* turns a precise amount left or right. This circular motion of the stepper is converted to back-and-forth motion by a worm gear attached to the head assembly. A flat motor spins the disk platter at a specific speed, so that information passes by the read–write heads at a known rate. With the heads positioned over the track requested by the logic board, the drive circuitry waits for the requested section to come around and then begins transferring information. Data can be transferred at rates of tens of megabytes per second, but because of the complex addressing which is mechanically performed, it takes on the order of 10 ms to reach the data. Compare that with the 60–80 ns access time of semiconductor RAM.

A file transferred to a disk consists of a directory and data. The data are all the programs and documents stored on your disk. The directory is an index that contains the name, location, and size of each file or subdirectory and the size and location of the available space on the disk. The reason why deleted files can sometimes be recovered is that when deleting, only parts of the directory entry are erased with the data remaining intact. If a subsequently placed file does not override data of the deleted file, it usually can be recovered.

---

[11]The magnetic coating of a disk, after manufacture, is a random magnetic surface, meaning that the tiny magnets that exist on the surface have their north–south poles arranged in a random pattern. During initializing or formatting of a disk, the read–write head writes the original tracks and sector information on the surface and checks to determine whether data can be written and read from each sector. If any sectors are found to be bad, that is, incapable of being used, then they are marked as being defective so their use can be avoided by the operating system.

Keyboards

Pressing and releasing keys actuates switches under the keys. A microcontroller inside the keyboard continuously looks for switch transitions, such as a key press or key release, and sends numeric transition codes to the logic board, identifying the keys you press and release. The system software translates the transition codes to character codes, such as ASCII (American Standard Code for Information Interchange).

Video Monitor

A video monitor is a descendent of a television set. It is only the additional receiving circuitry in a TV that distinguishes it from a monitor. However, a TV when used as a monitor has a resolution which is generally inferior to that of a computer monitor. The 4.2 MHz video bandwidth[12] of a TV restricts the screen display to a horizontal resolution of about 40 characters across the screen. Since a typical typed page has about 80 characters across, a TV monitor would appear blurred and barely readable if used in word-processing. It is clear that a computer monitor should have a bandwidth of at least 8.4 MHz so as to be able to display at least an 80-column page.

When evaluating a monitor, we need to know the monitor's ability to display color and its resolution. The monitor will display no more colors or finer resolution than its highest potential, no matter how much computer memory or processor power is available. For color display, monitors are ranked by their ability to display tens, hundreds, thousands, or more different color differentials. For example, if your video card is capable of 4-bit color, then only 16 colors can be displayed, but if it is 16-bit capable, 256 colors are possible. *Resolution* is expressed as the number of pixels (points of light) displayed across the screen by the number displayed up and down the screen. Numbers commonly seen for monitors are 640×480 or 1024×768 at a refresh rate of 60, 67, or 75 Hz. A refresh rate is the rate at which the image on a display is redrawn. A rate of 70 Hz or higher provides flicker-free displays. Refreshing is the process of constantly restoring information that fades away when left alone. In the case of a monitor, the phosphor dots or pixels on a monitor screen need to be constantly reactivated by an electron beam to remain illuminated. PC users will often find the combination of ability to display color and resolution expressed in rankings like VGA, hundreds, or thousands of colors with 640×480 resolution. The standards for what defines the various monitors, like VGA or SVGA, are not clearly defined, but SVGA monitors (1024×768 at

---

[12]Theoretically, a TV set that can display an image on the screen with a horizontal resolution of 500 pixels per line (commonly referred to as 500 lines of horizontal resolution) and vertical resolution of 525 raster lines at a rate of 30 screens (frames) per second displays $500 \times 525 \times 30 = 7.875$ million pixels per second. A sinusoidal video signal which, when displayed on the TV screen, results in one dark and one light pixel per period of the video signal has a bandwidth of $7.875/2 = 3.94$ million hertz, which is approximately equal to the bandwidth of the video circuitry in a TV set. In practice, however, only about 350 lines of horizontal resolution are realized in a TV set. A typical character (such as a letter of the alphabet) on a monitor screen is composed of an 8×10 pixel array. Allowing 2 pixels between characters, we obtain $350/(8 + 2) = 35$ characters per line for a TV monitor, clearly not adequate for an 80-column display.

72 Hz) will most likely meet the needs of today's user. Monitor size is a fairly obvious feature. However, larger monitors should have higher resolution levels, otherwise the screen display will appear granular.[13] For example, a 19-inch monitor has a recommended resolution of 1280 × 1024 pixels.

Finally, the remaining criteria are the dot pitch size and whether the monitor is interlaced or noninterlaced. *Dot pitch* is the distance in millimeters between the individual dots or pixels on a monitor screen. All other things equal, the less space between pixels, the better, since it allows for more pixels to be displayed, giving better resolution. A dot pitch in the range of 0.28 mm is acceptable. Interlaced monitors refresh the screen in a pattern of every other line and take two passes over the screen for a complete refresh. These monitors "jump" more and are more difficult on the eyes. Noninterlaced monitors refresh the screen in one pass without skipping any lines and are preferable.

Video RAM (VRAM)

VRAM is the memory used to display video. It determines how many colors the computer can display as well as the size of the monitor it can support. If your computer is expandable in this way, more VRAM can be installed for high-powered graphics work (for which the 1 MB of VRAM that usually comes with a computer is inadequate). Whereas in older systems it was adequate to display 16 colors, the newer graphical user interfaces require at least 256 colors. A screen image can occupy between 22 K and 2 MB of memory, depending on size and color. Some computers store screen images in the logic board's main memory, leaving less RAM for other uses. It is more efficient to store screen images in VRAM exclusively, freeing up the main RAM for programs and data storage. Currently SDRAM is the standard and for a resolution of 1280× 1024 a card with at least 16 MB is needed.

Modems

Inside the computer, fast communication between the CPU and peripherals, such as the hard disk, is by parallel data transport in which parallel words are sent over flat, multi-wire ribbon cables which rarely exceed several feet in length. For data transmission over long distances, it is more efficient in terms of cost to transmit the data bit after bit, which requires only a single conductor pair. Any single line such as a telephone line, coaxial cable, or fiberoptic cable uses serial data transport. A modem (modulator/demodulator) is a device that allows your computer to communicate with another remote computer by using ordinary telephone lines. Anywhere one can make a telephone call, one can send a computer message by modem. It is a slow connection—typical rates are 28.8, 33.6, or 56 kilobits per second (kbps), which are adequate for e-mail and bulletin boards but marginal for downloading graphics, sound, and video files that can be many megabytes

---

[13]Higher-resolution monitors can display more pixels than lower-resolution monitors. High resolution is more fine-grained. Lower resolution yields coarser results and/or a smaller image. Low-resolution images use less memory to display and take up less space on your hard disk.

in size. For comparison, a fast serial connection would be a direct connection to the Internet with an Ethernet card installed in your computer and a coaxial cable; connection speeds are then 10 megabits per second (Mbps) or higher. An advantage of modem connections, on the other hand, is their low cost because they are slow serial connections. Serial transmission of data sends one bit at a time which, for slow sending speeds, is suitable for the single wire pair of an ordinary telephone connection. Telephone lines are slow-speed lines restricted to a bandwidth of only 4 kilohertz (kHz). ISDN and ADSL are specially installed, costly telephone lines which are faster (ISDN at 128 kbps; ADSL up to several Mbps, depending on the distance from the central office) but are still much slower than a direct connection to the Internet by coaxial cable or the still faster fiberoptic cable.[14]

Serial interfaces are integrated circuits that convert parallel words from the data bus into serial words (and vice versa). Serial words can be used to transfer bits from one place to another or from one computer to another using a transmission line. Serial interface ICs are now part of every microcomputer. These ICs can be used to physically wire together several computers which are in close proximity to each other in a local area network (LAN). The wiring can be telephone wire or other inexpensive cable as long as the LAN is confined to distances on the same floor or building. A computer can then exchange messages and files serially with nearby computers that are in the LAN. For larger distances, these ICs are used by modems for the transmission of computer data over telephone lines. Before this can be done, two things must occur. First, the interface must have a standard for the multipin connector and for the signal characteristics when interfacing a computer with a modem. The industry, in 1962, adopted the RS-232 protocol for asynchronous (no common clock) serial data communications between terminal devices (printers, etc.), computers, and communications equipment (modems, etc.). This standard is now widely used between computers, terminals, keyboards, printers, plotters, digitizers, and other devices that can tolerate slow communication speeds, that is, maximum data rates in the kilobyte per second range. Second, a device must be present that converts the electrical pulses, or bits, from the computer into continuous signals, or tones, suitable for transmission over ordinary analog telephone lines. This function is performed by a modem. Another modem at the receiving end converts the analog signal, or tones, from the telephone into digital pulses the computer can understand.[15] The need for conversion arises because the sharp-edged digital pulses produced by the serial interface have a frequency content which is much too high to be transmitted over the narrow band telephone lines. As stated in the previous paragraph, telephone bandwidth has a maximum of 4 kHz, and as can be seen in Fig. 9.4*c* a microsecond

---

[14]This topic is treated in greater depth in the next chapter.

[15]Most modems are analog products. They are designed to send sound along the same sort of twisted-pair wires that your voice telephone uses. A modem sends and receives data as combinations of tones. At the slow speed of 300 bps, the originating modem sends binary 0's as 1070 Hz tones and 1's as 1270 Hz tones, and at the same time receives 0's as 2025 Hz tones and 1's as 2225 Hz tones from the other modem. At transmission speeds above 1200 bps, modems encode digital data by varying the phase of tones (how they overlap) instead of their frequency, encoding 2 bits of data per phase shift. Additional material on modems can also be found in Example 9.19.

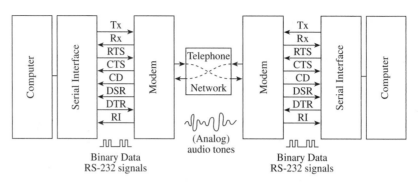

FIGURE 8.7   A modem is used to transmit computer data over telephone lines. The RS-232 standard signals are Tx, transmitted data (to modem); Rx, received data (to computer); RTS, request to send (to modem); CTS, clear to send (to computer); CD, carrier detect (to computer); DSR, data set ready (to computer); DTR, data terminal ready (to modem); and RI, ring indicator (to computer).

(μs) pulse contains frequencies up to 1 MHz, far exceeding telephone line bandwidth. Figure 8.7 shows a telephone communication path between two computers. The RS-232 interface provides the various signaling lines which are connected to the telephone modem. Notice that the transmit and receive lines are crossed in the telephone system because during normal operation the mouthpiece transmitter at one end is connected to the earpiece receiver at the other end. The RS-232 standard defines all the signals on the connector, a commonly used subset of which is shown in Fig. 8.7. Even this truncated listing is excessive as most cabling uses only three lines: Tx, Rx, and SG (signal ground) with data, control, and status information all being exchanged on these lines.

There are many factors that can affect modem performance, including the brand of modem, the time of day one is calling, where one calls from, and the age difference between the communicating modems. To communicate effectively, a modem must be compatible with the modems at the other end of the telephone line. For example, a 56 kbps modem and a 28.8 kbps modem can only communicate at 28.8 kbps at best. Even slight hardware differences between two modems can cause connection problems. Poor performance can also be due to heavy traffic over physical telephone lines, such as during the day when lines in general are more heavily used. Certain telephone subsystems tend to generate excessive line noise that may also cause connection problems.

There is also some jargon associated with modems. The more important are V.90 technology, which allows modems to receive data at up to 56 kbps; V.34, meaning data up to 28.8 kbps; V.32bis, meaning data up to 14.4 kbps (V.32bis modems fall back to the next lower speed when line quality is impaired and continue to fall back further as necessary, but also fall forward when line conditions improve); and V.42bis implies compression/error correction. Error correction checks for mistakes in transmission due to line noise and transparently resends data if necessary. Data compression looks at data being sent "on-the-fly" and recodes it to transmit more efficiently. V.42 is the international standard for error correction. V.42bis (which includes V.42) is a standard for data compression.

Finally, in addition to the software that comes with your particular modem, which is adequate for text-based access (e-mail), it is useful to have PPP (Point to Point Protocol) software which makes a modem-to-modem telephone link look like a direct Internet connection, albeit a slow one. With PPP one can run a variety of networking software (Netscape, Explorer, Fetch, etc.) that will not otherwise work with terminal programs such as ProComm. For additional information on connecting to the Internet see Example 9.19. The older SLIP (Serial Line Internet Protocol) is still in use and is supported.

8.3.9 Instrumentation Buses

A computer system can also be extended, as shown in Fig. 8.3, by expanding its bus to include external peripherals such as laboratory equipment designed to feed data to a computer. At the end of Section 8.3.7 on "The Three Buses" some standard system expansion buses such as the PCI bus were discussed. An external bus standard, the RS-232, initially designed to connect modems to computers, already introduced in the previous section, is a bit-serial, asynchronous connection using a 25-wire cable. The RS-232 serial interface port transmits and receives serial data 1 bit at a time. It has been updated as the RS-232C standard and is now used to connect other peripherals such as printers. Most popular PCs provide for RS-232C interfaces.

Manufacturers of laboratory equipment designed to be connected to a computer use a byte-serial interface designated as IEEE-488. It is a general-purpose, parallel instrumentation bus consisting of 16 wires, featuring 8 data lines and 8 control lines. The 8 data lines give this bus a byte-wide data path ("byte-serial" means transmitting and receiving 1 byte at a time). The control lines are used to connect the various instruments together. Each instrument connected to the bus can be in one of four different states: it can be idle, acting as a talker, acting as a listener, or controlling communication between talkers and listeners. IEEE-488 interface electronics are widely available on PCs, allowing laboratory instruments to be connected to this bus. The specification states that the bus can support at most 15 instruments with no more than 4 m of cable between individual instruments and not more than 20 m of overall cable length, with a maximum data rate of 1 megabyte per second (MBps). The intent was to provide interconnection for instruments within a laboratory room of typical dimensions.

SCSI, a parallel bus (although not as true an instrumentation bus as the IEEE-488), was initially developed to interface hard disk and tape backup units to a host computer but has since become a more general-purpose standard. Unlike RS-232C, SCSI can support multiple processors and up to eight peripheral devices, and unlike the IEEE-488 bus, it is not restricted to a single host processor. A 50-wire connector is standard in the SCSI I/O bus, which includes 9 data lines (8 data plus parity) and 9 control lines. Data transfers can take place up to 4 MBps in synchronous mode, decreasing to 1.5 MBps in asynchronous mode. The SCSI protocol is moderately sophisticated. Protocol refers to the set of rules or conventions governing the exchange of information between computer systems: specifically, the set of rules agreed upon as to how data are to be transferred over the bus. Thus, as we have seen, there can be many different ways to transfer data between the CPU and the peripherals, in addition to being classified as synchronous or asynchronous, depending on whether or not the transfer bears a relationship to the system clock. SCSI data transfer can be divided into three primary steps: arbitration, selection, and information. The CPU first checks if the bus is free. If it is, it then takes control of it. During selection, the CPU flags the target peripheral with which it desires to communicate. This is followed by the target responding with the type of information transfer in which it is prepared to engage: data in/out, message in/out, command request, or status acknowledgment. The CPU and peripherals communicate using a set of high-level command bytes, transmitted in packets. Macintosh computers were one of the first to use the SCSI bus.

The separate lines in a multiwire bus are termed *traces*. One should not readily assume that these bus standards merely involve assignments of particular traces to specific processor functions. Signal traces must be treated as transmission lines which posses distributed capacitance (farads/meter), distributed inductance (henries/meter), and characteristic impedance. This comes about because we cannot treat signals on such lines as propagating with infinite speed and producing the same instantaneous voltages and currents everywhere on the line—as is commonly assumed in electric circuits. The fact is that signals propagate down the line at speeds less than the speed of light, requiring approximately 1.5 ns to travel 1 ft. As the wire length and the signal frequency increases, it becomes more and more essential not to assume zero propagation time but to take the finite propagation time into account. Such transmission lines must be properly terminated (matched), otherwise transmitted signals will be reflected by a receiving device. The reflected signal can severely interfere with the transmission of information in the forward direction. For example, plugging an I/O board of impedance 20Ω into a bus that acts like a transmission line with a characteristic impedance Z_o equal to 100Ω will produce a reflected signal at the board location with voltage equal to $(100-20)/(100+20) = 67\%$ of the incident signal voltage. Such a large reflection will introduce significant errors, unless the wire lengths are very short. The matching problem is accentuated for fast computers which carry data in excess of gigabit rates (a gigabit per second digital signal has a gigahertz fundamental frequency with a nanosecond period). Transmission lines and how to properly terminate them is addressed in Section 9.5.5.

8.4   THE CPU

In this section we will examine the CPU in greater detail. Typically, the CPU chip is the largest on the motherboard, usually square in shape, and about 1 in. by 1 in. as illustrated in Fig. 8.8a. Figure 8.8b shows the main components that make up the CPU of Fig. 8.3. In Fig. 8.3 the CPU box was intentionally left blank as the emphasis in that figure was on the bus structure and the I/O devices connected to it. Registers are high-speed memory used in a CPU for temporary storage of small amounts of data (or intermittent results) during processing and vary in number for different CPUs. In addition to the ALU (arithmetic and logic unit), registers, and the control unit, a CPU has its own bus system (the internal bus) which connects these units together and also connects to the external bus. The bus system shown in Fig. 8.3 is therefore an external

a

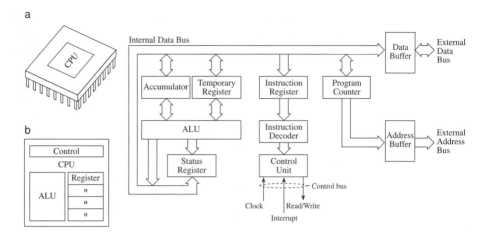

b

FIGURE 8.8   (a) Sketch of a typical CPU chip. (b) The main components of a CPU. (c) A simplified diagram of a microprocessor. (Note that the distinction between a CPU and a microprocessor is vague. Normally, a CPU is considered to be a subset of a microprocessor; also a microprocessor is assumed to have more RAM available to it.)

bus (also known as the I/O bus) and is typically wider than the internal CPU bus, which only needs to carry operations that are always well-defined (see footnote 9 on p. 292).

As already pointed out in Figure 8.4, *a microprocessor is essentially a programmable sequential logic circuit*, with all operations occurring under the control of a system clock. The CPU clock regulates the speed of the processor and it synchronizes all the parts of the PC. The general mode of operation follows a rather simple pattern. Instructions are fetched one at a time from memory and passed to the control unit for subsequent decoding and execution. The fetch–decode–execute sequence, however, is fundamental to computer operation. Hence, the CPU contains storage elements (registers) and computational circuitry within the ALU which at minimum should be able to add two binary words. To perform these functions, the CPU needs instruction-decoding as well as control and timing circuitry. These units, interconnected by the internal bus, are shown in Fig. 8.8c.

Summarizing, we can state that the primary functions of a CPU are

(a) fetch, decode, and execute program instructions
(b) transfer data to and from memory and to and from I/O (input/output) devices
(c) provide timing and control signals for the entire system and respond to external interrupts

When executing a program, the CPU successively transfers the instruction codes from the program memory[16] to an internal memory register and executes the instruc-

---

[16]The software program is read beforehand into RAM. Fetching different instructions will take varying numbers of system clock cycles. Memory reads and writes may take several clock cycles to execute. A

tions. In Fig. 8.8c this is accomplished by sequentially loading the *instruction register* with instruction words from the program memory. The translation of the coded words, which are strings of 0 and 1's, to a real operation by the CPU is performed by the *instruction decoder* and the *control unit*. The instruction decoder interprets the data coming from the data bus and directs the control unit to generate the proper signal for the internal logic operations such as read and write. The control unit—which is not a single block as shown in the figure but is actually distributed throughout the entire system—coordinates, by means of the clock, actions of both the CPU and all the peripheral circuits in the system.

Keeping track of the location of the next instruction in the program is the task of the *program counter*, which is a special register in the CPU. This register is incremented by 1 for every word of the instruction after each instruction is "fetched" from memory, placed into the instruction register, and executed. The program counter always points to the memory location where the next instruction is to be found (its output is the address of the location where the next instruction code is stored) and is updated automatically as part of the instruction fetch–decode–execute cycle. The program counter deviates from this routine only if it receives a jump or branch instruction, at which time it will point to a nonsequential address. Also, whenever the CPU is interrupted by an external device, the contents of the program counter will be overwritten with the starting address of the appropriate interrupt service routine. Much of the power of a computer comes from its ability to execute jump and interrupt instructions. Unlike the data bus, which must be bidirectional because data can go from CPU to memory as well as from memory to CPU, the address bus is unidirectional because the CPU always tells the memory (not vice versa) which memory location data are to be read *from* or written *to*.

Basic operations are executed by the arithmetic and logic unit. Even the simplest ALU has an adder and a shifter (shifting a number to the right or left is equivalent to multiplying or dividing by a power of the base). For example, if the ALU is directed to use the instruction ADD to add two binary numbers 0000 1010 and 0000 0101, the first number is placed in the accumulator and the second number is placed in the *temporary register*. The ALU adds the numbers and places the sum 0000 1111 in the accumulator and waits for further instructions. Typically the ALU can perform additional functions such as subtraction, counting, and logic operations such as AND, OR, and XOR. The results of all ALU operations are fed back via the internal data bus and stored in one of the accumulator registers. Of critical importance to the programmer is the status register which contains CPU status information. The *status register* is a group of flip-flops (or 1-bit flags) that can be set or reset based on the conditions created by the last ALU operation. One flip-flop could indicate positive or negative results, another zero or nonzero accumulator contents, and another register overflow. Such flags (also known as *status bits*) are used in the execution of conditional branching instructions. Based on the condition created by the last ALU operation, particular flag bits will determine if the

---

computer with a word length of 8 bits is not restricted to operands within the range 0 to 255. Longer operations simply take two or three such cycles. More can be accomplished during a single instruction fetch–decode–execute cycle with a 16-bit processor, and still more with a 32-bit one.

CPU proceeds to the next instruction or jumps to a different location. The temporary, accumulator, and status registers are frequently considered to be part of the ALU.

Two buffers are shown bridging the internal and external buses in Figure 8.8c. The data bus is bidirectional so the CPU, memory, or any peripheral device (which are all connected to the bus at all times) can be senders or receivers of data on this bus. However, only one device at a time can "talk." To avoid conflict, data from any device to the bus must be transferred through a tristate buffer (similar to the tristate logic of Fig. 8.6b) which acts as open or closed switches, thereby enabling only one output at a time. For example, when data are to be transferred from the CPU to memory, control signals enable (closed switch) the tristate buffers on the CPU and disable them (open switch) on the memory. The data from the CPU thus appear on the bus and can be stored by the memory. To transfer data from memory to CPU, the conditions of the tristate buffers are reversed. The tristate buffer thus has the enabling outputs of logical 0 and 1 and the disabling high-impedance state when it effectively disconnects the output from the bus. Typically, tristate buffers also include memory registers which are used to temporarily hold data which are being transferred from one device to another. This can be used to compensate for the different rates at which hardware devices process data. For example, a buffer is used to hold data waiting to be printed as a printer is not able to keep pace with the characters reaching the printer from a CPU. The CPU is thus freed to do other tasks since it can process data at a much faster rate. Similarly buffering is used when the CPU and peripheral devices have different electrical characteristics as, for example, when a CPU which operates at 5 V must interact with peripheral devices with many different voltage levels.

8.5   HEXADECIMAL NUMBERS AND MEMORY ADDRESSING

Microprocessors are generally programmed in a high-level language (e.g., FORTRON or C); a compiler program converts the high-level instructions to machine language, which is subsequently executed. Assembly language, more efficient but also more cumbersome, is used for small programs or to improve the efficiency of portions of programs compiled from high-level language; an assembler program converts the instructions to machine language. Rarely is machine language used to program a microprocessor because the instruction sets, which are long strings of 0 and 1's, become unwieldy for humans.

8.5.1   Hex Numbers

Since the basic word length in microprocessors is an 8-bit word called a *byte* (a poor acronym for "by eight"), it is convenient to express machine language instructions in the hexadecimal number system rather than in the binary system. Hex numbers are to the base 16—like binary numbers are to the base 2 and decimal numbers to the base 10.

The footnote to Section 7.4 on p. 248 gives examples of binary and decimal numbers. Similarly, we can state a hexadecimal number as

$$\cdots + x \cdot 16^3 + x \cdot 16^2 + x \cdot 16^1 + x \cdot 16^0 \tag{8.1}$$

where each x is one of 16 numbers: $0, 1, 2, 3, 4, 5, 6, 7, 8, 9, A, B, C, D, E, F$ (to avoid using double digits for the last 6 numbers, the letters A to F are used).

It is common to denote a hex number by the letter H (either at the end or as a subscript). Thus $0H = 0_H = 0_{16}$ is equal to 0 in binary (0_2) and 0 in decimal (0_{10}). Similarly $FH = F_H = F_{16} = 15_{10} = 1111_2$. The binary number 11011, which is used in the Section 7.4 footnote, can be expressed as a hex number by using (8.1); this gives

$$27_{10} = 11011_2 = 1 \cdot 16^1 + B \cdot 16^0 = 1BH = 1B_H \tag{8.2}$$

Hence binary 11011 is equal to $1B$ in hex.

In the following table we give the equivalent numbers in decimal, binary, and hexadecimal:

Decimal	Binary	Hex
0	0000	0
1	0001	1
2	0010	2
3	0011	3
4	0100	4
5	0101	5
6	0110	6
7	0111	7
8	1000	8
9	1001	9
10	1010	A
11	1011	B
12	1100	C
13	1101	D
14	1110	E
15	1111	F

This table shows that the 16 digits of the hex number system correspond to the full range of all 4-bit binary numbers. This means that 1 byte can be written as two hexadecimal digits, as is illustrated in the following table:

Byte	Hex number
00000000	00
00111110	3E
10010001	91
11011011	DB
11111111	FF

EXAMPLE 8.1   Express the following binary numbers in hex and decimal: (a) 10001, (b) 1010011110, and (c) 1111111111111111.

To express each binary number as a hex number, we first arrange the number in groups of four binary digits and identify each group with the equivalent hex number using the above tables. Thus

(a)

$$10001 = 00010001 = 0001\ 0001 = 11 \text{ or } 11_{16}$$

To find the decimal equivalent, we can use the hex number, which gives

$$1 \cdot 16^1 + 1 \cdot 16^0 = 17 \text{ or } 17_{10}$$

or the binary number, which gives

$$1 \cdot 2^4 + 0 \cdot 2^3 + 0 \cdot 2^2 + 0 \cdot 2^1 + 1 \cdot 2^0 = 17$$

(b) Similarly

$$1010011110 = 0010\ 1001\ 1110 \quad = \quad 29E \text{ or } 29E_{16}$$
$$2 \cdot 16^2 + 9 \cdot 16^1 + 14 \cdot 16^0 \quad = \quad 670 \text{ or } 670_{10}$$

$$1 \cdot 2^9 + 0 \cdot 2^8 + 1 \cdot 2^7 + 0 \cdot 2^6 + 0 \cdot 2^5 + 1 \cdot 2^4 + 1 \cdot 2^3 + 1 \cdot 2^2 + 1 \cdot 2^1 + 0 \cdot 2^0 = 670$$

(c) Similarly

$$1111\ 1111\ 1111\ 1111 = FFFF \text{ or } FFFF_{16}$$
$$15 \cdot 16^3 + 15 \cdot 16^2 + 15 \cdot 16^1 + 15 \cdot 16^0 = 65535 \text{ or } 65535_{10}$$
$$1 \cdot 2^{15} + \cdots + 1 \cdot 2^0 = 65535$$

8.5.2   Memory Addressing

From the above, we see that the largest 8-bit word (1111 1111) is *FF* in hex, the largest 12-bit word is *FFF*, and the largest 16-bit word is *FFFF*. A 2-byte word (16-bit word) can have $2^{16} = 65,536$ combinations. Hence, when used in addressing, a 2-byte address can locate 65,536 memory cells. The first location will have the address 0000 and the last *FFFF*, giving a total of 65,536 addresses (65,536 is obtained by adding the 0 location to the number $FFFF = 65535$, similar to the decimal digits 0 to 9, which are 10 distinct digits).

The digital codes for representing alphanumeric characters (ASCII) such as the letters of the alphabet, numbers, and punctuation marks require only 8 bits. In other words, 1 byte can hold one ASCII character such as the letter B, a comma, or a percentage sign; or it can represent a number from 0 to 255. Therefore, computers are structured so that the smallest addressable memory cell stores 1 byte of data, which can conveniently be represented by two hexadecimal digits.[17] Because a byte contains so little information, the processing and storage capacities of computer hardware are usually given in kilobytes (1024 bytes) or megabytes (1,048,576 bytes).

By addressable, we mean a cell has a unique address which locates its contents. If an address consists of 8 bits, the addressing capability or the largest number of cells which can be uniquely selected is $2^8 = 256$. This range is called the address space of the computer, and it does not need to have actual memory cells for all of these locations. A small computer could have an address space of 64kB; however, if the application only calls for 4kB, for example, only 4kB of memory need be included and the other 60kB of address space is not used. Sixty-four kilobytes (64K) memory,[18] which has $2^{16} = 65,536$ memory locations, each of which contains one byte of data or instruction denoted by a two-digit hex number, is illustrated in Fig. 8.9a. Each location has a 2-byte (16-bit) address, which is also given in hex form by the four-digit hex number. As shown, the addresses range from 0000_{16} to $FFFF_{16}$ with a particular memory location such as $FFFE$ containing byte 7A. If this is a RAM-type of memory, the control unit determines if information can either be read from or written to each memory location.

EXAMPLE 8.2 An Intel 8088 microprocessor may be interfaced with 64 K of memory. In a typical configuration the memory is divided into three blocks. For example, the first 8 K of memory could make up the first block and would be used for ROM purposes, the next block of 2 K would be used for RAM purposes, and the remaining 54 K of memory would be used as RAM for applications which the computer is intended to run.

Assume the memory structure is like that shown in Fig. 8.9a with the system bus consisting of a 16-bit address bus, A_0 through A_{15}, and an 8-bit bidirectional data bus, D_0 through D_7. ROM is enabled when an address is in the range of 0000_H to $1FFF_H$, which exactly addresses the first 8K of memory ($1FFF_H$ is equal to 8191_{10}). The address lines for ROM have therefore the range of values

	$A_{15} - A_{12}$	$A_{11} - A_8$	$A_7 - A_4$	$A_3 - A_0$
0000_H	0000	0000	0000	0000
$1FFF_H$	0001	1111	1111	1111

---

[17]Memory capacity is expressed in kilobytes (kB), which means thousands of bytes. However, the power of 2 closest to a thousand is $2^{10} = 1024$; hence 1 kB is actually 1024 bytes. Similarly, 64 kB $\approx 2^{16} = 65,536$ bytes, and one megabyte (1 MB) is actually 1 MB = 1 kB $\cdot$ 1 kB $\approx 2^{20} = 1,048,576$ bytes. The reason that a 1000-byte memory is not practical is that memory locations are accessed via the address bus, which is simply a collection of parallel wires running between the CPU and memory. The number of parallel wires varies with the word size, which is expressed by an integral number of bits. Hence a 10-bit address bus can access 1024 bytes, and a 30-bit bus could access $2^{30} = 1,073,741,824$ bytes ≈ 1 GB. Note that in "computerese" capital K is used to represent one kilobyte as 1 K, one megabyte as 1 MB, and one gigabyte as 1 GB.

[18]If it is a read/write type of memory (RAM), each bit of memory is similar to the 1-bit read/write cell shown in Fig. 7.31. A 4×4 RAM cell is shown in Fig. 7.32. Similarly, a 64 kB RAM module would be configured as 65,536 addressable cells, with each cell 8 bits wide. The addressing would be done with a 16-bit (16 parallel wires) bus, feeding a 16×216 decoder as shown in Fig. 7.33. A 64 kB memory module would therefore have $2^{16} \cdot 8 = 524,288$ 1-bit cells.

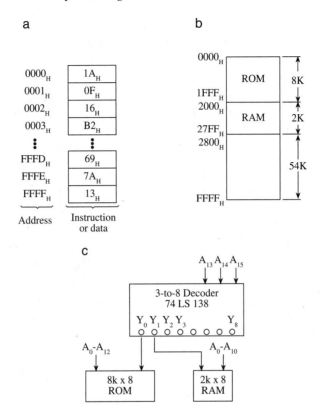

FIGURE 8.9 (a) Sixty-four kilobytes memory. Each one of the 65,536 memory locations can contain 1 byte of data or instruction. (b) Memory map for 64K of RAM. (c) A 3-to-8 decoder.

where bits $A_{15} - A_{12}$ select the first 8 K block and bits $A_{11} - A_0$ select locations within the first block.

The next 2 K RAM block is selected when the address is in the range 2000_H through $27FF_H$ ($27FF_H - 2000_H = 2047_{10}$ bytes $= 2$ K). The address lines for RAM therefore have the range of values

	$A_{15} - A_{12}$	$A_{11} - A_8$	$A_7 - A_4$	$A_3 - A_0$
2000_H	0010	0000	0000	0000
$27FF_H$	0010	0111	1111	1111

where bits $A_{15} - A_{12}$ select the second 2 K block and bits $A_{11} - A_0$ select locations within the second block. The addressing for the remaining 54 K space would be done similarly.

Figure 8.9*b* shows the memory map of the 64K memory and Fig. 8.9*c* shows the 74*LS*138, which is a 3-to-8 decoder. The select inputs of the decoder are driven by A_{15}, A_{14}, and A_{13}. When these are LOW, which they are when the address is in the range from 0000 to 1FFF, output Y_0 selects or enables the 8 K ROM. When A_{15} and A_{14} are LOW and A_{13} is HIGH, which they are when the address is in the range from 27FF to 2800, output Y_1 selects the 2 K RAM, which is then driven by $A_0 - A_{10}$. ■

Figure 8.9a shows a sequence of bytes of instruction or data with a 2-byte address. Because a computer is primarily a serial processor, it must fetch instruction and data sequentially, i.e., one after the other. The instruction format in a computer breaks the instruction into parts. The first part is an operation (load, move, add, subtract, etc.) on some data. The second part of the instruction specifies the location of the data. The first, action part of the instruction is called the operation, or *OP CODE*, which is an abbreviation for *operation code*, and the second part is called the *operand*, which is what is to be operated on by the operation called out in the instruction. In the program memory, the operation and the operand are generally located in separate locations (except for immediate data). The next example demonstrates a typical computer operation.

EXAMPLE 8.3   In this classic example, we will show the steps that take place in a computer running a word-processing program when key "A" is depressed and the letter A subsequently appears on the monitor screen.

After the computer is booted, the instructions for word processing are loaded into RAM, that is, into the program memory in Fig. 8.10. The data memory stores intermediary steps such as which key was just depressed. Actions that must take place are:

(1) press the "A" key
(2) Store the letter A in memory
(3) Display letter A on the monitor screen

Figure 8.10 shows the steps for the computer to execute the INPUT–STORE–OUTPUT instructions which are already loaded in the first six memory locations. Note that only three instructions are listed in the program memory:

(1) INPUT data from Input Port 1
(2) STORE data from Port 1 in data memory location 200
(3) OUTPUT data to Output Port 10

The remaining three memory locations are the data addresses. For example, the first instruction in memory location 100 contains the INPUT operation while memory location 101 contains the operand stating from where the information will be inputted. Recall that the microprocessor determines all operations and data transfers while it follows the fetch–decode–execute sequence outlined in Section 8.4. The student should try to identify how the CPU always follows the fetch–decode–execute sequence in each of the three parts of the program. The steps to execute this program will now be detailed.

The first part will be to carry out the INPUT part of the instructions.

(1) The first step is for the CPU to place address 100 on the address bus and use a control line to enable the "read" input on the program memory. "Read" enabled means that information stored in program memory can be copied—see

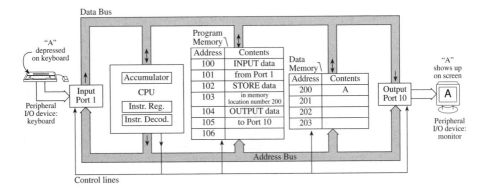

FIGURE 8.10 Computer operation as it executes instructions to display the letter A when key "A" is depressed.

Fig. 7.32. (In this step information flows from address bus to program memory.)

(2) "INPUT data," the first instruction, is placed on the data bus by the program memory. The CPU accepts this coded message off the data bus and places it in the instruction register to be subsequently decoded by the Instruction Decoder to mean that the CPU needs the operand to the instruction "INPUT data." (In this step information flows from program memory to data bus to CPU.)

(3) The CPU places address 101 on the address bus. The "read" input of the program memory is again enabled by the control line. (In this step information flows from address bus to program memory.)

(4) The "operand from port 1," which is located at address 101 in the program memory, is placed on the data bus by the program memory. The CPU accepts this coded message off the data bus and places it in the instruction register where it is subsequently decoded by the "instruction decoder." The instruction now reads, "INPUT data from Port 1." (In this step information flows from program memory to data bus to CPU.)

(5) The CPU now carries out the instruction "INPUT data from Port 1"; it opens Port 1 by using the address bus and control line to the Input unit. (In this step information flows over the address bus from CPU to Port 1.)

(6) The coded form for "A" is placed on the data bus and transferred to and stored in the accumulator register. (In this step information flows over the data bus from Port 1 to the accumulator.)

This completes the first part. The second part is to carry out the STORE instruction.

(7) After the program counter is incremented by 1, the CPU addresses location 102 on the address bus. The CPU, using the control lines, enables the "read" input on the program memory. (In this step information flows from address bus to program memory.)

(8) The program memory reads out the instruction "STORE data" on the data bus, which is then placed into the instruction register by the CPU to be decoded. (In this step information flows from program memory to data bus to CPU.)

(9) After decoding the "STORE data" instruction, the CPU decides that the operand is needed. The CPU places 103, which is the next memory location, on the address bus and uses the control lines to enable the "read" input of the program memory. (In this step information flows from address bus to program memory.)

(10) The program memory places the code for "in memory location 200" onto the data bus. This operand is accepted by the CPU and stored in the instruction register for decoding. The CPU now decodes the entire instruction "STORE data in memory location 200." (In this step information flows from program memory to data bus to CPU.)

(11) To execute the decoded instruction, the CPU places address 200 on the address bus and uses the control lines to enable the "write" input of the data memory. "Write" enabled means that data can be copied to memory—see Fig. 7.32. (In this step information flows from address bus to data memory.)

(12) The coded form for letter "A," which is still stored in the accumulator, is now placed on the data bus by the CPU. The letter "A" is thereby written into location 200 in data memory. (In this step information flows from data bus to data memory.)

This completes the second part. The third part is to carry out the OUTPUT instruction. It should be noted that an instruction such as "STORE data in memory location xxx" transfers data from the accumulator to address location xxx in RAM. This data are now contained in both RAM and the accumulator. The contents of the accumulator are not destroyed when data are stored.

(13) As the program counter increments, the CPU fetches the next instruction. The CPU sends out address 104 on the address bus and, using the control line, enables the "read" input of the program memory. (In this step information flows from address bus to program memory.)

(14) The program memory reads out the instruction code "OUTPUT data" onto the data bus; the CPU accepts this coded message and places it in the instruction register. (In this step information flows from program memory to data bus to CPU.)

(15) The CPU interprets (decodes) the instruction and determines that it needs the operand to the "OUTPUT data" instruction. The CPU sends out address 105 on the address bus and uses the control line to enable the "read" input of the program memory. (In this step information flows from address bus to program memory.)

(16) The program memory places the operand "to Port 10," which was located at address 105 in program memory, onto the data bus. This coded message (the address for Port 10) is accepted by the CPU and is placed in the instruction register. (In this step information flows from program memory to data bus to CPU.)

(17) The instruction decoder in the CPU now decodes the entire instruction "OUT-

PUT data to Port 10." The CPU activates Port 10 using the address bus and control lines to the OUTPUT unit. (In this step information flows from address bus to OUTPUT unit.)

(18) The CPU places the code for "A," which is still stored in the accumulator, on the data bus. The "A" is now transmitted out of Port 10 to the monitor screen. (In this step information flows from OUTPUT unit to monitor.)   ∎

8.5.3   Cache Memory

The speed of a computer is limited not only by how quickly the CPU can process data but also by how quickly it can access the data it needs to process. Overall speed of a computer is determined by the speed of the bus and the speed of the processor, both measured in MHz. Each computer uses a memory system that provides information to the CPU at approximately the rate at which the CPU can process it. But what happens when one upgrades to a faster CPU?[19] The memory system continues to provide information at the original rate, causing the new CPU to starve for data from a memory system designed for a less demanding processor. *RAM cache*, or simply *cache*, can alleviate this problem. A cache is a small amount of high-speed memory between main memory and the CPU designed to hold frequently used information for fast access. The cache stores information from main memory in the hope that the CPU will need it next. The result is that the newer CPU can reach maximum speed. Another example is MMX (multimedia extensions), which is a new technology that uses a large-scale cache, thus reducing the amount of code that multimedia applications need.

Thus, generally speaking, a bottleneck in computers is the bus system. While executing a program, the CPU frequently needs the same information over and over again which must be fetched repeatedly from RAM memory using the I/O bus.[20] If such much needed information could be stored more accessibly, as, for example, in memory that is placed directly on the CPU chip, access time could be greatly reduced. Cache memory is such memory that is built into the processor. Cache memory is the fastest type available due to the smaller capacity of the cache which reduces the time to locate data within, its proximity to the processor, and its high internal speed.[21] As the processor reaches out to the memory for data, it grabs data in 8 K or 16 K blocks. Before reaching out to memory a second time for new data, it first checks the cache to see if the data are already in cache. If so, the cache returns the data to the processor. This significantly reduces the number of times the processor has to reach into the slower main memory (RAM). When data are found in cache it is called a "hit." For example, 16 K

---

[19]Computer progress is often stated in terms of *Moore's law*, coined by Intel's chairman Gordon Moore: The density of transistors on chips doubles every 18 months while their cost drops by half. Note that the 8088 carried about 10,000 transistors, the 80286 about 100,000 transistors, the 386 about 500,000, the 486 about 1 million, the Pentium about 3 million, and the Pentium Pro about 5 million.

[20]The trend in bus design is to use separate buses for memory (called the data bus, memory bus, or main bus), for graphics (AGP), and for external devices (SCSI, IDE).

[21]The CPU and its internal processor bus frequently run two (referred to as "double clocking") to four times faster than the external I/O bus.

of internal cache provides a "hit rate" of about 95%, which can considerably increase computer speed.[22] Level 1 cache is 8 K, 16 K, or 32 K; is built into processor; is not upgradable; and is the fastest. Level 2 (L2) cache is found in 64 K, 128 K, 256 K, and 512 K amounts; is separate from the processor and is added onto the system board; is upgradable; and is the next fastest. Generally, Level 2 cache has a rate of diminishing returns beyond 256 K.

8.6   OPERATING SYSTEMS

There are many ways to characterize an operating system (OS). For example, in a tutorial session on basic technology given by IBM, computers were classified as needing

- Electricity (power)

- A method of instruction (input)

- Defined instructions (programs)

- A device to do the work (processor)

- A place to do the work (memory)

- A place to save the work (output)

The input could be keyboard, mouse, disk drive, scanner, CD-ROM, or LAN. Programs are the operating system and application programs such as Word Perfect and Excel. Popular processors are the Pentium and Power PC. Memory is RAM and cache. Output is the display, printer, disk drive, and LAN. A computer system was then compared to a kitchen in which the processor was the cook; the application program the recipe; the input device the refrigerator and the cupboards; the input data the water, flour, milk, eggs, and sugar; the memory the table and counter space; the bus the path between refrigerator, cupboards, oven, table, and counter space; the output device the oven; and the output data the cake. In this scenario, the operating system would be the instructions where and how to find the recipe, the oven, and the ingredients; how to light the stove and adjust its temperature; how to fetch and mix the ingredients; and how to place the cake in the oven and how to take it out again when it is baked. In other words, the operating system organizes and facilitates the cooking chores.

From the above list, operating systems are part of the programs category. We can state that operating systems such as Apple's Mac OS or Microsoft's DOS and Windows

---

[22]To state it in another way, the increase in throughput is determined by the "hit ratio"—the fraction of times the processor finds the desired information in the cache. The reason for performance increases with the addition of cache is that the microprocessor can keep its pipeline full, allowing for faster and more efficient processing. The microprocessor first checks its internal cache, then L2 cache, and finally main memory (DRAM, or dynamic RAM) for instructions. Because cache memory is faster than DRAM, it can be accessed more quickly, thus helping keep the pipeline full.

systems are the critical layer of technology between a computer's microprocessor and its software. It is what allows—or prevents—a computer from running a particular software application.

8.6.1   Controllers and Drivers

Early computers had no operating system. All instructions had to be manually inserted. As computers rapidly evolved and included a large range of peripherals such as a hard drive, floppy drive, monitor, printer, etc., it was realized that freeing the programmer of routine housekeeping chores for these peripherals could increase programmer productivity. In addition to relieving the burden on programmers, it was realized that CPU performance could be significantly improved if the CPU could be freed from housekeeping routines by adding hardware and circuitry to monitor and control peripherals. For example, the routines which are needed for the hard drive to run properly involve file management, formatting, and initializing.[23] Instead of having the CPU control such a hard disk and floppy disk drive directly, and have so much of its system resources tied up in disk overhead, it makes more sense to make use of a dedicated peripheral support chip (actually several chips mounted on a small board), thus freeing the CPU for other, more urgent tasks. Such a board, which contains the drive electronics and is mounted directly on the hard drive and on the floppy disk drive, is referred to as a disk controller. In general, a controller can be defined as hardware required by the computer to operate a peripheral device. Figure 8.11*b* shows typical interface signals required between a CPU, disk controller, and disk drive. The transferring of files between memory and disk can often be speeded up by including an additional DMA (direct memory access) controller, as indicated in the figure. DMA is a method by which data are transferred between a peripheral device and RAM without intervention by the CPU. Each controller requires software code, called a driver, to interface properly between the peripheral and the CPU. That is why drivers must be updated when adding a new printer or hard drive to your computer. Typically, your operating system contains a list of drivers, which should include those necessary for your new printer or hard drive. If not, software that comes with new peripherals should include new drivers that can be installed. A similar situation exists for every peripheral device. A monitor, for example, needs a video card (equivalent to a controller) mounted directly on the motherboard and a driver to be used

---

[23]File management includes opening and closing files, copying, deleting, changing file names, reading and writing files, and so on. Formatting includes writing address marks, track and sector IDs, and gaps on a disk. Initializing includes placing system information on a disk such as bitmaps onto the outermost track #00, followed by identification of all bad tracks. Figure 8.11*a* shows a view of the organization of the concentric tracks and the pie-shaped sectors of a disk. The magnetic read/write head can determine which track it is on by counting tracks from a known location, and sector identities are encoded in a header written on the disk at the front of each sector. Saved data on a disk have a directory entry which is an index that contains the name, location, and size of each piece of datum, and then the data themselves. A 1.2 MB floppy disk, for example, has 80 tracks, 15 sectors per track, and 512 bytes per sector on each side of the disk. It spins at 360 rpm. Hard disks contain many such disks which are rigid (called *platters*), spin at 3600 rpm or higher, and have a higher storage density than standard floppy disks.

a

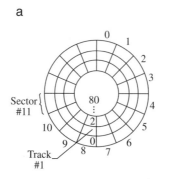

b

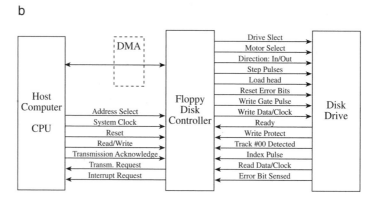

FIGURE 8.11 (a) Disk drives arrange recorded information on the disk in concentric bands called *tracks*; each track is divided into consecutive sectors. (b) Disk drive interface signals.

as part of the operating system. In general we can state that operating systems and advanced hardware for interfacing to peripherals evolved from a need for more efficiency in the use of computer systems.

Thus, it should be clear by now that the operating system is actually many files which are dedicated to a smooth interaction between the CPU, its peripherals, and a user. In fact the OS is a remarkably complex set of instructions that allocate tasks to be performed by the CPU, memory, and peripheral devices when the computer is executing a program. The CPU is directed by the operating system in the storage, loading, and execution of programs and in such particular tasks as accessing files, controlling monitors and memory storage devices, and interpreting keyboard commands. Operating systems can be very simple for minisystems to very sophisticated or complex for large general-purpose systems. The size and complexity of the operating system depend on the type of functions the computer system is to perform. For example, in the operation of large mainframe computers, the performance of the operating system is often as important as the performance of the computer hardware. Indeed the operating system software may be nearly as expensive as the computer hardware.

DOS

DOS (Disk Operating System) was one of the first operating systems for PCs. Its early incarnation was a rudimentary software system which basically managed files stored on disk. It maintained a file directory, kept track of the amount of free space left on disk, and could create new files as well as delete and rename existing files. As already made clear, DOS itself is a file which is loaded in at startup time by a bootstrap program in ROM. A *bootstrap program* is an initialization program which is capable of loading the OS into RAM. It appears that such a process is able to "pull itself up by its own boot-straps," and hence the term "booting up" when starting up an inactive computer. Once DOS is loaded into RAM, the user can type commands on a keyboard, instructing the OS to perform desired functions. Typing commands is now considered a cumbersome way to address a PC operating system. Although limited, DOS evolved into a sophis-ticated system which requires configuring, with most of this information stored in two files. The CONFIG.SYS file is a text file that contains commands to configure computer hardware components such as memory, keyboard, mouse, printer, etc., so DOS and the application programs can use them. This file must be modified whenever new hardware is added or when customizing the way DOS uses hardware, memory, and files. The AUTOEXEC.BAT file is a batch file that DOS runs immediately after carrying out the commands in the CONFIG.SYS file. The AUTOEXEC.BAT file contains any commands that need to be carried out when the system is started, for example, commands that define the port the printer is connected to, clear the screen of startup messages, or run a favorite menu program. These files are typically located in the root directory of the startup disk which usually is called drive C. Unless specified otherwise, DOS carries out the commands in both the CONFIG.SYS and the AUTOEXEC.BAT files each time the computer is started. The CONFIG.SYS file can be edited to add and change commands that configure the system. Because the settings in the CONFIG.SYS file control basic components of the system, such as memory and peripheral devices, if a new setting in this file is incorrect, the system might be unable to start correctly.

Macintosh Operating System

The Mac OS was the first widely used operating system to use a graphical user inter-face to perform routine operations. In the Mac OS, the user started programs, selected commands, called up files, etc., by using a device called a mouse to point to symbols (icons). Since the GUI's windows, pull-down menus, and dialog boxes appeared the same in all application programs, it allowed common tasks to be performed in the same manner in all programs. It was realized immediately that this had many advantages over interfaces such as DOS, in which the user typed text-based commands on a keyboard to perform routine tasks.

Macintosh computers have three layers of software:

(1) Application programs (like a word processor or spreadsheet)
(2) User-interface software
(3) Operating-system software

The last two form what is popularly known as the Mac OS (strictly speaking, user-interface software is not part of the operating system). On the Macintosh one interacts with application programs through the user-interface software. This approach, pioneered by Apple, gives Macintosh computers a consistent look-and-feel among applications, which makes all applications easier to learn and use. Other computers forced users to interact differently with different programs, because every application incorporated its own user interface. Now non-Macintosh computers have developed similar user-interface software, the most popular being Microsoft's Windows.

The user-interface software is called the Macintosh Toolbox. The Toolbox contains software modules called managers, which handle pull-down menus, windows, buttons, scroll bars, fonts, and other standard parts of the Mac's graphical user interface. The Toolbox also contains QuickDraw, the program that draws everything that appears on the Mac's video display. In typical use this software operates in the background and a user normally is not aware of its presence. Many of the tools in the Toolbox are stored in ROM; others are stored on a disk in the System Folder and then copied to RAM when the Mac is started.

The Toolbox and the application programs access the Mac's hardware through a third layer of software, the operating system, which provides routines that allow basic low-level tasks to be performed. The Toolbox is a level above the operating system and as stated above implements the standard Macintosh user interface for all applications. The Toolbox calls the operating system to do low-level operations such as controlling disk drives, the keyboard, the mouse, the video display, and all other parts of the system. The Toolbox thereby insulates your applications from low-level details of interacting with the computer's peripherals (e.g., data-storage hardware). The Mac is unique among popular computers because its users do not even know the Macintosh Operating System exists. On computers that use other operating systems, such as MS-DOS or Unix, users type commands to start programs, organize disk contents, and change system settings. On the Mac, many of these things are done with the Finder and control panels by gesturing with the mouse. The Finder is an application program that is launched automatically when the computer starts and with which the user interacts most of the time. It should not be confused with the Mac OS: the Finder is a shell that displays and manages the Desktop and its files and hard drives (in that sense it is similar to the Program Manager and Explorer in Windows and to Command.com in DOS). The Finder uses the Mac OS to do this, and it helps one to visualize the Mac's files and resources. It is sufficiently sophisticated to allow multiple applications to run and is about as close to true multitasking as one can get without a time-sliced OS *kernel*.[24] Each application has its own window. On the Macintosh desktop the multiple windows appear overlaid, one on top of the other. The currently running program has its window on top with the others laying like several sheets of paper on the desktop. Even though the Finder is not really part of the Macintosh operating system, it is such an important piece of the Macintosh graphic

---

[24] A kernel is the set of programs in an operating system that implement the most primitive of that system's functions.

user interface that it is sometimes difficult to tell where the Finder ends and the system software begins.

The System Folder contains all the system software. A disk that contains a System Folder is a startup disk. In addition to the Finder, this folder contains the Control Panels and Extensions that let the user change various Macintosh features. Control Panels are small programs which add to or modify the operating system to give the Macintosh a certain feel. For example, memory allocation, sound, and desktop patterns are governed by Control Panels. Numerous small programs such as screen savers, RAM doublers, etc., can be purchased and added as Control Panels or as Extensions; this is the current method for adding capabilities to the system software. Extensions are files that contain executable code and like Control Panels are loaded into memory at startup time. They include drivers, which make it possible for the computer to use a certain printer or other device, and programs that add features to the Finder or the system software.

The software routines that make up the Macintosh Toolbox and the Macintosh Operating System reside mainly in read-only memory (ROM), provided by special chips contained in every Macintosh computer, unlike DOS or other literal OS's (such as UNIX) which are more disk resident. This has allowed Apple Computer, Inc., to maintain complete control over the Mac OS and forced developers to give most Macintosh applications the same look and feel. When an application program calls a Toolbox routine, the Mac OS intercepts the call and executes the appropriate code contained in ROM. Instead of executing the ROM-based code, the OS might choose to load some substitute code into the computer's random-access memory (RAM); then, when a computer application calls the routine, the OS intercepts the call and executes that RAM-based code. RAM-based code that substitutes for ROM-based code is called a patch. Patches are also used to correct known bugs in the operating system.

Windows

Microsoft's Window 3.1 added a graphical users interface to DOS. It is an application program that sits on top of DOS and allows the user to access DOS not by typing DOS commands but by clicking on icons, similar to a Macintosh computer. Since it was an emulation program (a program that simulates the function of another software or hardware product) it suffered from all the shortcomings that DOS initially had. It did allow the user to flip between DOS and the Windows GUI, with DOS using the Command.com file which holds all the DOS commands to be typed in and Windows using the Program Manager to organize and display the windows and application icons on the desktop.

Microsoft followed up with Windows 95/98, which was a good improvement over Windows 3.1 because this operating system, much like the Mac OS, was not an emulation program but was designed from the beginning to interact with applications programs through graphical user-interface software and by the use of a mouse. Windows 95/98 has much the same feel as the Mac OS. It has a Control Panel and Explorer (Start Button is part of Explorer) which is similar to the Finder for the Macintosh. Both operating systems support true Plug and Play, which is the ability to attach a new peripheral with minimum setting of the files (drivers) that are needed to have the new peripheral work properly. In comparison, DOS required detailed settings of files such as AUTOEXEC.BAT and CONFIG.SYS whenever a new device was added to the computer. Windows 2000 is a further refinement of the Windows 95/98 family.

In 1993, Microsoft expanded Windows to Windows NT, which has gained acceptance as an operating system for workstations and servers. It is a powerful multitasking system that can handle UNIX in addition to MS-DOS.

The UNIX Operating System

Modern personal computers use relatively simple operating systems with the main function of managing the disk drives and the user's files, providing access to software such as word processing and spreadsheets, and supporting keyboard input and screen display. In that sense an OS is basically a collection of resource managers. On the other hand, it is clear that computers could not be as useful as they are today without an efficient operating system. A trend in operating system development is to make operating systems increasingly machine-independent. For example, users of the popular and portable UNIX operating system need not be concerned about the particular computer platform on which it is running. Hence, the user benefits by not having to learn a new operating system each time a new computer is purchased. In comparison to the operating systems discussed above (with the exception of Windows NT), UNIX is a complex and powerful operating system primarily for workstations and for networked computers and servers.

The UNIX system has essentially three main layers:

(1) The hardware
(2) The operating system kernel
(3) The user-level programs

A kernel is the set of programs in an operating system that implement the most primitive of that system's functions. In UNIX, the kernel hides the system's hardware underneath an abstract, high-level programming interface. It is responsible for implementing many of the facilities that users and user-level programs take for granted. The kernel also contains device drivers which manage specific pieces of hardware. As already pointed out above, the rest of the kernel is to a large degree device-independent. The kernel is written mostly in the language C, which was specifically written for the development of UNIX.

It is a popular myth that UNIX kernels "configure themselves." Whenever a new machine is added or the operating system upgraded, the kernel should be reconfigured, a job that is usually the province of the network administrator. Even though most versions of UNIX come with a "generic" kernel, already configured, it nevertheless should be reconfigured for reasons of efficiency. Generic kernels are designed for any kind of hardware and hence come with a variety of device drivers and kernel options. Efficiency can be increased simply by eliminating unneeded drivers and options. To edit configuration files or write scripts, UNIX uses text editors such as "vi" and "emacs" with which one must become familiar.

Most of us, at one time or other, will have to use some basic UNIX commands. For example, when logging onto a network to read one's e-mail, one might use the "elm" command to bring up the interactive mailer program. In UNIX, there is an excellent on-line instruction manual which can be accessed using the "man" command. To illustrate, at the UNIX prompt ($), typing the command "man" followed by a command for which one needs information, such as "man man," will bring up a message, "man-display reference manual pages; find reference pages by keyword." Similarly, entering "man ls" will bring up, "ls-list the contents of a directory." A short list of user commands which an occasional UNIX user needs follows:

command	result
telnet/rlogin	log into a remote system
cd	change directory
ls	list names of all files in current directory
rm	remove file
rmdir	remove directory
mv	move or rename file or dir
more	browse through a text file
cat	reads each filename in sequence and displays it on screen
who	who is logged in on the system
finger	display information about users
last	indicated last login by user or terminal
kill	terminate any running process
ps	display the status of current processes
ftp	file transfer protocol
lp/lpr	send a job to the printer
chmod	change the permissions of a file
vi/emacs	standard UNIX visual editor
man	displays reference manual pages
cd plonus	go to directory plonus
cd	goes to login/home directory
cd ..	up one directory
ls -l	long list
ps -ef	list all processes
man man	explains man command

8.6.2 Operating System Stability

What makes an operating system *stable*, i.e., not prone to crashes? It is protected memory, virtual memory, and object-oriented design. A modern and sophisticated operating system such as UNIX should possess the following features:

Preemptive multitasking: Multiple applications run smoothly. No single application is allowed to take over the entire system. In preemptive multitasking, the OS—

not the individual programs—allocates CPU time across multiple programs. The OS manages preemptive multitasking by giving each program the CPU's attention for a fraction of a second (called a *time slice*). Typically the programs follow one another in sequential order. Because all this happens within fractions of a second, programs appear to the user to be running simultaneously.

Protected memory: Applications that crash cannot crash other applications or the rest of the computer. With unprotected memory, all programs run in the same memory space, so when one program goes down they all do, like dominoes on your desktop. In an operating system with protected memory, programs run in their own memory space, meaning that the OS reserves a portion of memory to act on each program's data instructions.

Symmetric multiprocessing: Numerous independent tasks can execute at the same time (requires two or more processors to be installed).

Multithreaded user interface: Holding the mouse button down does not freeze the rest of the computer. Windows continue to be updated when moved. Processor efficiency is boosted by multithreading in which a program or a task within a program is divided into threads, or processes, that can run simultaneously. An example is the ability to print at the same time one is scrolling, rather than waiting for the printing to finish before one can move on. Compare that with multitasking, in which entire programs run in parallel.

Virtual memory: A true virtual memory system provides an application with as much memory as it needs, automatically, first by parceling out RAM and then by using space on a hard disk as additional RAM.

Object-oriented design: A benefit of an object-oriented application-programming interface is that not as many APIs (application-programming interfaces) are needed, making it easier for programmers to develop programs than doing so for the Mac OS and Windows. Fewer APIs makes such programs smaller, faster-launching, and—due to their simpler design—more stable.

8.6.3   The Network Computer

In four decades, computers have gone through four major generational shifts—first mainframe computers, then minicomputers, then personal computers, and now the handheld computers and networked computers focused on the Internet. However, thus far NC's have not proved successful.

The Net computer (NC) is a desktop machine with no disk drive or stored programs of its own. NCs, like terminals, rely on remote computers called servers for software. Software and files are executed on these servers with which the NC can communicate over computer networks. The same commands and controls that one uses to retrieve material from one's computer disk drive are used to get information from the Internet, once one is connected. This is analogous to the older mainframe–terminal technology except that the mainframe is replaced by the Internet and its servers—a sort of distributed "mainframe." However, this approach is much more sophisticated than that of the "dumb terminal" days of the 1960s and 1970s when no one had their own disks or software. The NC's main attractions are lower cost (usually less than $500), less likely to be made obsolete by new technology, elimination of the need to periodically update operating systems and application software (these would reside on the remote servers), and, because the server stores and executes programs, with the results appearing on the NC, only less than 4 MB of RAM memory are needed on the user's computer. Finally, in a corporation it provides centralized control over computing resources.

The network computer became a reality with the arrival of *Java*, SUN Microsystem's open, cross-platform scripting language designed to run on multiple operating systems. Java is designed specifically for the Internet and network-centric applications. Java breaks down the wall of software incompatibility and works as an operating system for NCs, giving users the option of downloading and running applications written for any system. The object-oriented language operates independently of any other operating system or microprocessor; that is, it would work equally well on a NC, PC, and Macintosh machine.[25] A computer language such as Java allows software to be broken down into small programs called *applets* that can be downloaded as needed by a network user. For example, a word processor would begin by downloading a small program to handle text, and another applet would be downloaded to format the text and run spelling checks. Other of these small programs would be downloaded as each function is needed. When the writing is done, the network computer would pluck a program for saving the text just produced and then by use of another applet store the actual document on an Internet server rather than on a hard drive or disk. To the user, this effort would be the same as if the software was contained on the desktop computer, rather than the Internet. As each new Java applet loads into the NC's limited 4 MB memory, the software that had been there before would be overwritten and reloaded when needed next.

At present the future of the network computer is uncertain as standard computers are now available in the same price range that made the NC initially promising. An even cheaper implementation of the NC, the $100 WebTV devices, can bring word processing, e-mail, and Web access to ordinary household TV sets.[26]

---

[25]The object-oriented programming language Java is designed to run on any system in a network. The premise underlying Java's development is that all computers can be described generically, that is, in terms of general functions like display and input capability. When a computer is described in this way, it is called a virtual machine. Java's developers realized that if all computers can be described in terms of general functions, then programs that are written using only general functions could run on all computers. Turning the general functions into the specific tasks would be a simple job for the computer on which the program ran. For a computer to run a Java program, then, it need only have a so-called *interpreter* to translate the Java code into instructions it can execute directly. Unlike current Windows or Macintosh applications, Java programs do not care what the computer's underlying operating system software or processor is. It thus frees the user from a particular type of system software.

[26]A telephone line plugs into the built-in 56 kbps modem of WebTV, which allows Web access using the TV as monitor. With the addition of a keyboard, WebTV's e-mail, which has built-in editing, including search, copy, and paste, can be used in place of a word processor by e-mailing yourself. And, as WebTV has a printer port, hard copies can be made.

8.7   SUMMARY

The material covered in this chapter serves two purposes. It shows the digital computer to be a crowning achievement of electronics. Even though electronics in its beginning was strictly analog, with the advent of high-speed circuits, digital techniques quickly evolved and are now dominant with a wide range of digital devices available. The second reason for the choice of chapter material and the way it is arranged is to give the student a basic knowledge about the digital computer so he or she is able to interact intelligently with others who are experts in this area. Therefore, the chapter is structured to focus on those parts of a computer with which the user interacts. To be reasonable, one cannot expect coverage on digital computers beyond basics in a small chapter such as this, when there are many thousands of detailed volumes devoted to this subject.

The chapter began with a brief history of the computer and the concept of the stored program. Programming languages were shown to be tools for communicating with a computer. The components of a computer were then identified. At the heart is the microprocessor which contains logic and control units and a small amount of on-chip memory. RAM memory, which is external to the CPU, is the short-term memory where the program and data are stored when executing an application program such as a word processor. It is volatile and before the computer is shut off, the work performed must be stored on a permanent medium such as a magnetic hard disk. I/O devices or peripherals (printers, modems, etc.) extend the capabilities of computers, and powerful systems can be created this way. The bus system, which uses a ribbon containing many parallel conducting wires, solves a major difficulty of too many internal connections even in a small computer. It does this by having all subsystems connected at all times but time-multiplexing the flow of information to allow different units to use the same bus at different times. The chapter closed with a study of operating systems, which act as a computer's manager that controls the basic functions of a PC. There are two primary categories of software: operating systems and applications. Operating-system software is designed to perform system and computer management utility and/or "housekeeping" functions such as directing the central processor in the loading, storage, and execution of programs and in accessing files, controlling monitors, controlling memory storage devices, and interpreting keyboard commands. Examples of operating systems include MS-DOS, Windows 95/98/2000/NT, Mac OS, Xenix, SCO-Unix, Solaris, SunOS, HP-UX, IRIX, Unix, Novell, and Linux.

Problems

1. Explain the difference between computer science and computational science.
2. What is the difference between a microprocessor and a microcontroller?
3. Discuss the difference between an assembler and a compiler.
4. Why do assembly language instructions execute faster than high-level language instructions?
5. Is programming in a high-level language dependent on the type on microprocessor used?

6. Is a compiler microprocessor-dependent?

7. Name the three main components of a computer system. Choose from auxiliary memory, auxiliary storage, CPU, floppy disk, I/O, memory, printer, and tape.

8. How many distinct 8-bit binary words exist?

9. If a microprocessor operating at 300 MHz takes four clock cycles to complete an instruction, calculate the total time it takes to complete the instruction.
   Ans: 13.3ns.

10. State typical access times for RAM memory and for hard disk memory. How much faster is RAM memory than disk memory?

11. A 16×8 memory array is shown in Fig. 8.5. Sketch an 8×8 memory array.

12. Consider the data bus and the address bus. Which is bidirectional and which is unidirectional? Explain why each has its particular directional character.

13. Explain the function of tristate logic in the bus system.

14. Why is a television set inadequate when used as computer monitor?

15. Compare the bandwidth of a television set to the minimum bandwidth required for a computer monitor.

16. If a modem can operate at a speed of 56 kbps, how long will it take to download a 4 MB file.
    Ans: 9.5 min.

17. A standard for asynchronous serial communication between a computer and a modem is RS-232. It defines a 25-pin connector. Communication between computer and modem can be with as few lines as three. Name these lines.
    Ans: Tx (transmit), Rx (receive), SG (signal ground).

18. Characterize the IEEE-488 interface.

19. Characterize the SCSI interface.

20. An interface is the point of meeting between a computer and an external entity, whether an operator, a peripheral device, or a communications medium. An interface may be physical, involving a connector, or logical, involving software. Explain the difference between physical and logical and give examples.

21. With respect to interfacing generally, what is meant by the term 'handshaking"?

22. Discuss the advantages and disadvantages of interrupts.

23. For each instruction in program memory, what kind of a sequence does the CPU follow?

24. Explain why program execution in a sequential machine follows the fetch–decode–execute cycle.

25. How can the fetch–decode–execute sequence be sped up?

26. Which bus does the microprocessor use when accessing a specific memory location?

27. Which bus is used when transferring coded information from microprocessor to data memory?

28. If the microprocessor decodes an instruction which says to store data in memory, the data would come from the accumulator or temporary register?

29. Which register keeps track of where one is in the program?

30. Which register keeps track of what type of instruction is being executed?

31. Which register keeps track of the result of the current calculation or data manipulation?

32. Which register keeps track of the values of the operands of whatever operation is currently under process?

33. Translation of the binary commands into specific sequences of action is accomplished by a _____?

34. The collection of registers, instruction decoder, ALU, and I/O control logic is known as the _____?

35. Electronics (such as A/D and D/A converters, buffers, and drivers) that match particular input and output devices to the input–output pins of the microprocessor chip are called _____?

36. Data, address, and control multibit transmission lines are called _____?

37. The larger the number of bits (the wider the bus), the faster the computer can become because fewer calls to memory are required for each operation. T or F?

38. Describe the difference between a data bus and an I/O bus.

39. Is a computer with a word length of 8 bits restricted to operands within the range 0 to 255?

40. If we have 256 memory locations, how many address bus lines are needed? How many memory locations can be selected using a 16-bit address bus?

41. It is generally agreed that a machine should be called a computer only if it contains three essential parts of a computer: CPU, memory, and I/O devices. T or F?

42. A buffer is an area of storage used to temporarily hold data being transferred from one device to another. T or F?

43. A tristate buffer allows only one device on the data bus to "talk." T or F?

44. A buffer can be used to compensate for the different rates at which hardware devices process data. T or F?

45. A buffered computer provides for simultaneous input/output and process operations. T or F?

46. Why is buffering important between a CPU and a peripheral device?

47. What is meant by the term *port*?

48. Convert the hexadecimal numbers that follow to their binary equivalents: 45, E2, 8B.
   Ans: 01000101, 11100010, 10001011.

49. Convert the hexadecimal numbers that follow to their binary equivalents: D7, F, 6.

50. The group of operations that a microprocessor can perform is called its *instruction set*. T or F?

51. Computer instructions consist of two parts called the "operation" and the "operand." T or F?

52. In program memory, the operation and the operand are located in separate locations. T or F?

53. What function does a software driver perform in a computer's operating system?

54. What is a controller?

55. Explain the term *kernel*.

56. Compare the access time of a floppy disk with that of a hard disk.

CHAPTER 9

Digital Systems

9.1 INTRODUCTION

The invention of the transistor in 1948 ranks as a seminal event of technology. It influenced the electronics industry profoundly, giving us portable TVs, pocket radios, personal computers, and the virtual demise of vacuum tubes. But the real electronics revolution came in the 1960s with the integration of transistors and other semiconductor devices into monolithic circuits. Now integrated circuits (called chips) can be found in everything from wristwatches to automobile engines and are the driving force behind the boom in personal computers. The latest advance, though, involves the integration of digital communication networks with the computer which has given birth to the global Internet.

9.2 DIGITAL COMMUNICATION AND THE COMPUTER

Underlying the rapid progress in technology is the transformation from analog to digital technology. Modern digital technology has evolved along two parallel but independent paths—*logic* and *transmission*. Binary logic proved to be a particularly efficient way to mechanize computation. The digital nature of computers is familiar to everyone. Digital computers were born in *Boolean algebra* and the language of 0 and 1's. With the invention of the transistor, which is almost an ideal on–off switch that mimics the 0 and 1's of computer language, computer technology rapidly developed to its current advanced state. However, in communications, the transformation to digital transmission was much slower. When communicating by telephone or broadcasting by radio or television, the use of continuous (analog) signals is natural. But it was years after the publication in 1948 of Claude Shannon's fundamental papers on communication theory, which put forth that noise-free transmission is possible if digital signals in place of analog ones are used, that the switch from amplitude and frequency modulation to pulse-code modulation began in the 1960s. In this scheme of communication, messages are converted into binary digits (0 and 1's) which can then be transmitted in the form of closely spaced electrical pulses. Such digital encoding permits signals to be sent over large distances without deterioration as it allows signals to be regenerated again and again without distortion and without contamination by noise. Even today, successful analog implementations in music, video, and television are being replaced by digital technology.

The common language that is used between the computer and the ever-growing digital communication networks brought forth rapid, interactive changes that resulted in another leap in technology, completing the digital revolution. By using computers that can store and access quickly large amounts of information at the nodes of a world-wide digital communications network, the Internet was born. Computers function as servers at the nodes of a network and provide a wealth of information for any user. The combination of computers and communication networks, which include telephone lines, coaxial cables, fiberoptic lines, and wireless technologies, is changing the face of entertainment, business, and information access. Such a system is frequently referred to by the media as the information super highway.

Fundamental building blocks for digital systems were considered in the previous chapter. We classified logic gates as elementary blocks which in turn could be arranged into larger blocks such as registers, counters, adders, etc., which are the fundamental components of modern digital equipment.

In this chapter, digital systems such as those exemplified in elementary communication systems and in computers will be considered. Fundamental notions that define the transmission of digital information in communication systems will be introduced. We will consider the rate at which an analog signal must be sampled when changing it to a digital signal (the *Nyquist criterion*); channel capacity, which is the maximum information that a channel can carry; and bandwidth, which relates to the amount of information in a signal. The previous chapter was devoted to the digital computer. We introduced the microprocessor, which is a small computer on a chip. Included on the chip is a minimal amount of memory and a minimal amount of control functions which are sufficient to allow the microprocessor to serve as a minimal computer in calculators, video game machines, and thousands of control applications ranging from automobile ignition control to smart instruments, programmable home thermostats, and appliances. Adding more memory to a microprocessor (which is now referred to as a CPU or central processing unit) in the form of RAM and ROM chips and additional input/output (I/O) circuitry, we obtain a microcomputer, of which the personal computer (PC) is the best example. Larger and more powerful computers are obtained by adding even more memory and running the CPU at a faster clock speed.

9.3   INFORMATION

In any interval or period, the quantity of information I_o can be defined as

$$I_o = \log_2 S \text{ bits} \tag{9.1}$$

where S is the number of distinct states that can be distinguished by a receiver in the interval.[1] Choosing log to the base 2 gives the quantity of information in units of binary digits or bits. For example, turning a light on or turning it off corresponds to 1 bit of information ($\log_2 2 = 1$) as there are two distinguishable states. If someone states that a light is on, he has conveyed one bit of information.

Let us first examine a simple information system such as a traffic light and progress to more complex ones such as television.

9.3.1 Traffic Light

Consider a display consisting of three lights. Figure 9.1a shows the 8 possible states of the three on–off lights. Such a display, for example, can be used to regulate traffic or show the status of assembly on a production line. As already discussed in Section 7.2, three binary lines can "speak" 8 ($= 2^3$) words.

Similarly here, the number of 3-bit words (or information states) the display can show is 2^3. Technically speaking, we say such a display uses a 3-bit code. Thus the quantity of information I_o per code is

$$I_o = \log_2 8 = 3 \text{ bits} \tag{9.2}$$

Each time the lights change, we receive 3 bits of information. Regardless of the speed with which the display changes, in other words, regardless if the interval between changes is short or long, the information that we receive remains constant at 3 bits per interval.

The output of the three-light display appears digital, as the on–off states of the lights correspond to 0 and 1's. To do digital processing on the output, we change the output to digital form by assigning binary numbers (called *code words*) to each output state of the display. For example, Fig. 9.1a shows the assignment of the binary numbers 0 to 7 to the eight states of the display. The output is now ready for processing by a digital computer because the use of code words that are expressed as binary numbers (a sequence of 0 and 1's) is also the language of the digital computer.

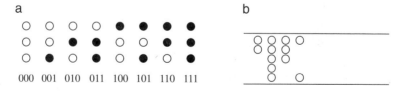

FIGURE 9.1 (a) The possible states of three on–off lights. (b) A teletype ribbon showing the five digit position code represented by holes punched across the ribbon.

---

[1] The choice of logarithm makes the quantity of information in independent messages additive. The choice of logarithm to the base 2 makes the unit of information the binary digit (bit), whereas a choice of base 10 would make the unit the decimal digit.

9.3.2 Teletype

One of the early transmission systems for information such as news and stock market quotations was the *teletype* machine, which at the receiving end produced long paper ribbons called *ticker tape* which had up to five holes punched across the ribbon as shown in Fig. 9.1*b*. The position of holes across represented the five digits of the code which provided $2^5 = 32$ information states. These states were used for the 26 letters in the alphabet and for control operations (carriage return, space, line feed, and numbers shift). The quantity of information I_o per code was therefore

$$I_o = \log_2 32 = 5 \text{ bits} \tag{9.3}$$

Modern examples are the fax machine for transmitting pictures over telephone lines and the modem for transmitting text over telephone lines using ASCII code (American Standard Code for Information Interchange). Each character in ASCII is represented by seven data bits, giving a total of $2^7 = 128$ different characters to represent lowercase and uppercase letters, numbers, punctuation marks, control characters, etc.

9.3.3 Speech Signal

Next we consider a continuous signal such as speech. Figure 9.2*a* shows a typical speech signal plotted on a log scale as a function of time. The vertical log scale reflects the fact that the human ear perceives sound levels logarithmically. Ordinary speech has a dynamic range of approximately 30 dB,[2] which means that the ratio of the loudest sound to the softest sound is 1000 to 1, as can be seen from Fig. 9.2*a*. Combining this with the fact that human hearing is such that it takes a doubling of power (3 dB) to give a noticeably louder sound, we can divide (quantize) the dynamic range into 3 dB segments, which gives us 10 intervals for the dynamic range, as shown in Fig. 9.2*a*. The 10 distinguishable states in the speech signal imply that the quantity of information at any moment of time t is

$$I_o = \log_2 10 = 3.32 \text{ bits} \tag{9.4}$$

We need to clarify resolution along the vertical scale, which is the resolution of the analog sound's amplitude. Using 10 quantizing steps is sufficient to give recognizable digital speech. However, a much higher resolution is needed to create high-fidelity digital sound. Commercial audio systems use 8 and 16 bits. For example, the resolution

---

[2]A decibel (dB) was already defined in Section 5.4 as the smallest change in sound level which the human ear can distinguish. The average ear, on the other hand, can only detect a 3 dB change in sound level, that is, a doubling of power (which means that when increasing the sound level by increasing the power of an audio amplifier from 50 to 100 W, or for that matter, from 1 to 2 W, a listener will be barely aware that a change in sound level has taken place; to an audiophile just about to invest in expensive audio equipment, this is usually a surprising fact). When used as a measure of change in power P, dB is defined as $10 \log 10 P_2/P_1$. Therefore, a 1000 to 1 range translates into $10 \log 10 P_2/P_1 = 10 \log_{10} 10^3 = 30$ dB. Under more controlled (quiet) conditions a 1000 to 0.1 dynamic range is possible which translates to a 40 dB range. Classical music, especially, has even a larger dynamic range, whereas background music (elevator music) or rock-and-roll is characterized by a narrow range.

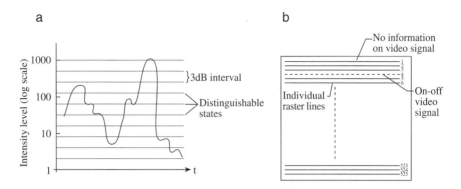

FIGURE 9.2 (a) A typical speech signal showing a 30 dB dynamic range. Along the vertical scale the signal is divided into 10 information states (bit depth). (b) Screen of a picture tube showing 525 lines (called raster lines) which the electron beam, sweeping from side to side, displays. All lines are unmodulated, except for one as indicated.

obtainable with 16-bit sound is $2^{16} = 65,536$ steps or levels. As bit depth (how many steps the amplitude can be divided into) affects sound clarity, greater bit depth allows more accurate mapping of the analog sound's amplitude.

9.3.4 Television Signal

A television picture in a black and white TV is produced by an electron beam moving across the screen of a picture tube (the technical name is CRT or cathode ray tube). A coating of phosphorus (actually an array of discrete phosphorus dots) on the inside of the picture tube screen lights up at those points that are struck by the electron beam. The electron beam moves repeatedly from side to side, tracing out 525 visible horizontal lines on the screen, which is called a frame. As the beam moves across the screen, its intensity also varies, producing 525 lines of varying brightness, as shown in Fig. 9.2*b*, which the eye interprets as a picture. It is the video signal, a component of the received television station signal, which controls the intensity variations of the electron beam. Another component of the station signal synchronizes the beam in the picture tube to a scanning beam inside a television camera which is photographing a scene to be transmitted. To capture moving scenes, television cameras produce 30 frames per second.

Intensity variations along a horizontal line cannot be arbitrarily fast. Let us say that we want to display alternating light and dark spots as, for example, a single row of a checker board. The circuitry[3] of a TV set limits the variations in brightness to about

---

[3]If the station signal were to vary faster, the beam would blur or smear the variations along a line. We say that the TV set circuitry, which is designed to have a maximum bandwidth of 6 MHz, cannot follow more rapid variations than 500 pixels per line. Faster-changing signals which require a larger bandwidth are simply smeared. For example, a TV set, when used as a monitor, can display about 40 columns of type. A typical typed page, on the other hand, requires about 80 columns, implying that computer monitors must have at least 12 MHz of bandwidth, with high-resolution monitors having much more bandwidth than that. To

500 alternating black and white spots along one horizontal line, also referred to as 500 pixels (picture elements) or as 500 lines of horizontal resolution. Thus the best that a TV set can do is display 250 dark and 250 light spots along a single line. For each pixel the eye can perceive about 10 gradations of light intensity (brightness). The brightness of a pixel depends on the strength of the electron beam at the pixel location. If we take a pixel, which is 1/500 of a line, as the basic information interval, we have as its quantity of information (for black and white TV)

$$I' = \log_2 10 = 3.32 \text{ bits per pixel} \tag{9.5}$$

We could also consider a single line as our basic information interval. Given that we have 10 states in each pixel and 500 pixels per line, the number of possible states in each line is 10^{500}. We can now define the quantity of information in one line as

$$I' = \log_2 10^{500} = 1661 \text{ bits per line} \tag{9.6}$$

Similarly, the number of possible states per frame is $10^{500 \times 525}$, which gives as the quantity of information per frame

$$I' = \log_2 10^{262500} = 872,006 \text{ bits per frame} \tag{9.7}$$

If we consider that 30 frames per second are received, we obtain $10^{500 \times 525 \times 30}$ possible states, which gives

$$I' = \log_2 10^{7875000} = 26.2 \cdot 10^6 \text{ bits per second} \tag{9.8}$$

In conclusion, we observe that the quantity of information must always be identified by the length of the interval.

9.4   INFORMATION RATE

Typically the displays considered in the previous section change with time. We can express the rate of information change in units of bits per day, bits per hour, or what is adopted as a standard, bits per second. The rate of transmission of information I is defined as the product of the number n of identifiable intervals per second during the message and the quantity of information I_o which is in each interval, that is, $I = nI_o$ or

$$I = n \log_2 S \text{ bits per second (bps)} \tag{9.9}$$

where S is the number of states in the interval, as defined by (9.1).

Note that rate is independent of interval length and is expressed in bits per second. Furthermore, convention is that the stated rate is always the maximum possible rate

---

verify the above 6 MHz bandwidth for TV sets, we observe that ideally a TV screen should be able to display $500 \cdot 525 \cdot 30 = 7.8 \cdot 10^6$ pixels per second. Equating the time interval between two pixels (one light and one dark) with the period of a sinusoidal signal, we obtain 3.9 MHz, which is reasonably close to the 4.2 MHz video signal component of a 6 MHz TV signal.

for a system. For example, if the traffic light display board in Fig. 9.1a is designed to change once every 15 s, then during an hour it will have changed 240 times, giving $240 \cdot 3 = 720$ bits for the maximum possible quantity of information. It is irrelevant that the rate is 0 at times when the display is shut down, or has a rate less than 720 bits per hour when at times the display malfunctions. The rate that characterizes this system is always the maximum rate of 720 bits per hour.

9.4.1   Traffic Light

The information rate of the traffic light system in Fig. 9.1a, if it changes once every 15 s, is

$$I = \frac{1}{15} \text{ intervals/s} \cdot 3 \text{ bits/interval} = \frac{1}{5} \text{ bits/s} \tag{9.10}$$

9.4.2   Teletype

Teletype machines ran typically at 60 words per minute when reading the punched ribbons. If the average word contains 5 letters followed by a space, we need 6 codes per second when receiving 1 word per second. The information rate is therefore, using (9.9),

$$\begin{aligned} I &= 6 \frac{\text{codes}}{\text{word}} \cdot 1 \frac{\text{word}}{\text{s}} \cdot \log_2 2^5 \frac{\text{bits}}{\text{code}} \\ &= 6 \log_2 2^5 = 6 \log_2 32 = 30 \text{ bits/s} \end{aligned} \tag{9.11}$$

where the code rate is equal to $n = 6$ codes/s.

9.4.3   Speech Signal

A speech signal, as in Fig. 9.2a, is a continuous mix of many frequencies. The predominant frequencies in speech are in the range of 100 Hz to 3000 Hz. Technically speaking, we say that speech has a frequency bandwidth of 2900 Hz. Speech sounds quite natural with frequencies restricted to this range—note the good quality of speech on telephones which are restricted to the above bandwidth.

The example of a speech signal or any other that is a continuous (analog) signal is fundamentally different from the previous examples of the traffic light and the teletype machine that display discrete (digital) signals. There we talked about information per code, meaning information per interval during which the code is displayed. We assigned a binary number, shown in Fig. 9.1a, to each discrete state of the lights. This number remained unchanged during the interval that the lights did not change. We had a digital input (state of the three lights) and a digital output (the binary numbers). Output in digital form is required for processing by a computer. Similarly, for an analog signal such as a speech waveform, we would like to abstract a corresponding digital signal for processing in digital form. It appears that this should not be any more difficult than in the previous case. All that needs to be done is assign a binary number to each of the 10 states of the speech signal in Fig. 9.2a, and we have our digital form. What is different though is that the magnitude of the analog signal is continuously changing with time, requiring continuously changing binary numbers. We are now presented with a dilemma: a number needs a finite time for its display, but if the signal changes significantly during the display time, the number has little meaning. Clearly, we must choose a display time short enough so the signal remains approximately constant in that time interval. This means that the rapidity with which the analog signal is changing determines the sampling intervals. Fortunately, a rigorous sampling theory exists which puts these notions on a precise foundation and allows a design engineer to choose the best sampling rate.

Conversion to Digital: The Sampling Process

Figure 9.3a shows a speech signal $v(t)$ in which the horizontal time axis has been divided into sampling intervals of duration t_s. The sampling rate, which is the number of sampling intervals per second, is then a frequency f_s equal to $1/t_s$. Clearly, the fastest changing parts of the signal determine the length of t_s. To obtain an estimate of what t_s should be, consider the fastest wiggle that can be observed in the signal and then imagine the wiggle to be a short segment of a sinusoid (shown dotted in Fig. 9.3a). It is reasonable to associate the frequency of this sinusoid with the highest frequency f_h that is present in the signal. For the wiggly part of the signal to stay approximately constant during a sampling period t_s, t_s must be shorter than the period of the wiggle or, conversely, the sampling rate must be faster than the highest frequency in the speech signal.

The question, "What is the appropriate sampling rate?" is precisely answered by the *Nyquist sampling criterion* which states that no information is lost by the sampling process if the sampling frequency f_s satisfies the criterion[4]

$$f_s \geq 2f_h \tag{9.12}$$

where f_h is the highest frequency present in the waveform being sampled. Thus, if the signal is sampled twice during each period of the highest frequency in the signal, we have converted the signal to digital form for further processing (by a computer, for example) and for transmission. The Nyquist criterion guarantees us that the sampled signal v_s which is a sequence of discrete values, $v(0), v(t_s), v(2t_s), v(3t_s), \ldots$, contains all the information of the original analog signal and in fact the original signal

---

[4] Since the highest frequency f_h and the bandwidth B of a speech signal are approximately equal, (9.12) is often stated as $f_s \geq 2B$. Bandwidth is a range of frequencies, typically defined as $B = f_h - f_l$, where the subscripts h and l denote the highest and lowest frequencies. Since speech covers the range from the lowest frequency (approximated by 0) to about 3000 Hz, the bandwidth of speech is $B = 3000$ Hz and we can state that $f_h = B$. Speech is also referred to as being at *baseband*. Baseband systems transmit the basic or original signal with no frequency translation. Since a baseband signal such as speech is limited to a short distance over which it can be heard, it can be placed on a carrier which has the desirable property of traveling long distances with little attenuation. For example, in Chapter 5 we studied AM radio which takes a baseband signal such as speech or music and superimposes it (called modulation) on a high-frequency signal (the carrier) which can travel far into space.

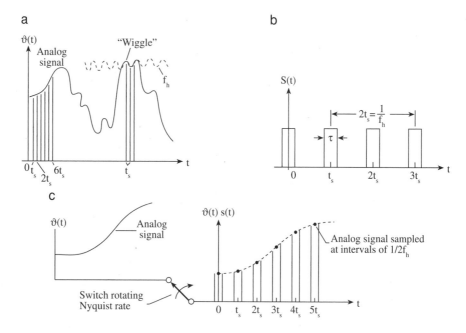

FIGURE 9.3   (a) A speech signal $v(t)$ divided into sampling intervals whose length is determined by the highest frequency present in the signal. (b) A sampling function is a periodic train of pulses. (c) A switch rotating at the Nyquist rate gives sampled values of the signal at periodic intervals.

$v(t)$ can be uniquely extracted from $v_s(t)$. This rather amazing and not at all obvious result—which implies that valuable information is not lost in the sampling process—is demonstrated whenever we use digital audio and video. The concepts presented in this paragraph are known as the *sampling theorem*.

How is such sampling performed in practice? Figure 9.3b shows a sampling function, call it $s(t)$, which can be produced by a high-speed electronic switch.[5] If $s(t)$ is multiplied by an analog signal $v(t)$, we obtain the sampled signal $v_s(t) = v(t)s(t)$. Thus the sampling operation is simply multiplication by $s(t)$, where $s(t)$ is nothing more than a periodic pulse train as shown in Fig. 9.3b. Note that the width τ of the sampling pulses is not really important as long as τ is much smaller than the sampling period t_s (if τ becomes vanishingly small, we refer to it as an *ideal sampling train*).

Reconstructing the Analog Signal

How do we know that the sampled signal contains all of the information of the original analog signal? To answer this question, we must look at the frequency content of

---

[5]For purposes of illustration, Fig. 9.3c shows a rotating mechanical switch which periodically connects the input signal to the output and thereby provides sampled values. If the switch rotates at the Nyquist rate or faster, it will make contact for a brief time (assumed to be τ seconds) and by that means allow a small portion of the analog signal to pass through, repeating the process every t_s seconds.

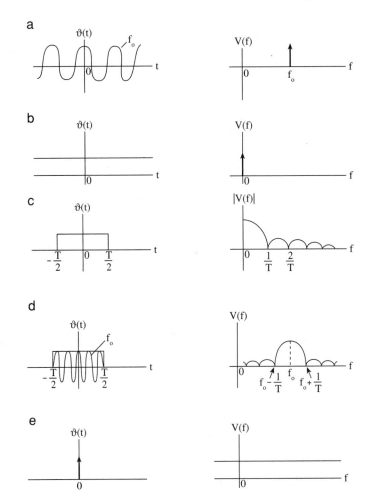

FIGURE 9.4   Time-domain signals and their spectra in the frequency domain, obtained by Fourier-transforming the time signals.

our analog signal. The analog signal can be described uniquely in the time domain as $v(t)$ or in the frequency domain as $V(f)$, where $V(f)$ gives the amplitudes of all the frequencies that are present in the analog signal. The Fourier transform relates $v(t)$ and $V(f)$ in the two domains.

Detour: Signals and Their Spectra

Before we go on, let us familiarize ourselves with some common signals and their spectra. In Section 5.6 we already introduced the Fourier series which related a periodic time signal to its *discrete* spectrum—"discrete" meaning that only specific frequencies were present. Similarly we can show that an aperiodic signal such as single pulse in the time domain will give a spectrum that is *continuous*—"continuous" meaning that all frequencies are present. Figure 9.4 gives a table of familiar time-domain functions and their spectra. For example, Fig. 9.4a, a sinusoid $\cos 2\pi f_o t$, has obviously only a single frequency f_o for its spectrum. Similarly a DC voltage, Figure 9.4b, has a single frequency $f = 0$ for its spectrum. These are two examples of uniform motion in the time domain which are single-frequency phenomena. Nonuniform and abrupt motions, on the other hand, result in a spread of frequencies, as Figs. 9.4c to e illustrate. Figure 9.4c, a square pulse such when a battery is turned on and off, results in a continuous spectrum referred to as a sinc function. A highly useful signal is obtained by starting with a single-frequency sinusoid and turning it on and off as shown in Figure 9.4d. The frequency spectrum of such a signal is the frequency spectrum of a pulse, shape $V(f)$ in 9.4c, but shifted to the frequency f_o. In other words we have placed the baseband information of a pulse on a carrier which can transmit the baseband information over long distances solely determined by the properties of an electromagnetic wave of frequency f_o. A series of pulses representing binary numbers is transmitted that way. A final figure to become familiar with is a very sharp and narrow pulse in time (a lightning stroke, a sharp noise spike, etc.) which yields a frequency spectrum, as in Fig. 9.4e, which is *uniform*—meaning that all frequencies are present in equal strength. We can now make a general observation: a gradually changing signal in t has a narrow spread of frequencies whereas an abruptly changing signal produces a wide spread of frequencies. Note that a square pulse of length T seconds will have most of its frequencies concentrated in a bandwidth of $1/T$ Hertz—implying that a very abrupt pulse for which T is very short will be, so to speak, all over the place in the frequency domain. This inverse relationship between t and f is fundamental to the Fourier transform. ∎

   Let us return to the reconstruction of a sampled signal. Say the spectrum of an analog signal $v(t)$ to be sampled looks like that in Fig. 9.5a: it has a strong DC component ($f = 0$) with the higher frequencies tapering off linearly to zero at B, which we call the bandwidth of the analog signal (in the time domain a signal $v(t) = (\sin t/t)^2 = \sin c^2 t$ would produce such a spectrum). The highest frequency f in this band-limited signal is therefore $f = f_h = B$ and no frequencies beyond B exist in the signal.

   The reconstruction of the original analog signal from the sampled signal is surprisingly simple once we realize that the spectrum of the sampled signal is just like the original spectrum except that it repeats itself periodically as shown in Fig. 9.5b. This repetition of the spectrum in the frequency domain is a consequence of sampling the analog signal periodically in the time domain.[6] The period of the new spectrum is equal to the sampling rate f_s. It appears that the sampling operation has preserved the message spectrum because the spectrum of the original signal and the sampled-signal spectrum for which $f \leq B$ are identical—as long as the sampling is done at the Nyquist rate or faster (oversampling). But when undersampling, Fig. 9.5d shows that the replicas overlap, which alters and distorts the baseband, making recovery impossible. The recovery

---

[6]The fact that the spectrum after sampling is the sum of $V(f)$ and an infinite number of frequency-shifted replicas of it is a property of the Fourier transform: multiplication of two time functions, as in $v(t) \cdot s(t)$, is equivalent to the convolution of their respective Fourier transforms $V(f)$ and $S(f)$.

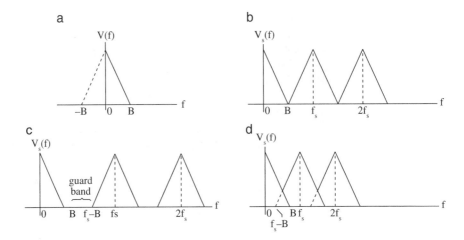

FIGURE 9.5   (a) The spectrum of an analog signal. (b) The spectrum of the same signal but uniformly sampled at the rate $f_s = 2B$. (c) The spectrum when oversampled at $f_s > 2B$. (d) The spectrum when undersampled at $f_s < 2B$.

of the original signal, not at all obvious in the time domain, is obvious in the frequency domain: merely eliminate the frequency content in the sampled-signal spectra above the baseband frequency B. In other words, eliminate the extra replicas which were created by the sampling process, and one is left with the original signal. It is that simple. In a real situation once a sampled signal is received, it is passed through a low-pass filter which eliminates all frequencies above $f = B$, thus reproducing the original signal in real time at the receiving end.

We can use a simple RC filter, covered in Sections 2.3 and 5.5, to perform the low-pass filtering. The RC filter shown in Fig. 2.6 (repeated in Fig. 9.7b), has a cutoff frequency

$$f_h = B = \frac{1}{2\pi RC} \tag{9.13}$$

and its bandwidth is sketched in Fig. 9.6a.[7] Passing an analog signal that is sampled at $f_s = 2B$ through such a filter would "strip off" all high frequencies above B, as suggested in Fig. 9.6b, leaving the original analog signal, Fig. 9.6c. It should also be evident that oversampling, even by a small amount can be very beneficial as it creates guard bands between the replicas. Guard bands, as is evident in Fig. 9.5c, are dead zones of frequencies in the range $B < f < f_s - B$. The existence of guard bands makes the design of the RC filter less critical because now the cutoff frequency (9.13)

---

[7]For the sake of simplicity, we are using an elementary RC filter which attenuates the high frequencies slower than the slope of the filter in Fig. 9.6a suggests. To obtain the sharper cutoff characteristics $H(f)$ shown in Fig. 9.6a, filters more complex (and more expensive) than the simple RC filter are used in practice. An ideal filter would have a slope which is vertical (see Fig. 9.10b), that is, pass all frequencies up to B equally and completely attenuate all frequencies larger than B. One reason for oversampling is that shallow-slope filters become usable instead of steep-slope filters which are expensive, complicated, and difficult to design.

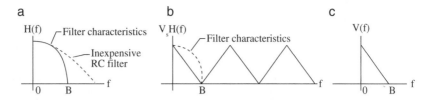

FIGURE 9.6 Recovery of the original signal from the sampled signal. A low-pass filter (a) eliminates all replicas in the spectrum of the sampled signal, (b) allowing only the spectrum of the original signal (c) to pass.

can be anywhere in the guard band for the filter to be effective. Note that when sampling is at precisely the Nyquist frequency, the guard band is zero, as shown in Fig. 9.5*b*.

There are two mechanisms that can cause severe distortion. One occurs when the analog signal possesses frequencies beyond a bandwidth that contains most of the signal.[8] For example, even though speech as shown in Fig. 9.7*a* is concentrated at frequencies less than 3000 Hz, implying that a sampling frequency twice that should be adequate for sampling the speech signal, speech does have frequency content beyond 3000 Hz. It is the frequency content beyond 3000 Hz that will cause distortion if sampling is at 2×3000 Hz because these frequencies are in effect undersampled and therefore will show up in the baseband, making distortion-free recovery impossible—this is identical to the distortion suggested in Fig. 9.5*d*. A distortion mechanism in which replica frequencies of a sampled signal overlap the baseband is called *aliasing*, meaning that high-frequency components take on the identity of a lower frequency in the spectrum of the sampled signal. To avoid such distortion, frequencies above 3000 Hz in the original signal must be removed once it is decided to sample at a 6000 Hz rate. This is called prefiltering of the original signal and is accomplished by passing the signal through a low-pass RC filter before sampling the signal. Of course, to avoid aliasing we can always sample at a faster rate. The second distortion mechanism occurs when the signal is undersampled, *causing* aliasing. The remedy is to sample at a faster rate.

EXAMPLE 9.1 The frequency spectrum of a typical speech signal is shown in Fig. 9.7*a*. It shows that even though speech can have frequencies as high as 10 kHz, much of the spectrum is concentrated within 100 to 700 Hz, with it sounding quite natural when the bandwidth is restricted to 3 kHz. Show the necessary design steps to transmit this as a digital voice signal over telephone lines.

Telephone speech signals are generally restricted to 3 kHz of bandwidth. To prepare this signal for sampling, which must be done at least at a 6 kHz rate, we will first low-pass filter the speech signal by passing it through a RC filter of the type shown in Fig. 9.7*b*. Such prefiltering will eliminate aliasing errors. If we choose the cutoff

---

[8]Real signals are rarely band-limited, i.e., confined to a band of frequencies with no frequencies outside that band. The reason is that real signals are time-limited, that is, they have a beginning and an end and therefore cannot be band-limited simultaneously. An examination of our table of signals and their spectra, Fig. 9.4, confirms this as it shows that only unlimited signals in time (Figs. 9.4*a* and *b*) can be band-limited. It shows the remaining time-limited signals to have unlimited bandwidth.

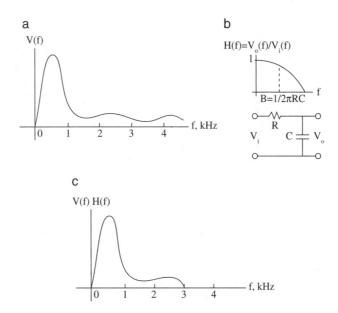

FIGURE 9.7   (a) A typical frequency spectrum of a speech signal. (b) A simple low-pass filter. Generally more elaborate filters with sharper cutoff characteristics are used. (c) Speech spectrum after passing the speech signal through a low-pass filter.

frequency of the filter as 3 kHz, then from (9.13), the resistance is calculated as $R = 1060\ \Omega$ if capacitance is chosen to have a value of $C = 0.05\ \mu F$. The spectrum of a prefiltered speech signal looks like that shown in Fig. 9.7c. Generally one samples at a higher rate than 6 kHz to ensure separation of baseband and replica spectra (i.e., create guard bands) and to compensate for poor filtering characteristics of nonideal filters which do not have sharp cutoff characteristics. A standard for telephone systems is to sample at 8 kHz.   ∎

EXAMPLE 9.2   The device that makes conversions of analog signals to a digital form is the *analog-to-digital converter* (ADC). Every ADC requires a certain amount of time to perform the A/D conversion. A typical ADC device is characterized by a maximum conversion time of 40 μs. What is the highest-frequency signal that could be sampled on the basis of the Nyquist criterion?

The highest rate of data conversion using this device is

$$f_{max} = \frac{1}{40\ \mu s} = \frac{1}{40 \cdot 10^{-6}s} = 25\ \text{kHz}$$

If this is to be the highest sampling frequency, then according to the Nyquist criterion, the highest signal frequency that can be represented without distortion is

$$\frac{1}{2}f_{max} = \frac{25\ \text{kHz}}{2} = 12.5\ \text{kHz}$$ ∎

EXAMPLE 9.3 The rudder position of an airplane is telemetered by a signal which has a bandwidth of 0–30 Hz. What is the shortest time for this signal to change to a distinctly new state?

Referring to the fastest-changing part of a signal such as the "wiggle" in Fig. 9.3*a*, we see that when the signal changes from a min to a max, that change takes a time of a half-cosine. This time is a measure of the speed with which the rudder can change to a new position. Since the shortest time for a signal change is determined by the highest frequency in the signal bandwidth, we calculate that

$$\frac{1}{2} \cdot \frac{1}{30 \text{ Hz}} = 0.017 \text{ s}$$

is the shortest time.

An acceptable answer to this problem can also be obtained by equating the shortest time with the 10–90% rise time of a pulse considered in (5.32) of Section 5.6. The rise time can be considered as the minimum time needed to change from one state to another. Thus $t_r = 0.35/f_h = 0.35/30 \text{ Hz} = 0.012$ s, which is somewhat shorter than the half-cosine time—not unexpected as t_r is taken between 10% and 90% of the rise to a new state. ∎

EXAMPLE 9.4 A temperature transducer is a device that converts temperature to an analog (continuous) electrical signal. A particular temperature transducer is listed as having a time constant τ of 0.4 s. (a) What is the signal bandwidth that can be associated with such a transducer? (b) If the tranducer signal is to be converted to a digital signal by sampling, how frequently must the transducer output be sampled?

(a) Specifying a device by a time constant implies it is a linear, first-order system such as an RC low-pass filter considered previously in (9.13). Any transducer should be able to follow slow variations, but eventually due to inertia of its parts, will cease to produce output when the input varies too quickly. To use a time constant and its associated bandwidth to describe when a transducer fails to follow input variations, consider the corner frequency of a low-pass filter, which is the frequency at which the output has decreased by 3 dB. Using (9.13) we have that $f_h = B = 1/2\pi RC = 1/2\pi\tau$, where the time constant is $\tau = RC$. Hence, the signal bandwidth B is given by $B = 1/2\pi\tau = 1/(2\pi \cdot 0.4 \text{ s}) = 0.398$ Hz. Should the temperature vary faster than this frequency, the tranducer would not be able to follow and hence could not generate a corresponding electrical signal that is higher in frequency than 0.398 Hz.

Note: to refresh the relation between time constant and bandwidth, the student should review Eqs. (6.27, p. 222), (5.32, p. 181), (5.22, p. 173), and (2.14, p. 59), and Section 1.8.2, "Time Constant," which follows Eq. (1.53, p. 39).

(b) Once we decide on a sampling frequency for the analog signal, we must make sure that no frequencies greater than half the sampling frequency are present in the analog signal. We can do this by passing the analog signal through a low-pass filter. Ideally, the filter should have a very sharp cutoff at half the sampling frequency. The penalty for using a filter with a gradual roll-off such as the simple RC filter is a higher sampling frequency.

It was shown that the highest signal frequency available from the transducer is 0.398 Hz. If we were sure that no higher frequencies are present, then sampling at $2 \cdot 0.398 = 0.796$ Hz would be sufficient. In practice, however, we would first pass the transducer output through a low-pass filter with a cutoff frequency of, say, 0.45 Hz. Then instead of sampling at the required minimum rate of $2 \cdot 0.45 = 0.9$ Hz we should slightly over-sample at 1 Hz to account for presence of any low-amplitude higher frequencies that might get past the filter because the filter roll-off might not be sufficiently steep. Sampling at 1 Hz (that is, taking one sample every second) should give a sufficient safety cushion for good performance.   ■

9.4.4 Information Rate of Speech

Now that we have established what the sampling rate for a continuous signal must be, we can find the information rate of a speech signal. For transmission of speech by telephone, telephone circuits limit the maximum frequency to about 3000 Hz. The Nyquist criterion tells us that the slowest that we can sample such a signal is at 6000 Hz; sampling at this rate guarantees that the original signal can be reconstructed. Thus in such a system, with 10 distinguishable volume levels as shown in Fig. 9.2a, the information rate is

$$I = 2 \times 3000 \times \log_2 10 = 19{,}920 \text{ bits/s} \approx 20 \text{ kbps} \qquad (9.14)$$

EXAMPLE 9.5   For comparison, let us look at commercial telephone lines and at compact CDs.

For digital transmission of voice in the telephone network, the sampling rate is 8000 Hz, and 8 bits (or $2^8 = 256$ resolution or quantizing levels) per sample are used (to use 256 quantizing levels implies that the noise on the telephone line must be sufficiently low to allow the division of the voice amplitude into 256 steps). The sampling rate corresponds to a maximum voice frequency of 4000 Hz but the voice signal is filtered so that its spectrum is in the range of 300 Hz to 3400 Hz, allowing for a small guard band. Thus, to send a voice signal as a bit stream one must send 8 bits, 8000 times per second, which corresponds to a bit rate of 64 kbps.

For the example of a compact disc, the sampling rate is about 41 kHz, which corresponds to a maximum frequency of 20 kHz for the audio signal. The amplitude of each sample of the signal is encoded into 16 bits (or $2^{16} = 65{,}536$ quantization levels), giving it a bit rate of $16 \cdot 41 = 656$ kbps for each of the left and right audio channels. The total rate is therefore 1.3 Mbps. A 70-min CD stores at least $70 \cdot 60 \cdot (1.3 \cdot 10^6) = 5460$ Mbits or $5460/8 = 682$ Mbytes, which makes CD-ROMs a convenient medium to distribute digital information such as software programs.   ■

EXAMPLE 9.6   Compare music CDs' 44 kHz, 16-bit sound and 22 kHz, 8-bit sound.

We can ask how much resolution is needed to create convincing digital sound. When converting an analog signal to digital we speak of resolution in two dimensions: along the horizontal axis we sample the signal at discrete intervals and along the vertical axis

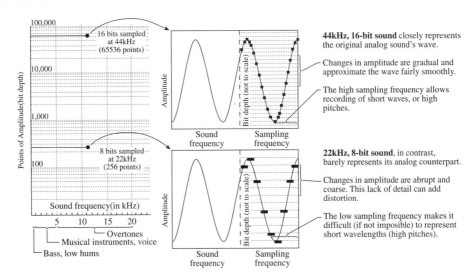

FIGURE 9.8   Comparison of high-resolution 16-bit sound sampled at 44 kHz to 8-bit, 22 kHz sound.

we quantize the amplitude. As Fig. 9.8 illustrates, the important factors in producing a convincing digital reproduction of natural sound are sampling frequency (how often the change in amplitude is recorded) and bit depth (how many steps the amplitude can be divided into, also known as quantization). The sampling frequency must be at least twice the rate of the analog sound's frequency (if sampling and sound frequency were the same, one would keep sampling the sound wave at the same point in its cycle, producing a straight line, or silence). Sampling at 44 kHz can capture sounds with frequencies up to 22 kHz, which can be considered the full range of human hearing. Bit depth affects sound clarity. Greater bit depth allows more accurate mapping of the analog sound's amplitude. ∎

9.4.5   Information Rate of Television Signal

A picture is created on a television screen by an electron beam sweeping out a frame of 525 horizontal lines every 1/30 of a second. If each line can display 500 pixels, the rate of pixel display is

$$n = 500 \times 525 \times 30 = 7.875 \cdot 10^6 \text{ pixels/s} \qquad (9.15)$$

Since we assumed that there are 10 gradations of light intensity which the eye can perceive at any point in the received picture, we have for the information rate of television

$$I = n \log_2 10 = 7.875 \cdot 10^6 \times 3.32 = 26.2 \text{ Mbits/s} \qquad (9.16)$$

The information rate of a television signal is therefore 26.2 megabits per second. TV signals[9] are assigned a bandwidth of 6 MHz per channel, of which the video portion takes up 4.2 MHz. Currently TV signals are amplitude-modulated (AM) analog signals. A question naturally arises: if TV signals were digitized, how much bandwidth would be required to transmit such a signal, and would it be possible to send it over a 4.2 MHz bandwidth? If we use simple on–off coding (pulse-code modulation or PCM) which achieves at best 2 bits/cycle (or 2 bps/Hz), then a bandwidth of 26.2 Mbps/2 bps/Hz = 13.1 MHz is needed. To send a 26.2 Mbps digital signal over a 4.2 MHz channel would require ingenious coding with an efficiency of 26.2/4.2 = 6.2 bits/cycle. Another possibility would be a band-compression technique which would compress the 26.2 Mbps digital signal to a 9.2 Mbps signal. A 9.2 Mbps signal could then be transmitted in a 4.2 MHz-wide channel.

The quality of a TV picture is limited by the resolution in the vertical and horizontal direction. For each direction, resolution is expressed in terms of the maximum numbers of lines alternating between black and white that can be resolved.

Vertical Direction

Resolution in this direction is primarily specified by the number of horizontal scanning lines (also called raster lines) when system specifications were adopted. In the United States, the NTSC (National Television System Committee) standard specifies 525 horizontal lines per frame. In Europe it is 625 lines, and the HDTV (high-definition television) standard has 1125 lines. In practice, however, the 525 lines of the NTSC format must be reduced by the retrace lines, leaving about 483 lines, and not all of these can be active in image formation because in general the raster does not align perfectly with the rows of the image. Arbitrary raster alignment typically reduces the 483 line resolution by a factor of 0.7.

Horizontal Direction

Horizontal resolution is determined by the 4.2 MHz bandwidth (B) allotted to the video signal. It is expressed in terms of the maximum number of lines that can be resolved in a TV picture along the horizontal direction. If we assume the incoming video signal is a sinusoid, varying at a maximum frequency limited only by the bandwidth B, i.e., $f_{max} = B$, the resulting TV picture would be a sequence of alternating dark and light spots (pixels) along each horizontal scanning line.[10] The spacing between spots would be a half-period of the sinusoid, in other words, two spots per period of f_{max}. This gives

---

[9]For example, channel 2 is assigned the frequency spectrum of 54–60 MHz. The frequency spectrum in MHz of the VHF TV channels is **2**, 54–60; **3**, 60–66; **4**, 66–72; **5**, 76–82; **6**, 82–88; **7**, 174–180; **8**, 180–186; **9**, 186–192; **10**, 192–198; **11**, 198–204; **12**, 204–210; **13**, 210–216. The UHF channels 14 and up are in the range of 470–806 MHz.

[10]If the spots of the 525 scanning lines line up vertically, we would see vertical lines fill a picture frame. Therefore, it is common practice to speak of lines of resolution in the horizontal direction.

for the horizontal resolution R_h

$$R_h = 2B(T - T_r) \tag{9.17}$$

where T is the duration of one scanning line (equal to 1/(525 lines per frame $\times$ 30 frames per second) = 63.5 μs per line) and $T_r = 10$ μs is the retrace time per line. Using these figures we can calculate a horizontal resolution of 449 lines. This is rarely achieved in practice and a resolution of 320–350 lines is common in typical TV sets. For comparison, the resolution of VCRs is only about 240 lines.

EXAMPLE 9.7 The characteristics of a HDTV are 1125 horizontal lines per frame, 30 frames per second, and a 27 MHz bandwidth. Find the line frequency, horizontal resolution, and number of pixels per frame.

$T_{line} = (1/30)s/1125$ lines= 29.63 μs/line, which gives for the line frequency $f_{line} = 1/T_{line} = 33{,}749$ lines per second.

Horizontal resolution, using (9.17), is $R_h = 2 \cdot 27$ MHz $(29.63$–$2.96)\mu s = 1440$ pixels or lines of horizontal resolution. In the above calculation 10% of the time needed to trace one horizontal line was used as the retrace time.

The maximum number of pixels per frame is therefore $1440 \cdot 1125 = 1.62$ million. In practice this number would be less as the number of resolvable horizontal lines might only be around 800 (1125 is the number of scanning lines). ∎

EXAMPLE 9.8 Calculate the time it would take to fax a magazine page, 30×50 cm in size with a resolution of 40 lines/cm, over a voice telephone channel with a bandwidth of $B = 3$ kHz. Assume zero time for horizontal and vertical retrace and that all scanning lines are active.

The number of pixels per page are $(30 \cdot 40) \cdot (50 \cdot 40) = 2.4$ million. Assuming that we can transmit 2 pixels per period of the highest frequency in the bandwidth, the rate of pixel transmission is then $f = (3 \cdot 10^3$ periods/s) $\cdot(2$ pixels/period$) = 6 \cdot 10^3$ pixels/s. The time for transmitting a page is then given by $T_{page} = 2.4 \cdot 10^6/6 \cdot 10^3 = 400$ s $= 6.7$ min. ∎

9.5 DIGITAL COMMUNICATION NETWORKS

The majority of today's communication systems are digital. It is much cheaper to store, manipulate, and transmit digital signals than analog signals because advances in electronics have made digital circuits more reliable and cheaper than analog circuits.

One of the advantages of digital communication is noise-free transmission of messages over long distances. A transmitted signal continues to deteriorate with distance because noise is continuously adding along the path of the channel as suggested in Fig. 9.9. Analog signals become progressively more contaminated with noise, and because they are complex, cannot be regenerated to their original shape at the receiving end. Digital signals, on the other hand, are simple on–off pulses which also become contaminated and distorted as they travel, but as long as one can still recognize the presence or absence of a pulse at the receiving end the original digital message can be

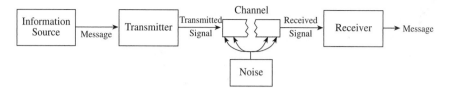

FIGURE 9.9   Block diagram of a communication system. Noise which will distort the message, is added along the communication channel which can be free space or transmission lines such as twisted copper wire pair or coaxial or fiberoptic cable.

completely restored. Communication or transmission channels respond to the different frequencies of a signal much like a low-pass RC filter. Transmission lines like a coaxial cable transmit low frequencies well but gradually begin to attenuate higher frequencies, which is exactly how a simple low-pass RC filter behaves (see Fig. 9.14).

Another advantage, not available to analog signals, is *time-domain multiplexing*. Its use is a natural in digital systems. It is a technique for packing many digital messages in a single channel, making for very efficient transmission of information. Before we continue with multiplexing, let us consider bandwidth first.

9.5.1   Bandwidth

There is no universally satisfying definition for bandwidth. Bandwidth gives important information about the signal in the frequency domain. We have already used it in connection with analog signals. There we stated that bandwidth of a continuous signal is simply equal to the difference between the highest and lowest frequencies contained in the signal. For digital signals, though, the discontinuous steps of the digital pulses introduce infinite frequencies and use of bandwidth is more challenging—sometimes bandwidth is even confused with the bits per second information rate which gives important information about the digital signal in the time domain. Bandwidth is used in two ways: to characterize a signal and to characterize a system. Let us first consider typical analog signals such as speech.[11]

9.5.2   Bandwidth of Signals

If we state that a signal has a bandwidth B, we mean that the signal contains a range of frequencies

$$B = \Delta f = f_2 - f_1 \tag{9.18}$$

---

[11]Technically speaking we do not need to differentiate between analog and digital signals. After all, we need only to look at the spectral content of each signal, the extent of which determines bandwidth. Typical analog signals, though, usually have a finite bandwidth, whereas digital signals usually have unlimited bandwidth, as a look at Fig. 9.4 shows. But also for digital signals it is useful to specify a finite bandwidth. We define bandwidth for digital signals as a range of frequencies that contains most of the energy of the signal. For example, we normally state that the bandwidth B of a T-long pulse is $B = 1/T$, because as shown in Fig. 9.4c, most (90%) of the energy of the pulse is contained between the frequencies of 0 and $1/T$. Note that Fig. 9.4 shows signal spectra; the energy or power spectrum is the square of the signal spectrum.

where f_2 is the highest, and f_1 the lowest, significant frequency of the signal. Figure 9.7c shows the bandwidth of a telephone speech signal, where $B = 3$ kHz with $f_2 = 3$ kHz and $f_1 \approx 0$. We say this signal is at baseband.[12] Baseband signals are produced by sources of information. In music or speech directed at a microphone or a scene filmed by a video camera, the microphone and the camera will produce baseband signals. For transmission purposes baseband signals are shifted to higher frequencies by the modulation process and are then referred to as passband signals. For example, if a telephone speech signal were to be broadcasted on the AM band by, say, station WGN, the speech signal would be used to modulate the 1 MHz station signal (also referred to as a carrier signal) and the composite signal would be sent out over the airwaves. The range of frequencies of the transmitted signal would be 1 MHz$- f_2$ to 1 MHz $+ f_2$ (assuming $f_1 = 0$), giving us a bandwidth of $B = 2f_2 = 6$ kHz. The bandwidth of the transmitted signal is thus twice the baseband signal. We call this double-sideband transmission and it is a peculiarity of AM modulation which introduces this redundancy in the transmitted signal, putting the baseband information in the upper (1 MHz $+ f_2 - 1$ MHz $= 3$ kHz) and lower sideband (1 MHz $-$ (1 MHz $- f_2$) $= 3$ kHz). Except for this wasteful doubling of the transmitted frequency spectrum, AM is perhaps the most efficient modulation technique in terms of frequency spectrum needs. For more detail on AM see Fig. 9.19.

It appears that a strong relationship between bandwidth and information exists. Using the speech signal in Fig. 9.7 as an example, we can state that bandwidth is a range of frequencies required to convey the information of speech. As a corollary, we can state that a signal with zero bandwidth has zero information.[13] Zero-bandwidth signals are single-frequency signals. In practice such signals are observed quite often. They are the station signals remaining when a broadcast—music or speech—momentarily stops, leaving the station carrier signal as the only signal being broadcast. For WGN, for example, it would be the 1 MHz signal. At such time no information is conveyed except for 1 bit of information that is the answer to the question, "Is the station on or off?" Summarizing, we can generally state that more bandwidth is required for signals with more information as the following examples indicate: telephone speech requires 3 kHz, high-fidelity music 15 kHz, and television 6 MHz.

9.5.3 Bandwidth of Systems

Bandwidth is also used to characterize a system. A system can be as simple as a low-pass filter or an amplifier, or as complicated as an entire satellite communication link. Bandwidth, when referring to a system or device, usually means the ability to pass,

---

[12] See also footnote 4.

[13] This is also nicely demonstrated by Figs. 9.4a and b, which show the sinusoid and the DC signal as two zero-bandwidth signals. Switching a DC signal on and off (Fig. 9.4c) generates an information bandwidth with frequencies primarily between 0 and $\approx 1/T$, where T is the duration of the pulse. Using such a pulse to switch a frequency f_o on and off, as in Fig. 9.4d, is what a transmitting station does when it modulates a carrier signal f_o by a pulse and sends the composite signal out to be received and interpreted as a binary signal.

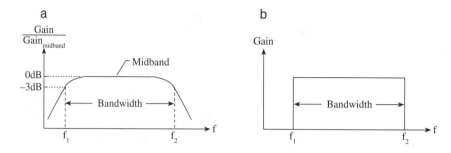

FIGURE 9.10 (a) Bandwidth of an amplifier is defined between the −3 dB points. (b) An ideal band-pass curve that is not physically realizable.

amplify, or somehow process a band of frequencies. Whereas bandwidth for a signal can be subjective—as, for example, the speech signal in Fig. 9.7a where the bandwidth of maximum energy could be specified as the range between 100 and 600 Hz, or the bandwidth for telephone-quality speech as between 100 and 3000 Hz—bandwidth for a system is usually defined between the −3 dB points. Figure 9.10a shows a typical band-pass curve for the amplifier of Fig. 5.9. Bandwidth is defined as that range of frequencies for which the amplifier gain is within 3 dB of midband gain. The frequency response of low-pass filters, high-pass filters, band-pass filters, and amplifiers can have steeper sides than those shown in Fig. 9.10a, but frequency response cannot have a discontinuous shape with a flat top and vertical sides as shown in Fig. 9.10b, referred to as an *ideal band-pass curve*, because devices are made with components such as resistors, capacitors, transistors, etc., for which voltage and current vary continuously (although a goal of elaborate and expensive filters is to approach the ideal band-pass curve).

We can now make a simple but powerful observation: when the bandwidth of a transmission system (consisting of amplifiers, filters, cables, etc.) is equal to or larger than the bandwidth of a signal that is to be transmitted over the system, that is,

$$B_{\text{signal}} \leq B_{\text{system}} \tag{9.19}$$

then the entire information content of the signal can be recovered at the receiving end. Conversely, when the transmission bandwidth is less than the signal bandwidth some degradation of the signal always results. B_{system} is also known as the *channel bandwidth*.

As examples of commercial utilization of the free-space spectrum, we can cite AM broadcasting which occupies a bandwidth of 1.05 MHz in the range of 550 kHz to 1600 kHz. Each AM station uses a 5 kHz audio signal to amplitude-modulate a carrier signal which results in a double-sided 10 kHz signal which is broadcasted. Since the bandwidth of an individual AM radio station is 10 kHz, we could frequency-divide (see Fig. 9.21a) the AM band and have potentially 1050/10 = 105 stations broadcasting simultaneously. Practically this is not possible as the 10 kHz bandwidth of each station does not occur abruptly but tapers off gradually, leaving some signal energy above and below the 10 kHz band of frequencies. It is the job of the FCC (Federal Communi-

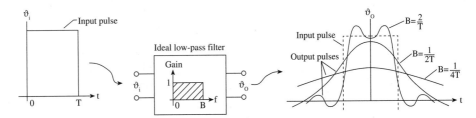

FIGURE 9.11   Passing an ideal square pulse of duration T through an ideal low-pass filter with a cutoff frequency B results in a distorted output pulse. Absence of high frequencies in the output manifests itself as a smoothing effect on the square input pulse: all sharp corners are rounded.

cation Commission) to allocate frequencies to stations in a region that will minimize interference between stations. This means that for interference-free reception in any given locality, the number of stations must be much less than 105. The FM radio spectrum occupies a 20 MHz band of frequencies between 88 and 108 MHz with a 200 kHz bandwidth allotted each station. The bandwidth of a TV station is 6 MHz, which if allotted to AM could carry 600 AM radio stations, or if assigned to telephone companies could carry 1500 telephone conversations. It is the scarcity of open free-space frequencies that explains why so much communication is done over wire and fiberoptic cables. In conclusion, when media people talk about bandwidth, they usually mean system bandwidth, specifically that of a transmission channel, trunk, or link, which would allow more signals and faster transmission of those signals with increased bandwidth.

9.5.4   Bandwidth of Digital Signals

Now let us consider digital signals and the bandwidth requirements for pulse transmission. We have to distinguish between the case of an exact reproduction at the receiving end of a transmitted square pulse (which represents a binary digit 1) and a distorted reproduction. An exact reproduction would require a transmission channel with infinite bandwidth as a square pulse has infinite bandwidth (see Fig. 9.4c). But if we only need to detect that a pulse has been sent, we can get by with a finite channel bandwidth. For example, if we were to calculate the effect of an ideal low-pass filter on a square pulse, we would find the output to be a distorted pulse which resembles the original pulse better and better with increasing bandwidth B of the filter.

Figure 9.11 demonstrates this effect and shows that a filter bandpass of $2/T$ allows excellent identification of the original pulse but that a bandwidth of $1/4T$ would give the smooth and shallow curve for the pulse, which because of noise would result in too many misidentifications. For many purposes the bandwidth

$$B = \frac{1}{2T} \tag{9.20}$$

yields a resolution with an acceptable error rate. Since a transmission channel such as coaxial cable has band-pass characteristics similar to that of a low-pass filter, a pulse propagating down a cable will be affected similarly. Thus as long as the shape of the

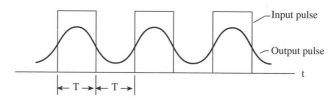

FIGURE 9.12   A bandwidth of $B = 1/2T$ is sufficient to allow resolution of square pulses. Shorter pulses or smaller separations between pulses would require a larger bandwidth.

received signal can still be identified as a pulse or the absence of a pulse, the sent message can be identified at the receiving end. This is at the heart of digital transmission when compared to analog transmission in which a signal becomes irreversibly distorted due to the addition of noise. In digital transmission, even though the individual pulses become badly distorted during propagation, as long as the distorted pulse that is received can be identified with the presence or absence of a pulse, the original message is preserved. Furthermore, the distorted pulses can be regenerated to their original shape at a repeater station and resent, thus making transmissions over long distances possible. Figure 9.12 shows a sequence of square pulses after low-pass filtering. Equation (9.20) gives the condition for distinguishing between output pulses spaced by T. A smaller spacing or a smaller bandwidth would result in overlap, making it difficult to identify separate pulses at the receiving end. Summarizing, we can state that the bandwidth required for a digital signal of R bits per second is $R/2$ Hz.

EXAMPLE 9.9   A transmission channel is used for the propagation of pulses. Find the minimum channel bandwidth required for pulse detection and resolution of a sequence of 5 μs pulses which are randomly spaced. The minimum and maximum spacing between pulses is 2 μs and 10 μs, respectively.

To resolve pulses of 5 μs duration would require a transmission bandwidth of $B = 1/2.5 \mu s = 100$ kHz. However, to resolve a spacing between pulses of 2 μs requires a larger bandwidth, given by $B = 1/2(2 \mu s) = 1/4 \mu s = 250$ kHz, which is the required minimum bandwidth.   ∎

EXAMPLE 9.10   Here we examine the relationship between bandwidth and information rate for simple coding (pulse code modulation or PCM).

A mechanically activated printer punches small holes in a paper tape, creating a coded message consisting of holes or the absence of holes and representing bits or the absence of bits of information. If the printer can punch up to 10 consecutive holes per second and if the tape is used for transmission at that rate, the bandwidth necessary for transmission is at least five cycles per second (5 Hz). If the printer can punch 2 million holes per second, and again if the tape is used for transmission at the equivalent rate, the source will be said to have a bandwidth of at least a million hertz. By the same token, in order to carry a signal of that bandwidth one needs a channel able to accommodate frequencies of up to a million hertz in order to transmit the information at the rate at which it is produced. Therefore, a source which produces information at the rate of N

bits per second (bps), coded as simple on–off pulses (PCM), has a minimum bandwidth of $N/2$ Hz.

For this case we can justify the bandwidth of 1 Hz for 2 bits of information by viewing the square wave in Fig. 5.10 as a stream of a pair of on–off pulses per period T of the square wave. The fundamental frequency of the square wave is $1/T$, and hence 2 bits/second per hertz. We can also arrive at the same conclusion by simply looking at Fig. 9.4a, which shows two distinguishable intervals per period $T = 1/f_o$. Hence, the number of distinguishable intervals in a band of frequencies from f_1 to f_2 is $2B$, where $B = f_2 - f_1$.

Using more efficient coding than PCM, higher rates than 2 bps/Hz can be obtained. For example, digitizing a 3 kHz telephone voice signal (Example 9.5) results in a 64 kbps digital signal. Using simple on–off pulses to represent the 64 kbps signal would require a channel bandwidth of at least 32 kHz to transmit the digitized voice signal. Thus PCM requires 8 ($= 32/4$) times as much bandwidth as the original analog signal. Modems, using error-correction and band-compression techniques, are able to send a 28.8 kbps signal over a 4 kHz telephone line, giving a 7.2 bps/Hz efficiency of bandwidth usage. The ratio of data bit rate to channel bandwidth, measured in bps/Hz, is called bandwidth efficiency or spectral efficiency.   ∎

9.5.5   Transmission Channels

Channels for transmission of information can be an ordinary telephone transmission line consisting of a twisted pair of copper wires, a coaxial line, a parallel wire line (300 Ω TV twin lead), a waveguide, a fiberoptic cable; free-space transmission channels such as radio, television, mobile, and satellite antennae to receiver antennae; and point-to-point antennae such as the dish antennae on microwave towers.

Transmission systems carry electrical signals. Hence, before a twisted pair of copper wires can carry a telephone conversation, the voice (acoustical signal) of a caller must first be converted by a microphone in the handset to an electrical signal, which in turn must be reconverted (by a speaker in the handset) to an acoustical signal at the receiving end for a listener's ear. The electrical signals on an open-wire line such as a twisted pair travel at the velocity of light, which is determined by the expression

$$v = \frac{1}{\sqrt{\varepsilon\mu}} \tag{9.21}$$

where ε and μ are the *permittivity of free space* (capacitance per unit length measured in farads/meter) and the *permeability of free space* (inductance per unit length measured in henries/meter), respectively. In free space $v = 3 \cdot 10^8$ m/s given that $\varepsilon = 9.854 \cdot 10^{-12}$ F/m and $\mu = 4\pi \cdot 10^{-7}$ H/m. The signal travels as an electromagnetic (EM) wave just outside the wires. It differs from a free-space EM wave (such as one launched by a TV, radio, or mobile antenna which spreads out in all directions), only in that it is bound to and is guided by the wires of the transmission line.

When do two connecting wires become a transmission line? It is when the capacitance and inductance of the wires act as distributed instead of lumped, which begins to happen when the circuit approaches dimensions of a wavelength (wavelength λ and frequency f are related by $\lambda = v/f$). At sufficiently high frequencies, when the length of connecting wires between any two devices such as two computers is on the order of a wavelength or larger, the voltages and currents between the two devices act as waves that can travel back and forth on the wires. Hence, a signal sent out by one device, propagates as a wave toward the receiving device and the wave is reflected unless the receiving device is properly terminated or matched. Of course, should we have a mismatch, the reflected wave can interfere with the incident wave, making communication unreliable or even impossible. This is important when networking computers, printers, and other peripherals which must be properly matched to avoid reflections.

Other phenomena associated with waves are also important for transmission lines. One is characteristic impedance Z_o of a transmission line,

$$Z_o = \sqrt{\frac{L}{C}} \tag{9.22}$$

which is the ratio of the voltage wave and current wave that propagate down the line. Another is the velocity of propagation on a transmission line which, by analogy to (9.21), is given by

$$v = \frac{1}{\sqrt{CL}} \tag{9.23}$$

where C and L are the distributed capacitance (farads/meter) and distributed inductance (henries/meter) of a particular transmission line. Another is the reflection coefficient r at the end of a transmission line which is terminated by a load impedance Z_L,

$$r = \frac{Z_L - Z_o}{Z_L + Z_o} \tag{9.24}$$

The reflection coefficient is a ratio of reflected to incident voltage, that is, $r = V_{\text{reflected}}/V_{\text{incident}}$. Hence, r gives the part of the incident voltage that is reflected by a terminating or receiving device. From (9.24) we see that to avoid the occurrence of a reflected voltage wave, we need to match the characteristic impedance of the transmission line to the load impedance; that is, if $Z_o = Z_L$, then $r = 0$ and the incident wave is completely absorbed by the load. In high-data-rate communication networks and in fast computers, the operating frequencies are sufficiently high that even short connections act as transmission lines which require matching. Hence, in high-speed computers if the physical dimensions of the computer can be kept small, reflections might be avoided.

EXAMPLE 9.11   (a) Find the power drained from a 12 V battery that is connected at time $t = 0$ to a very long RG 58 coaxial cable. Sketch the outgoing voltage and current waves before they reach the end of the cable. The setup is sketched in Fig. 9.13a.

(b) If a 1 km-long coaxial cable of the same type as in (a) is terminated by a 25 Ω load resistance, calculate the reflected voltage and power and sketch the voltage that exists on the coaxial cable after reflection took place.

Solution.   (a) The characteristic impedance of the RG 58 coax is 50 Ω, the inductance per unit length is 253 nH/m, and the capacitance per unit length is 100 pF/m.

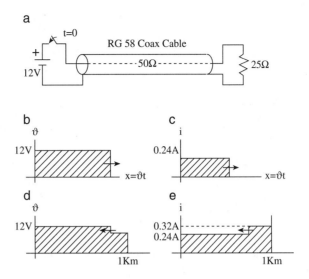

FIGURE 9.13   A coaxial transmission line (suddenly connected to a battery) showing incident and reflected waves of voltage and current.

Using (9.23) we can calculate the wave velocity as $2 \cdot 10^8$ m/s. When at time $t = 0$ the switch is closed, a 12 V voltage wave is launched by the battery and proceeds to travel down the line at a speed of $2 \cdot 10^8$ m/s. Similarly a coupled current wave of strength 12 V/50 Ω = 0.24 A accompanies the voltage wave. Figures 9.13b and c show these waves. The battery is therefore drained at the rate of 12 V $\cdot$0.24 A = 2.88 W.

Solution.   (b) The incident wave, upon reaching the end of the line, sees a sudden change in impedance from 50 Ω to 25 Ω. Such a discontinuity causes a reflection of the incident wave similar to any other situation that involves a discontinuity, such as a tennis ball hitting a wall. Not all of the incident voltage will be reflected. Of the 12 V incident on the load resistance, the reflected voltage is $V_r = rV_i = -12/3 = -4$ V, where the reflection coefficient r is given by (9.24) as $r = (25 - 50)/(25 + 50) = -1/3$. The power absorbed by the load resistance is equal to the incident power minus the reflected power, i.e.,

$$P_L = P_i - P_r \qquad (9.25)$$

$$\frac{V_L^2}{Z_L} = \frac{V_i^2}{Z_o} - \frac{V_r^2}{Z_o}$$

$$\frac{V_L^2}{Z_L} = \frac{V_i^2}{Z_o}(1 - r^2)$$

Since the voltage across the load is $V_L = V_i + V_r = V_i(1 + r) = 12(1 - 1/3) = 8$ V, the power absorbed by the load is $P_L = 8^2/25 = 2.56$ W, the incident power is $P_i =$

$12^2/50 = 2.88$ W, and the reflected power is $P_r = 2.88(1/3)^2 = 2.88 \cdot 1/9 = 0.32$ W. Figures 9.13*d* and *e* show the total voltage and current that exist at one particular time after reflection took place. You will notice that the reflected voltage decreases the total voltage on the line (the voltage reflection coefficient is negative) but that the reflected current increases the total current. The reason is that the current reflection coefficient is the negative of the voltage reflection coefficient, a fact that derives from conservation of energy at the load as follows: if the power at the load is expressed by (9.25), then

$$
\begin{aligned}
P_L &= V_L I_L = (V_i + V_r)(I_i + I_r) = V_i(1+r)I_i(1-r) \\
    &= V_i(1+r)(V_i/Z_o)(1-r) = V_i^2/Z_o(1-r^2)
\end{aligned}
$$

and the reflection coefficient of voltage and current have the same amplitude but are opposite in sign. In other words, conservation of energy gives us the term $(1-r^2)$, which factors as $(1-r^2) = (1+r)(1-r)$. Thus $r \equiv r_V = -r_I$.

The motivation for this rather lengthy and detailed example is to convince the reader that unless the length of interconnections is short between two computers that communicate with each other at high data rates, the associated high frequency of the high data rates makes any length of wires larger than a wavelength act as a transmission line. Unless properly terminated, the resulting reflection of the data streams can make communication between two such computers, or any other peripherals for that matter, impossible. ∎

EXAMPLE 9.12   The capacitance of a 10 m-long transmission line is measured and found to be 500 pF. If it takes 50 ns for a pulse to traverse the length of the line, find (a) the capacitance per unit length of the line, (b) the inductance of the line, (c) the characteristic impedance of the line, and (d) the terminating or load resistance to match the transmission line.

Solution.   (a) The capacitance per unit length is 500 pF/10 m = 50 pF/m. (b) Since the time to traverse the line is specified we can calculate the speed of the pulse as 10 m/50 ns = $2 \cdot 10^8$ m/s. If we use this result in (9.23), the inductance/length of the line can be calculated as $L = 1/v^2 C = 1/(2 \cdot 10^8)^2 \cdot 50 \cdot 10^{-12} = 0.5$ μH/m, which gives 5 μH for the inductance of the line. (c) Using (9.22), the characteristic impedance of the line is $Z_o = (L/C)^{1/2} = ((0.5 \cdot 10^{-6})/(50 \cdot 10^{-12}))^{1/2} = (1000)^{1/2} = 31.6$ Ω. (d) A matched load would absorb all of the energy in the incident wave. This reflectionless case takes place when $Z_L = Z_o = 31.6$ Ω.

Even though the one-dimensional guided waves on transmission lines appear to be different from free-space waves which can spread out in all three dimensions, the basic principles are the same for both types. Free space can be considered as a transmission line between transmitter and receiver antennae. Similar to transmission lines, free space has distributed capacitance and inductance (for vacuum, ε and μ are given as $9.854 \cdot 10^{-12}$ F/m and $4\pi \cdot 10^{-7}$ H/m, respectively), it has a characteristic impedance, sometimes called wave impedance, given by $Z_o = \sqrt{\mu/\varepsilon} = 377$ Ω, just like (9.22), and has a reflection coefficient r just like in (9.24).

Twisted-Wire Transmission Line

This is the classic line used in telephone circuits. The frequency response of such a line, which is shown in Fig. 9.14, looks like that of a low-pass RC filter (see Fig. 9.6a or 2.6b). Because the resistance of the wires increases with frequency, leading to increased I^2R wire losses, the telephone line loss increases with frequency, and hence the low-pass filter action.[14] Figure 9.14 shows that the frequency response of a 1 km-long, twisted line is down by −9 dB at 1 MHz and continues to decrease for higher frequencies. Let us assume that a simple, 1 km-long transmission system is needed. If a loss of −9 dB can be tolerated at the receiving end, a pair of twisted copper wires 1 km long would be the simplest implementation. The 1 km cable would have an effective bandwidth of 1 MHz which could be frequency-divided into 250 four-kilohertz-wide channels, each carrying analog voice signals. Of course a shorter cable would have a larger bandwidth and could carry more voice channels. Commercial telephone companies use twisted-wire lines for short distances. The bandwidth of each line is divided into many voice channels, each 4 kHz wide (the division is done at the central station). Even though voices sound quite natural when restricted to 3 kHz, a 4 kHz bandwidth is used to give some separation (guard bands) between channels so as to minimize cross talk. On the other hand, the newer DSL service tries to use the entire 1 MHz bandwidth to give fast service at Mbps rates over twisted copper lines.

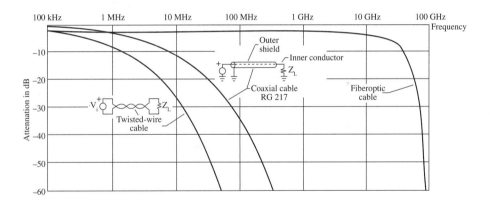

FIGURE 9.14 Frequency response of three types of cables, each 1 km long. In the two metal-conductor cables the resistance and hence the loss increase with frequency f as $f^{1/2}$. In the fiberoptic cable attenuation increases as f^2 because of pulse dispersion, where f is the frequency of the signal that modulates the light.

---

[14]The loss increases with frequency f as $\sqrt{f}$ because the resistance of the wires increases as $\sqrt{f}$. The increase in the wire resistance is caused by the *skin effect*, which forces the current in the wire to flow in a progressively thinner layer as frequency increases. This loss, which is proportional to $-\sqrt{f}$ and is shown in Fig. 9.14, determines the frequency response of the transmission line. The decrease by this curve with frequency gives a transmission line a characteristic of a low-pass filter.

ISDN, Modems, and Twisted-Wire Lines

In an ideal telephone network the signals from each customer would be digital and remain digital throughout the system until its final destination. In the present system, local lines which are pairs of twisted copper wires, originally installed for analog telephony, are used to transmit baseband voice signals from a customer or PBX (private branch exchange) to the local switching center (local exchange). There they are converted to digital signals for onward transmission over digital lines called trunks. Digital trunk lines interconnect exchanges. At the receiving exchange, the digital signal is converted back to an analog signal suitable for the local lines. When a customer uses a digital device such as a fax machine or computer, a modem (modulator/demodulator) must first be used to convert the digital signal to analog form. A modem codes the digital information into a stream of sinusoidal pulses which the local lines then treat as though they were analog voice signals. If local lines were digital, modems would be unnecessary and digital devices could be connected directly to the lines—this is the principle behind ISDN, which is discussed next. However, because of the number of installed links from private customers to local exchanges, they will remain analog for the foreseeable future.

ISDN (integrated services digital network)[15] is an effort to enable twisted copper lines to carry digital signals with a much higher bandwidth than the 4 kHz for which they were originally designed. Pulse code modulation codes the customer's analog devices to generate digital signals. The basic rate for ISDN access is two 64 kbps channels (B channels) plus one 16 kbps data channel (D channel), referred to as the $2B + D$ access. This gives a combined rate of 144 kbps, which is about at the limit for installed links between customers and exchanges which are twisted copper wires initially intended for analog service. To convert an ordinary twisted-pair copper wire line to ISDN service, the old-fashioned and slow analog switch must be changed to a faster digital switch at the local exchange.

DSL (digital subscriber line) is a service[15] designed for much higher rates (in the Mbps range) using all of the 1 MHz bandwidth of twisted copper lines. It uses packet switching, which is a technique for data transmission in which a message is divided into packets and transferred to their destination over channels that are dedicated to the connection only for the duration of the packet's transmission. DSL users can receive voice and data simultaneously.

Coaxial Cable

A 1 km-long coaxial line of the type shown in Fig. 9.14 has a larger band-pass than a twisted line by about a factor of 8, with a loss of only -3 dB at 1 MHz (-9 dB at 8 MHz). It is also shielded as the outer cylindrical conductor of a coaxial cable is normally grounded. This means that one can route a coax through a noisy environment with little of the noise signals being added to the information signal that propagates on the coax. The frequency response of both lines can be made flat up to about 100 MHz by using equalizer amplifiers which amplify the higher frequencies more than the low ones.

---

[15]ISDN and DSL are discussed in more detail in later sections.

Possibly the most popular coax cable for short interconnections between computers is the 50 Ω RG 58 cable; for television connections the common cable is the 75 Ω RG 59. Both cables are light, flexible, and comparatively thin but have rapidly increasing losses for longer lengths.[16]

Fiberoptic Cable

For really large bandwidth the optical fiber reigns supreme. A typical glass fiber is thinner than a human hair and is made of pure sand, which, unlike copper, is not a scarce resource. Light propagates in a fiber at a wavelength of 1.3 μm (230 THz = 230,000 GHz = $2.3 \cdot 10^{14}$ Hz) virtually loss-free (transmission losses can be as low as 0.2 dB/km). Because of this low attenuation, the distances between repeaters can be very long, up to a few hundred kilometers, and short transmission links do not need any repeaters between the transmit and receive terminals. The flat attenuation of 1 to 2 dB over a large frequency range shown in Fig. 9.14 is due to light loss. A natural way to place information on an optical fiber is by rapid on–off switching of the light that enters the fiber, making such a cable a natural for the transmission of digital signals. It is the frequency of the digital modulating signal that is used to label the horizontal axis in Fig. 9.14 for the fiberoptic cable. Modulating signals can be as high as 10% of the carrier frequency, making the theoretical bandwidth about $2 \cdot 10^{13}$ Hz for fiber optics, which is extremely high. Such high switching speeds are beyond our present technology, and even if they were possible, the steep roll-off shown in Fig. 9.14 limits the bandwidth of a fiber cable. The roll-off is determined by *pulse dispersion* on the optical cable, which gives an effective attenuation of 3 dB at about 30 GHz. If we use the -3 dB point to define the low-pass band of a fiberoptic cable, we still have a very large bandwidth of 30 GHz—which gives voice, video, and data speeds of at least 100 times faster than those for the standard copper wiring used in telecommunications. Note that this bandwidth represents only a small fraction of the promise of this technology.

The practical information-carrying rate of a fiber is limited by pulse dispersion, which is the tendency of a pulse which represents the binary digit 1 to lose its shape over long distances. Optical fibers transmit different frequencies at slightly different speeds. Since the frequency content of a short pulse is high (see Fig. 9.4d) and since the higher frequencies travel a bit faster than the lower ones, the propagating pulse tends to stretch or elongate in the direction of propagation—consequently a series of pulses become smeared and unrecognizable. This pulse widening due to dispersion grows as f^2 (where f is the modulating frequency of the information signal) and accounts for the steep attenuation in optical fiber shown in Fig. 9.14. Nevertheless the optical fiber has revolutionized long-distance communications and will continue to do so, especially with progress toward dispersion-free propagation in fibers, faster electronic switching, and eventually all-optical switching.

---

[16]For more data on transmission lines see Chapter 29 of *Reference Data for Radio Engineers*, H. W. Sams, Indianapolis, 1985.

Wireless Transmission

Whenever it is impractical to string cables between two points we resort to free-space communication. Familiar cases are those of AM, FM, and TV broadcasting which involves one localized transmitter broadcasting to many receivers, not unlike wireless (cellular) telephony which also resorts to broadcasting in order to reach a called party that can be anywhere in some region (cell). Similarly for mobile radio.

In free-space transmission we can also have point-to-point communication, although perhaps not quite as effective as stringing a cable between two points, which guarantees privacy and awareness of any unauthorized wire tampering. In order to concentrate radio waves in a sharply directed beam like that of a search light, the transmitting antenna must be large in terms of wavelength, which suggests that to have antennae of reasonable physical size we need to use a short wavelength (or high frequencies). Recall that frequency f is related to wavelength λ by

$$\lambda = \frac{v}{f} \tag{9.26}$$

where v is given by (9.21) and is the velocity of light, which in free space is $3 \cdot 10^8$ m/s. Typical point-to-point communication as between antennae mounted on microwave towers or between satellites and earth is at microwave frequencies (1 GHz to 300 GHz or 30 cm to 1 mm wavelengths) and uses dish antennae, perhaps a meter in diameter. For example, at 10 GHz, a 1 m-diameter dish is 33 wavelengths across, which gives it good directivity. *Rayleigh's criterion* in optics can be used to relate directivity to

$$\theta = 2.44 \frac{\lambda}{D} \tag{9.27}$$

where θ gives the angle of a conical beam emanating from an aperture antenna of diameter D as shown in Fig. 9.15a. Eighty-five percent of the radiated energy is contained within the angle θ. Even though this law has its basis in optics—which has a frequency range where most physical structures such as lenses are millions of wavelength across—it nevertheless gives usable results when structures are only a few wavelengths across. Because microwave systems are line-of-sight, microwave towers are spaced approximately every 40 km along their route.

EXAMPLE 9.13   A transmitting dish has a 1 m diameter aperture and radiates a beam at a frequency of 5 GHz. Find the beam width and the beam width area at 10 km.

Solution.   The conical beam angle, using (9.27), is $\theta = 2.44(6 \text{ cm})/1 \text{ m} = 0.146$ rad $= 8.4°$, which is a fairly narrow beam. That is, whereas an isotropic antenna would radiate the transmitter's power uniformly over 360°, this dish antenna concentrates this power in a beam of 8.4°.

The cross section of the beam at a distance of 10 km is calculated by using the arc length of a circular segment, that is, $s = r\theta$, where r is the radius of the circle. Thus $s = (10 \text{ km}) \cdot 0.146 = 1460$ m, which is much larger than any receiving dish. Hence since only a small fraction of the transmitted energy is intercepted by a receiving dish,

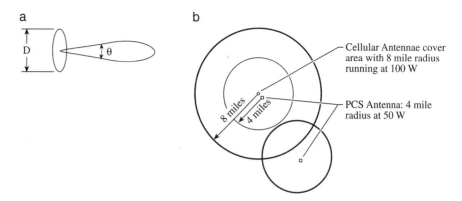

FIGURE 9.15   (a) A circular aperture antenna of diameter D can concentrate EM waves in a beam of width θ. Such antennae are used in free-space point-to-point communication to establish a link. (b) Cellular antennae typically cover an area with a 8 mile radius while PCS antennae have a 4 mile radius.

this link is not very efficient. To increase the efficiency, either the transmitting or the receiving dish must be larger or a higher frequency must be used. The beam width area is $\pi \, (s/2)^2 = 3.14 \cdot (0.73 \text{ km})^2 = 1.674 \text{ km}^2$.   ∎

EXAMPLE 9.14   Compare cellular telephony with the newer PCS (personal communication service) standard.

At the present time wireless telephony connected to the public switched telephone network consists of cordless telephones, cellular radio systems, and personal communication systems. Cellular technology was designed in the 1960s and 1970s by Bell Labs for use in cars and originally relied on basic analog radio signals. An area is divided into many contiguous cells with an antenna in each cell and all antennae controlled by a central computer. A car driving through the cells would be handed off from cell to cell almost seamlessly as the signal from a given cell began to weaken and a stronger signal from an adjacent cell appeared. Analog frequency modulation of the signals, which are in the 800 MHz frequency band, was used. Currently digital modulation and digital voice compression in conjunction with time-division multiple access (TDMA) exist, as well as a form of spread spectrum multiple access known as code-division multiple access (CDMA). PCS networks, introduced in the 1990s, use digital technology, which increases efficiency, enhances privacy and security, and allows for services beyond basic two-way communication; however, at present cellular and PCS technology are nearly indistinguishable to the typical wireless user. Frequencies for PCS are in the 2 GHz range, which makes for smaller antennae that are cheaper to manufacture and are easier to install in high places such as towers, tall buildings, and telephone posts. Recall that for an antenna to radiate efficiently it must be on the order of a wavelength λ, with $\lambda/4$ antennae most common as they give a uniform radiation pattern which is circular, as shown in Fig. 9.15b. Thus, the 800 MHz cellular antennae have a $\lambda/4$ length which is $\lambda/4 = (300/f_{\text{MHz}})/4 = (300/800)/4 = 0.094 \text{ m} = 9.4 \text{ cm}$ long, whereas PCS antennae would be only 3.75 cm long. However, typical cellular antennae sit atop a tower

FIGURE 9.16   Noise and its effect on the transmission of a signal.   (a) Input signal.   (b) Noise in the channel which adds to the signal.   (c) The resolution of the output signal is now limited by noise.

about 300 ft. high, run at 100 W of power, and reach a typical radius of 8 miles, whereas PCS antennae are on towers about 100 ft. high, run at 50 W of power, and reach a typical radius of 4 miles.   That means that PCS antennae, to avoid annoying dead spots, require about five times as many antennas as cellular networks to serve the same area. Comparison of the two radiation patterns are given in Fig. 9.15b.   An estimated 30,000 antennae have been built by the cellular industry across the United States.   For the same demand PCS would have to build in excess of 100,000 antennae, which presents a challenging problem as many communities object to a proliferation of antenna structures in their neighborhoods.   Installers avoid building new towers at all cost and try to install antennae on existing structures such as water towers, church steeples, electric utility poles, etc.   ■

9.5.6   Signal-to-Noise Ratio and Channel Bandwidth

In this section we will show that the amount of information that a channel can carry depends on the bandwidth of the channel and the magnitude of noise present in the channel.   Any physical channel such as coaxial cable has noise present, either man-made or created by the electron current that flows in the copper wires.   Even though electrons flow smoothly in the wires of a coax that carry a signal current, they do have a random motion component which gives rise to a noise signal.   Fiberoptic cables have other noise sources.   The amount of noise present in any channel limits the number of distinct amplitude levels that a signal propagating in the channel may have.   For example, if a varying analog signal has a maximum level of 10 V and the noise level is 5 V, the signal may have only few levels.   On the other hand, if the noise level is only 1 mV, the same signal can be divided into approximately $10 \text{ V}/1 \text{ mV} = 10^4$ levels. Thus when converting an analog signal to a digital signal, as in Fig. 9.3, the amount of noise present determines the maximum number of quantization steps that each sample of the analog amplitude may have.   Figure 9.16 shows pictorially how noise that has been added during transmission can degrade the signal and hence its resolution at the receiving end.

The signal-to-noise ratio (SNR) is the standard measure of the noise level in a system.   It is the ratio of received signal power P_s to noise power P_n, and since power is proportional to voltage squared, we can express SNR as

$$\text{SNR} = \frac{P_s}{P_n} = \left(\frac{v_s}{v_n}\right)^2 \tag{9.28}$$

where v_s is the received signal voltage and v_n is the noise voltage (noise voltages, because of their multitude of random amplitudes, are typically given as rms voltages). SNR is usually expressed in decibels (dB) as[17]

$$\text{SNR}_{dB} = 10\log_{10}\text{SNR} = 20\log_{10}\left(\frac{v_s}{v_n}\right) \tag{9.29}$$

Note that a power ratio of 10,000 and a voltage ratio of 100 are both 40 dB.

In the preceding paragraph we gave an example of reception in a very noisy environment which would result in a SNR of 4, or $\text{SNR}_{dB} = 20\log_{10}(10\text{ V}/5\text{ V}) = 20 \cdot (0.301) = 6.02$ dB. On the other hand, if the noise level is reduced to 1 mV, the resulting signal-to-noise ratio is $\text{SNR} = (10^4)^2 = 10^8$ or $\text{SNR}_{dB} = 20\log_{10}(10\text{ V}/1\text{ mV} = 10^4) = 80$ dB, which is an excellent signal-to-noise ratio environment. For comparison, AM radio operates typically with a SNR of 3000 ($\text{SNR}_{dB} = 10\log_{10}3000 = 35$ dB).

EXAMPLE 9.15 A music signal is characterized by a rms voltage of 2 V and a bandwidth B of 15 kHz. A noise process adds a rms noise voltage of 4 mV to the music. Find the information rate and the signal-to-noise ratio of the music.

According to Nyquist's sampling criterion (9.12), we sample the music at $f_s = 2f_h = 2B = 2 \cdot 15$ kHz $= 30$ kHz, which results in 30,000 samples per second. Let us use the analog noise that exists with the music to approximate the maximum number of quantization intervals as $v_s/v_n = 2$ V/4 mV$= 500$. The music signal voltage for each sample, on average, can then be subdivided into no more than 500 intervals (quantization intervals) because the noise would mask any finer division. Since $2^9 = 512$, we can use 9 bits per sample to number the 500 quantization levels. As there are 30,000 samples per second, the information rate to encode the music is $I = 9 \cdot 30,000 = 270$ kbits/s (kbps). If this music signal were to be sent over a transmission system as a binary signal (coded as simple on–off pulses—see Example 9.10), the system would need to have a bandwidth of 270 kbits/sec $\cdot$ cycle/(2 bits) $= 135$ kcycles/s $= 135$ kHz. That is a ninefold increase in bandwidth for the digitized music signal when compared to the original analog 15 kHz music signal. As we will show in the following section, to transmit the digitized signal requires a transmission medium with a channel capacity nine times the channel capacity needed for transmission of the analog signal.

The signal-to-noise ratio is $\text{SNR}_{dB} = 20\log_{10}(v_s/v_n) = 20 \cdot \log_{10}500 = 54$ dB. ■

---

[17]The decibel measure was introduced in Section 5.4.

Channel Capacity

Let us now show that transmission of information in a channel depends on SNR and channel bandwidth. As a matter of fact we may trade off one for the other when transmitting information. Assume that noise power P_n and signal power P_s exist in a channel. A message is sent as a sequence of information intervals. At the receiving end, during each information interval, we find ourselves trying to distinguish a signal amplitude of $\sqrt{P_n + P_s}$ volts in the presence of a noise amplitude of $\sqrt{P_n}$ volts. From our previous discussion we can state that the number of distinct amplitudes is limited by the ratio of these two voltages. Hence, in each interval the number of available information states, on average, will be

$$S = \frac{\sqrt{P_n + P_s}}{\sqrt{P_n}} = \sqrt{1 + \frac{P_s}{P_n}} = \sqrt{1 + \text{SNR}} \tag{9.30}$$

The information content of S resolvable or distinguishable states[18] is then, from (9.1),

$$
\begin{aligned}
I_o &= \log_2 S = \log_2 \sqrt{1 + \frac{P_s}{P_n}} \\
&= \frac{1}{2} \log_2 \left(1 + \frac{P_s}{P_n}\right) \\
&= \frac{1}{2} \log_2(1 + \text{SNR}) \text{ bits}
\end{aligned}
\tag{9.31}
$$

What this means is that at the receiving end we cannot distinguish more than I_o bits no matter how many bits were sent. Noise imposes this limitation. If the channel were noiseless, it would be possible to distinguish an infinite number of different amplitude levels for each sample, implying that an infinite number of bits per second could be transmitted.

The remaining question concerns how fast we can send these bits through the channel if the channel bandwidth is given as B_{ch}. In Example 9.10 we concluded that the maximum number of distinguishable intervals in a band of frequencies B is $2B$. Hence, the available transmission rate in bits per second, also known as information capacity or channel capacity, can be stated as the number of distinguishable intervals per second multiplied by the bits per interval,

$$
\begin{aligned}
C &= 2 B_{\text{ch}} \tfrac{1}{2} \log_2(1 + \text{SNR}) \\
&= B_{\text{ch}} \log_2(1 + \text{SNR}) \text{ bits/s}
\end{aligned}
\tag{9.32}
$$

This is one of the most famous theorems in information theory, formalized by Shannon and known as the *Hartley–Shannon Law*. The interpretation of channel capacity C is as

---

[18]From the discussion in Section 9.3 we see that (9.31) also gives the number of bits required to code S levels. For example, if we have 8 levels, it takes 3 bits to number the levels (recall that $8 = 2^3$). Hence, taking log to the base 2 of a quantity gives its bit depth.

follows: while it is possible to transmit information up to C bits per second (bps) with negligible error, the error increases rapidly when trying to transmit more than C bps. Furthermore, we may trade off signal-to-noise ratio for channel bandwidth: keeping the channel capacity C fixed, B_{ch} can be decreased if SNR is increased and conversely B_{ch} must be increased if SNR decreases. Thus, in a noisy environment in which SNR is low, using a broad bandwidth to send messages might be advantageous. If data need to be transmitted at a faster rate, (9.32) shows that bandwidth and/or SNR must be increased. To increase SNR by increasing P_s is costly as a linear increase in C requires an exponential increase in P_s/P_n, which explains the emphasis on increasing bandwidth in communication systems.

Recall that if an analog signal in which the highest frequency is f_h is to be converted to a digital signal, the sampling frequency must be $f_s = 2f_h$. If the analog signal in each sample is represented by S quantization levels and if the S levels in turn are represented by an n-bit binary number, the information rate of the analog signal is given by

$$I = 2f_h \log_2 S \text{ bits/s} \tag{9.33}$$

where the number of bits needed to represent the S levels is $n = \log_2 S$. Assume we want to transmit the digitized signal through a channel which has a capacity of transmitting C bps. Another theorem, *Shannon's fundamental theorem*, states that error-free transmission is possible if[19]

$$I \leq C \tag{9.34}$$

We have now presented the two most powerful theorems of information theory. The first, Eq. (9.32), states that there is a maximum to the rate at which communication systems can operate reliably, and the second, Eq. (9.34), states that for any code rate I less than or equal to the channel capacity C, error-free transmission of messages is possible. It should be pointed out that prior to Shannon's work it was believed that increasing the rate of information transmitted over a channel would increase errors. What was surprising is that Shannon proved error-free transmission provided that $I \leq B_{ch} \log_2(1+\text{SNR})$, that is, as long as the transmission rate was less than the channel capacity. It should also be pointed out that although Shannon showed error-free communication is possible over a noisy channel, his theorems give limits without any guide on how to design optimum coders or modulators to achieve those limits.

EXAMPLE 9.16   Compare the following two communication systems by finding their respective transmission rates (or channel capacity) C. System (a): $B_{ch} = 5$ kHz and $\text{SNR}_{dB} = 50$ dB. System (b): $B_{ch} = 15$ kHz and $\text{SNR}_{dB} = 17$ dB.

To use (9.32) for the transmission rate, we first need SNR, which from (9.29), for system (a), is SNR $= 10^{50/10} = 100,000$. Then $C = 5$ kHz $\cdot \log_2 100,000 = 5$ kHz $\cdot 16.7 = 83.3$ kbits/s. For system (b) we obtain $C = 15$ kHz $\cdot \log_2(10^{17/10}) = 15$ kHz $\cdot 5.7 = 85$ kbits/s. Hence the two systems, even though substantially different, have about the same channel capacity.   ∎

---

[19]Note that (9.34) is more general than (9.19), which is more useful for analog signals.

9.5.7   Noise Created by Digitization

To convert an analog signal like the one shown in Fig. 9.17 to a digital signal requires two steps. The first step, called sampling,[20] consists of measuring periodically the value of $v(t)$. These values, called samples, are denoted by the black dots in the figure. The second step is quantization and it consists of assigning a rounded-off numerical value to the sample and representing that value by a binary number,[21] that is, a binary with a finite number of bits. Clearly the more bits we have available the more accurately we can represent each value because the amplitudes of $v(t)$ can then be subdivided into finer intervals (called quantization intervals). For example, with 3 bits one can represent 8 intervals (2^3 combinations: 000, 001, 010, 011, 100, 101, 110, 111). If we have available 16 bits (the standard for audio digitization) we can divide each sample value in 65,536 intervals and thus represent the amplitude of $v(t)$ very accurately. In our example shown in Fig. 9.17, the digitization hardware decomposes the range of $v(t)$ values into a set of 8 quantization intervals, and associates a binary number with each interval. Each sample of the analog signal is now represented by a 3-bit binary

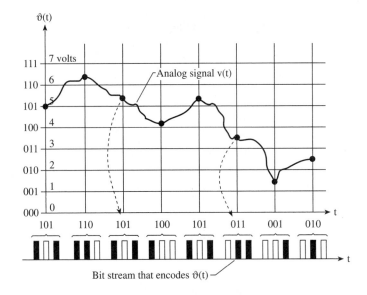

FIGURE 9.17   Sampling and quantization of the analog signal results in the digital bit stream shown.

---

[20]Nyquist's theorem (9.12) states that no information is lost provided that the sampling rate is at least twice the maximum frequency or bandwidth of the analog signal. This theorem states precisely that the faster a signal changes (the larger its bandwidth), the more frequently one must measure it in order to observe its variations.

[21]Recall that any sequence of discrete symbols or characters can be represented by a sequence of binary numbers. For example, the characters on a computer keyboard can be represented by a 7-bit sequence as there are less than 128 characters and $2^7 = 128$. Of course the binary string is seven times as long as the character string. Decoding the binary string, 7 bits at a time, we restore the character string. Representing the bits by a series of on–off pulses is pulse code modulation.

number (from then on the binary numbers represent $v(t)$ and it is the binary numbers that are processed and/or transmitted over a channel). The hardware, by using the two steps of sampling and quantization, replaces the analog signal $v(t)$ by the bit stream (shown on the bottom of Fig. 9.17) of binary numbers associated with the quantization intervals of the samples. The process of sampling and quantization together is known as digitization. Encoding with binary numbers which can be represented by simple on–off pulses is called pulse code modulation.

It is clear that sampling and quantization introduce errors. The more frequently we sample the better we can represent the variations of the analog signal, and the more quantizing intervals we use the more accurately we can represent the amplitude of the signal. As in every measuring operation, one must decide how precise to make each measurement. A certain amount of rounding off is inevitable. For instance, a sample value of 3.2 V in Fig. 9.17 might be encoded as the binary number equivalent to 3 (i.e., 011). The remainder, 0.2 V, is the quantizing error. Quantizing errors are small if the quantization intervals are small. These errors act just like analog noise that is always present to some extent in any analog signal. We call these errors quantization noise. In encoding speech or television pictures the code words must have enough bits so the noise due to quantizing errors cannot be detected by the listener or viewer, or at least does not add appreciably to the noise already present in the analog signal. With 16-bit encoding of music, quantization noise is essentially undetectable. In black-and-white television a 6-bit code (64 levels) begins to fool the eye into seeing nearly continuous gradations of shades of gray. In color television the use of too few bits results in a noisy picture marred by "snow." Similar to analog noise, we measure quantization noise by the ratio of signal power and quantization noise power. Thus if n bits are used to number the quantization intervals, there are then 2^n intervals and a typical error has a magnitude inversely proportional to the number of intervals, i.e., 2^{-n}. Similar to analog signals, where power is proportional to the square of the magnitude, we can state that the noise power is proportional to $(2^{-n})^2$, that is,

$$P_{\text{quant.noise}} \propto 2^{-2n} \tag{9.35}$$

The signal-to-noise ratio, which is $\text{SNR} = P_s / P_{\text{quant.noise}} = 1/2^{-2n} = 2^{2n}$, can now be expressed in terms of decibels as

$$\text{SNR}_{\text{dB}} = 10 \log_{10} \text{SNR} = 10 \log_{10} 2^{2n} = 20n \log 2 \approx 6n \tag{9.36}$$

This is a useful formula which, given a desired signal-to-noise ratio, states the number of bits n needed when digitizing an analog signal.

With the derivation of quantization noise we can now state that there are two major sources of noise. The performance of a digital system is influenced by channel noise, an always-present, additive noise (natural or man-made) introduced anywhere between the input and the output of a channel that affects equally analog and digital signals, and quantization noise which is introduced by the digitization hardware at the input of a channel and propagates along the channel to the receiver output. We have little control over channel noise—it is always there. We minimize it by choosing locations with less noise, using shielded cables, and if possible increasing the transmitted power so as to increase the SNR. This type of noise mainly introduces errors in the received signal: a symbol 1 can be confused with a symbol 0 and vice versa. Quantization noise, on the other hand, is under the control of the designer. We can always decrease this type of noise by dividing the signal range into finer quantization intervals and using more bits (i.e., larger binary numbers) to represent the intervals. However, we do not gain anything by reducing quantization noise much below channel noise as then channel noise sets the effective signal-to-noise ratio. For example, in telephone transmission, a SNR of 48 dB is acceptable. Using (9.36), $48 = 6n$, which gives $n = 8$ bits. Assuming channel noise is 48 dB, when using 8 bits for encoding, channel and quantization noise are comparable. If more than 8 bits are used, channel noise dominates the quality of received speech; if less bits are used quantization noise dominates. Finally, unlike channel noise, quantization noise is signal dependent, meaning that it vanishes when the message signal is turned off.

EXAMPLE 9.17   Compare bit rates and noise in some practical systems.

If the hardware which performs the digitization uses n bits per sample and if it samples the analog signal at a rate of f_s times per second, then the hardware produces a bit stream which has a rate of $n \cdot f_s$ bits per second. The frequency f_s of sampling must be at least twice as high as the highest frequency f_h in the signal, that is, $f_s \geq 2f_h$. The bit stream then has a rate of $2nf_h$ and a signal-to-noise ratio of $6n$ dB.

The telephone network can transmit voice signal frequencies up to 4 kHz and achieve a signal-to-noise ratio of 48 dB. Sampling must then be at $2 \cdot 4$ kHz $= 8$ kHz with $48/6 = 8$ bits per sample, giving the digitized voice signal a rate of 8 kHz$\cdot 8 = 64$ kbps.

Compact CD discs can achieve SNRs of 96 dB and high frequencies of $f_h = 20$ kHz. Sampling at 40 kHz and quantizing with $96/6 = 16$ bits per sample gives 640 kbps per stereo channel or 1.28 Mbps for both channels.

If a television video signal has a high frequency of 4.2 MHz (at baseband) and requires at least 48 dB for acceptable viewing, the bit stream must be $2 \cdot 4.2$ MHz $\cdot(48/6) = 67.2$ Mbps.   ∎

9.5.8   AM, FM, and Pulse Code Modulation (PCM)

Modulation is a technique of imparting baseband information on a carrier. Unlike baseband signals such as music and speech which are acoustic signals and have a limited propagation range, carrier signals are electromagnetic waves and can propagate long distances at the velocity of light.[22] Modulation, such as AM, is used to move a baseband signal anywhere in the frequency spectrum, which is important when multiplexing (combining) many signals. The carrier signal can have any waveshape, although in practice, it usually is a sine wave with amplitude and frequency that can be modulated by the baseband signal.

---

[22]Using a historical perspective, we can compare a mail-pouch message to a baseband signal message—both need to be delivered from point A to point B. The Pony Express did it for the pouch at a speed of about 10 miles per hour, while the carrier signal does it for the baseband signal at 186,000 miles per second.

Amplitude Modulation (AM)

Figure 9.18a shows a carrier signal which is a sinusoid with constant amplitude and frequency. It is typically a radio frequency (RF) wave and must be much higher in frequency (at least by a factor of 10) than the modulating baseband signal. The next figure, Fig. 9.18b, is the modulating signal—it could be a small part of a voice signal. An AM modulator is a multiplier. The amplitude-modulated signal, shown in Fig. 9.18c, is the product of multiplying a carrier and a modulating signal. It results in a new signal which has the same frequency as the carrier but its amplitude varies in synchronism with the modulating signal. The variation of the amplitude manifests itself as two new bands of frequencies about the carrier frequency. As shown in Fig. 9.19, it is these two bands that contain the information of the original baseband signal. AM therefore doubles the bandwidth of the original message. The amplitude-modulated signal is now ready to be transmitted, either over a transmission line or as a wireless signal.

Although not apparent from Fig. 9.18c, AM shifts the baseband frequencies to two higher bands of frequencies centered about the carrier frequency f_c. We can readily show this by considering a very simple baseband message, $A \cdot \cos \omega_m t$, which is a pure tone of frequency f_m and amplitude A such as produced by a whistle. Then, with the help of the trigonometric identity

$$2 \cos x \cdot \cos y = \cos(x - y) + \cos(x + y)$$

if we use the shifted[23] baseband signal $(1 + 0.8 \cos \omega_m t)$ to modulate the carrier signal $\cos \omega_c t$ we obtain

$$(1 + 0.8 \cos \omega_m t) \cos \omega_c t = 0.4 \cos(\omega_c - \omega_m)t + 0.4 \cos(\omega_c + \omega_m)t + \cos \omega_c t \tag{9.37}$$

Equation (9.37) states that the modulating process (the left side) results in three frequencies: a difference frequency between carrier and baseband signal, a sum frequency between carrier and baseband, and the carrier signal. This is pictured in Fig. 9.19a, a frequency-domain diagram. If in place of a single-frequency message we have a band of frequencies ranging from 0 to B as shown in Fig. 9.19b, the modulating process[24]

---

[23]Baseband signals are usually typical AC signals that have positive and negative excursions. If such a signal were to modulate a high-frequency carrier, the top and bottom envelopes (or the peaks) of the modulated signal shown in Fig. 9.18c would not be a faithful copies of the baseband signal as they would cross into each other. By giving the baseband signal a DC level, we make sure that the envelope of the modulated signal has the shape of the original baseband message. Here we have chosen a 1 V DC level and 0.8 V peak level for the AC signal. We must be careful not to overmodulate the carrier as then distortion would again result. For example, for the AC signal peak level larger than 1 V (when added to the 1 V DC level) the modulating signal would again have negative parts which would result in crossing of the bottom and top envelopes, causing distortion. Summarizing, a constant DC level is added to the modulating signal to prevent overmodulation of the carrier. See also p. 347 for an introduction to AM.

[24]The doubling of bandwidth with AM modulation can also be seen by examining the Fourier transforms in Figs. 9.4c and d. In Fig. 9.4c we have a T-long pulse which has a baseband of $1/T$ (the infinite bandwidth of a square pulse is approximated by a finite bandwidth which contains most of the energy of the pulse and is seen to extend from 0 to the frequency $1/T$). When such a pulse is used to multiply (modulate) a sinusoid (carrier) of frequency f_o, in Fig. 9.4d, the bandwidth is doubled and centered at f_o.

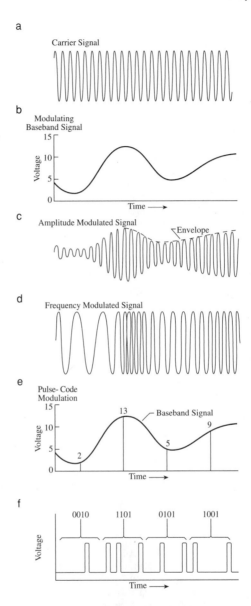

FIGURE 9.18   (a) The carrier is a pure sinusoidal radio frequency wave—an analog signal that can be modulated for transmission by AM, FM, and PCM. (b) The modulating signal which typically is a complex AC signal must be given a DC bias so as not to have a section with negative voltage. (c) An AM signal, which is a continuous variation in the amplitude of the carrier wave, is transmitted and the original signal (the envelope) is recovered or demodulated at the receiver and can be reproduced by a loudspeaker, for example. (d) In an FM signal the amplitude of the modulating signal varies the frequency of the carrier wave without affecting its amplitude (which makes FM more immune to noise). (e) When PCM is used to transmit an analog signal (b), the voltage is first sampled and the voltage values are then expressed as binary numbers. (f) The binary numbers are then transmitted as pulsed signals. At the receiver, the sequence of pulses is decoded to obtain the original signal.

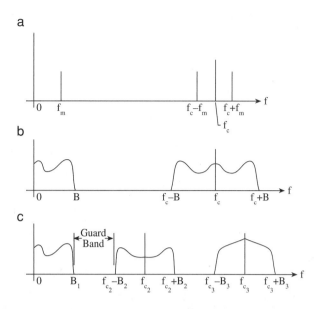

FIGURE 9.19   (a) The message band (baseband) contains a single tone of frequency f_m. AM shifts f_m to two higher frequencies. (b) AM shifts the entire message band to a higher frequency f_c. (c) Use of AM to separate three messages of bandwidth B_1, B_2, and B_3, referred to as frequency-division multiplexing.

results in shifting the original baseband spectrum to a double-sided spectrum centered about the carrier frequency f_c. The implications of this are as follows: say we have three telephone signals, each 4 kHz wide, that we need to combine and transmit simultaneously over a trunk (a wide-band channel). We can send one message at baseband, the second shifted to a higher frequency, and the third shifted to a still higher frequency. The shifting is done by amplitude-modulating a carrier signal f_{c2} by the second message and modulating carrier f_{c3} by the third message, as shown in Fig. 9.19c. As long as the message bands do not overlap one another the messages will be transmitted distortion-free; it is up to the designer how wide to make the guard bands between messages without wasting spectrum space. This process is referred to as frequency multiplexing and will be addressed further in the next section.

Frequency Modulation (FM)

In frequency modulation, the frequency rather than the amplitude of the carrier wave is made to vary in proportion to the varying amplitude of the modulating signal, as shown in Fig. 9.18d. A simple method to achieve FM is to vary the capacitance of a resonant LC circuit in a transmitter. Because the frequency of a radio wave is less vulnerable to noise than the amplitude, FM was originally introduced to reduce noise and improve the quality of radio reception. In order to accomplish this FM radio signals have bandwidth several times that of AM signals. Bandwidths six times or larger are common. For example, commercial stereo FM broadcasting (88–108 MHz) is assigned

a

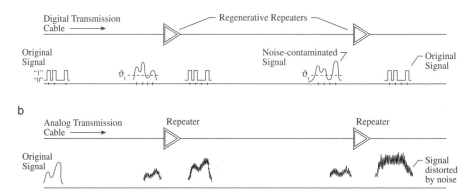

b

FIGURE 9.20   (a) Digital and (b) analog signals are both severely attenuated and distorted by noise when traveling long distances either on wires or in free space. In analog transmission, both noise and distortion are amplified by the repeaters (which are basically amplifiers), leading eventually to an unrecognizable signal. Analog amplification has no mechanism for separating noise from signal. In digital transmission, the contaminated signal at the input of each regenerative repeater is examined at each time position to see if the voltage is higher or lower than the reference voltage v_t. Using this information the repeater then regenerates the original digital signal.

a bandwidth of 200 kHz in which to broadcast 15 kHz of audio-music bandwidth. One speaks of FM trading bandwidth for noise. Also in AM if the amplitude of modulation is to be increased, the power must be increased proportionately. In FM the amplitude of the frequency modulation can be increased without increasing the power at all. In addition, since the amplitude of the FM signal remains constant, amplitude limiters can be set close to the FM signal amplitude and thus very effectively reduce impulse noise. AM was adopted for the transmission of the video part of a TV signal because AM is the least wasteful of the radio frequency spectrum, which is a precious commodity in a wireless environment. FM, though, because of its relative noise-free reception, is used to transmit the audio part of the television signal.

Pulse Code Modulation (PCM)

PCM is one of the most noise-resistant transmission methods. The main advantage of pulse code modulation is that the message in PCM is a train of pulses of equal height which can be regenerated almost perfectly any number of times along the propagation path because the information is not dependent on the precise height and shape of the pulses, but is more dependent on recognizing whether the pulse is there or is not there. A pulse which has deteriorated in shape and size still can be identified at the receiver as the presence of a pulse, that is, as the presence of a bit. Before the pulses in a pulse train become so distorted that error-free decision between 1's and 0's becomes difficult, the pulse train is simply regenerated to its original form and resent as if it were the original message. Figure 9.20 shows how regenerative repeaters extract the original digital signal from a noise-contaminated digital signal. Although the received signal has a very different appearance from that transmitted, it is possible to obtain a perfect replica of the original by means of a circuit which samples the distorted waveform at the midpoint of the digital symbol period, and generates a high or low voltage output depending on whether the sample is above or below an appropriate threshold voltage v_t. If it is above then it is decided that a "1" is present, and if below then a "0." Hence, a PCM repeater is not a conventional amplifier but rather a circuit designed to detect whether an electrical pulse is present or absent. A moderate amount of noise added to a pulse does not alter the basic fact that a pulse is present. Conversely, if a pulse is absent from a particular time slot, a considerable amount of noise must be present before there will be a false indication of a pulse. A device that can perform the A/D (analog/digital) and the D/A conversions is referred to as a *coder-decoder*, or simply a *codec*. Bandwidth requirements for the transmission of pulses were already determined in (9.20): essentially pulses of length T can be transmitted over a channel of $1/2T$ hertz. It can now be stated that in modern communications PCM has facilitated the convergence of telecommunications and computing: because simple on–off pulses that represent bits are transmitted, PCM messages can easily interact with the digital computer. This natural interaction between digital communication networks and computers gave rise to local area networks (LANs) and the Internet.

Pulse code modulation is a process which begins by low-pass filtering the analog signal to ensure that no frequencies above f_{max} are present. Such a filter is called an antialiasing filter. The next step is to sample the signal. A clock circuit generates pulses at the Nyquist sampling rate of at least $2f_{max}$, which are used by the sampler to produce $2f_{max}$ samples of the analog signal per second. This is followed by quantizing (into 2^n levels) and rounding off each sampled value. The result of sampling and quantizing— called digitization—is a series of varying-amplitude pulses at the sampling rate. Such a pulse-amplitude-modulated (PAM) signal, discrete in time and in amplitude, is shown in Fig. 9.16a. This signal is basically a step-modulated AM signal subject to degradation by noise as any AM signal is. To convert these step pulses to a digital signal, the quantized value of each step is expressed (coded) as a binary number. The bits of the binary number can then be represented by simple on–off pulses. The pulses are grouped according to the grouping of the bits in the binary number and are then transmitted in rapid[25] sequence. As the transmitted on–off pulses are of equal amplitude, they are immune to additive noise in the sense that only the presence or absence of a pulse needs to be determined at the receiver. This process is reversed at the receiver, which converts binary data into an analog signal by first decoding the received binary numbers into a "staircase" analog signal (Fig. 9.16a) which is then smoothed by passing it through a low-pass filter with bandwidth f_{max}.

EXAMPLE 9.18 Figure 9.17 shows an analog signal and the pulse-code-modulated bit stream which represents it. Since the sampling of the analog signal is fairly coarse, it is

---

[25]The rapidity of the pulses is determined by the quality of the hardware, with high-grade equipment operating at gigabit rates. An additional bit might be needed to signal the beginning and end of each group, allowing the receiver to identify and separate the bit groups.

sufficient to use 3 bits ($n = 3$) to code the voltage range, giving an accuracy of 1 part in 7. The resulting bit stream has a rate of $2 f_{max} \cdot n = 6 f_{max}$ and the signal-to-noise ratio is $6n = 18$ dB. The channel bandwidth required for transmission of the bit stream is $3 f_{max}$ using the criterion of 2 bits per cycle at the highest signal frequency (see (9.20)). Thus when transmitting a 3-bit coded PCM signal, the channel bandwidth must be three times larger than when sending the original analog signal.

Figure 9.18 f is another bit stream with a 4-bit grouping representing the analog signal of Fig. 9.18b. Here the bit stream is $8 f_{max}$ and the SNR is 24 dB. The bandwidth requirement is $2 f_{max} \cdot n/2 = f_{max} \cdot n = 4 f_{max}$, or four times the analog signal requirement. *Note*: for PCM the necessary transmission bandwidth is simply the analog signal bandwidth multiplied by the bit depth.   ■

9.5.9 Multiplexing

Sending more than one message over a single channel generally requires a technique referred to as multiplexing.[26] The need for multiplexing of messages is obvious: when building a single communication link between two cities, whether by laying down a bundle of twisted-pair telephone lines, a coaxial cable, fiberoptic cable, or a satellite link, one can expect that thousands of messages will be routed over such a link, also referred to as a trunk.

Frequency-Division Multiplexing

In the older system of analog transmission what is typically done is to divide the entire bandwidth of such a link, which can be viewed as one big channel, into smaller subchannels. For example, a single twisted-pair telephone cable, less than 1 km long, has a channel bandwidth of about 1 MHz which could be potentially divided into 250 telephone channels. Each of these subchannels would have a bandwidth of about 4 kHz, which is sufficient for voice conversation, but not music, which requires at least a 10 kHz bandwidth and would sound unnaturally tinny if received over telephone. We refer to division of a large bandwidth into smaller ones as frequency-division multiplexing. It is a method that is available for both analog and digital signals. The technique for separating messages in frequency space is AM modulation, which was already considered in the previous section on AM. In Fig. 9.19c we showed that when each of several messages modulates a single-frequency signal (a carrier), the messages are separated and

---

[26]To explain multiplexing, which is a technique to send different messages over the same link at the same time, many references use the following analogy: two pairs of people in a crowded room could carry on distinct conversations if one pair would speak in low-frequency tones and the other in high-frequency tones, with each pair being able to listen to only their tones, thus filtering out the other conversations, which would be an example of frequency-division multiplexing. Another way that the two conversations could take place would be for each pair to speak at different times only. Alternating conversations is an example of time-division multiplexing. Different pairs huddling in different parts of the room for the purpose of conversing would be an example of space-division multiplexing (using different links or circuits to carry conversations). Finally, if different pairs speak on different topics, one pair might be able to listen only to their topic, screening out other conversations, which would be an example of code-division multiplexing.

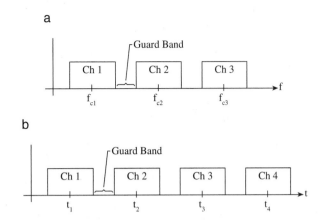

FIGURE 9.21 Finite-bandwidth signals are separated in the (a) frequency domain and (b) time domain by time-division multiplexing.

shifted to higher frequency bands which then can be transmitted together over a wide-band channel. Figure 9.21a shows three messages occupying three distinct frequency bands. The messages are separated by the difference in frequency of the different carrier signals f_{c1}, f_{c2}, and f_{c3}. Of course the wide-band channel must have a bandwidth equal to the sum of the individual channel bandwidths. At the receiver, a bank of filters, occupying the same frequency bands as those shown in Fig. 9.21a, is used to separate the different message signals which then can be demodulated. Another example of frequency multiplexing is AM and FM broadcasting. There, many stations transmit different frequency signals simultaneouly over a wide-band channel known as the AM band (540 kHz–1.6 MHz) and the FM band (88–108 MHz).

Time-Division Multiplexing

Time-division multiplexing (TDM), on the other hand, is only used with digital signals, which are a stream of pulses representing the 0 and 1's of a message. Since modern digital equipment can process 0 and 1's much faster than the 0 and 1's that come from a typical message, we can take several messages and interleave the 0 and 1's from the different messages and send the packet simultaneously over a single channel. Thus TDM is a method for combining many low-speed digital signals into a single high-speed pulse train. Multiplexing C channels, each sampled at S samples per second and coded using n bits per sample, gives a rate R for the pulse train of

$$R = C \cdot S \cdot n \text{ bits per second} \tag{9.38}$$

At the receiving end the messages are separated by a decoder. This is a very efficient way to send many messages simultaneously over the same channel.

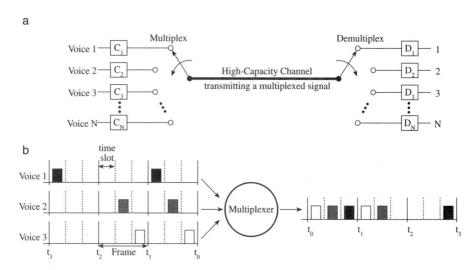

FIGURE 9.22   (a) Time-division multiplexing of N voice channels.   (b) Three low-speed digital voice signals are combined (by interleaving) into a higher-speed pulse train.

Figure 9.22a shows a multiplexer and demultiplexer, represented by a mechanically driven switch. N voice[27] channels are placed sequentially on a high-capacity channel and again separated at the receiving end by the demultiplexer. In TDM the high-capacity channel is divided into N "logical" channels and data in each of the N incoming voice channels are placed in a designated "logical" channel. The procedure is as follows: time on the high-capacity channel is divided into fixed length intervals called frames. Time in each frame is further subdivided into N fixed-length intervals usually referred to as slots: slot 1, slot 2, ... , slot N. A slot is 1 bit wide.[28] A "logical" channel occupies a single slot in every frame. For example, the first "logical" channel occupies slots $1, N + 1, 2N + 1, \ldots$; the second occupies slots $2, N + 2, 2N + 2, \ldots$; the third slots $3, N + 3, 2N + 3, \ldots$; and so forth. A given "logical" channel therefore occupies every Nth slot, giving us N "logical" channels in which to place the N incoming messages. At the receiving end of the high-capacity channel the bit stream is readily demultiplexed, with the demultiplexer detecting the framing pattern from which it determines the beginning of each frame, and hence each slot. An integrated-circuit codec (encoder/decoder) carries out antialiasing filtering, sampling, quantization, and coding of the transmitted signal as well as decoding and signal recovery on the receiving side. Figure 9.22b shows how TDM interleaves the three voice signals represented by the

---

[27]The fact that the signals are all shown as encoded voice is unimportant. Digitized signals of all types (data, graphics, video) can be interleaved in the same way. Only the pulse rate, and not the nature of the original signal, is a factor.

[28]A reference timing signal (called the *framing signal*) identifies the pulse position of voice channel 1; the subsequent pulse positions of channels 2, 3, ... , N are generated by an electronic counter that is synchronized to the framing signal. For ease of understanding we stated a slot to be 1 bit wide. However, it can be wider, often 1 byte wide as in 8-bit coding of telephone signals shown in Fig. 9.23.

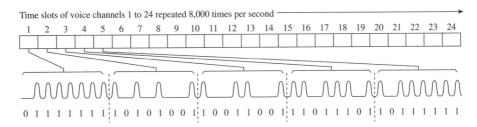

FIGURE 9.23 Time-division multiplexing of 24 telephone voice channels which are sampled in sequence 8000 times a second, with the amplitude of each sample represented in binary form.

dark, shaded, and clear pulses into a faster bit stream. The time frames, denoted by the vertical lines, have three slots, one for each voice signal. The first voice signal pulse occupies the first slot in each frame, the second signal pulse the second slot, and the third the third slot. At the receiving end the three pulse streams are separated by use of a reference timing signal (framing signal). The framing signal identifies the pulse position of voice 1; the voice 2 and voice 3 pulse positions are generated by an electronic counter that is synchronized to the framing signal.

T-1 Carrier System

To allow effective connection of digital equipment into a network that was originally devised to carry only analog voice signals, Bell Laboratories in 1962 designed the T-1 carrier system. This system is designed for digital transmission of voice signals by time-multiplexing 24 voice channels, each using a 64 kilobit per second data rate, into a 1.544 megabit per second pulse stream which is transmitted over the same line. Because the standard twisted pairs of copper wires used are very lossy, short-haul distances of only 30 miles or so can be achieved and only then by use of regenerative repeaters which are placed about every mile along the path to restore the digital signal. To replace all twisted pairs of copper wires currently in use by broadband and less lossy coax cables or fiberoptic cables is too expensive at this time but the trend is in that direction. Figure 9.23 shows the T-1 system in which 24 voice channels, with each voice channel first filtered to a bandwidth of 4000 Hz, are sampled in sequence 8000 times a second (or every 125 microseconds (μs)), and the amplitude of each sample is represented in binary form using an 8-bit binary number, as in PCM. A frame is the time interval in which all 24 samples are transmitted or alternatively the interval between samples of a single voice channel. Each time slot holds an 8-bit binary number which gives the value of a voice channel sample. A frame then has $24 \cdot 8 = 192$ bits. An extra bit (the 193rd bit) is added to each frame as the framing signal, resulting in the above-stated rate of 193 bits/frame $\cdot$ 8000 frames/s $= 1.544$ Mbps. This can also be obtained from (193 bits/frame)/(125 μs/frame) $= 1.544$ Mbps.

As was stated in Example 9.17, in telephony a sample of a telephone signal consisting of 8 bits is sent every 1/8000 of a second. But the transmission of this signal takes so little time that a large part of the 1/8000 of a second is not used. Applying time multiplexing, the T-1 system sends 24 signals simultaneouly by interleaving the binary values of the samples of the different messages and transmitting the values as a succession of 0 and 1's. The high-speed, broadband 1.544 Mbps T-1 service is available for home and office but is expensive, currently about $1000 per month.

If the bandwidth offered by T-1 is inadequate, T-3 service is available and offers 30 times the capacity of T-1. T-3 can handle digital data at 44.7 Mbps. (T-2 seems to have been passed by.)

9.5.10   ISDN

Analog service over a twisted pair of copper wires, which until now has provided the standard telephone service between homes and the analog switch in the local exchange (central office), is known as POTS (plain old telephone service). The home–central office connection, which provides the access to high-speed networks, is also known as the *last mile*, the local loop, or the subscriber loop, and typically is a bottleneck as it is the slowest link in a network. Various techniques are under investigation, or currently being implemented, to enable the standard twisted copper wires to carry digital signals with a much higher bandwidth than the 4 kHz for which they were originally designed.[29] One of these techniques,[30] initiated in the early 1980s and in service now, has become a world-wide standard: it is *ISDN*, which stands for integrated services digital network. It is a relatively slow, narrowband digital service which includes digital voice telephone, fax, e-mail, and digital video. Modern communication networks carry voice calls and data in digital form; in that sense ISDN can be called modern except for its relatively slow speed. Because of its slow speed, ISDN is falling out of favor in the marketplace and other, speedier techniques such as DSL and cable modems have been developed to be used in the "last mile" to the home (or the first mile to the Internet). ISDN sends digital signals over standard telephone copper wires. Clarity and high-speed information transfer are the advantages of a digital network. Clarity is the result of digital signals being more robust against noise that deteriorates analog transmissions. High speed is the result of digital technology using efficient signal processing, modulation, and coding techniques to achieve a much higher information transfer rate than analog techniques. Because of intense, ongoing research, there is promise of even greater speeds as digital compression and other sophisticated techniques become available. To be more precise, we can state the advantages of digital technology as follows:

(i) Digital transmission can be regenerated provided that the noise is not too high.

---

[29]It appears that V.90 modems have reached a maximum speed of 56 kbps using the 4kHz bandwidth of POTS.

[30]Another example—out of many—is ADSL (asymmetrical digital subscriber line), which will be considered more fully in a later section; it is a leading technology for providing high-speed service over traditional copper line facilities, typically running at 1Mbps and beyond. ISDN, ADSL, Ethernet, 1–5 Mbps cable modems, and ordinary 56 K modems provide the access part, the "last mile," to high-speed data networks which are being built with optical fibers. These fiber networks connect switching stations, cities, and countries and run at speeds in the gigabit per second range and are the high-speed backbone of the Internet. One can speculate that in the future when the cost of optical fiber links decreases sufficiently, even the "last mile," currently a severe bottleneck, will be high-speed fiber optics.

(ii) Data and voice can be multiplexed into one bitstream.

(iii) Sophisticated coding and processing techniques can be used to guarantee reliability and increase the information delivery rate.

Even in its most basic configuration, ISDN can move 128 kbps, which is more than
twice as fast as the standard computer modem (56 kbps). Data-compression software
can give still higher speeds, allowing faster downloads of software from the Internet
or graphics-rich Web pages. Thus, the same twisted-pair copper telephone line that
could traditionally carry only one voice, one computer (by use of modem), or one fax
"conversation" can now carry as many as three separate "conversations" at the same
time, through the same line. However, it is a costly (about $100 per month) and hard-to-
install telephone connection that requires the purchase and configuration of a terminal
adapter—the ISDN equivalent of a modem. Keep in mind, though, that each ISDN
circuit is equivalent to three communication channels which can be combined for digital
data, used separately for voice, fax, or modem,[31] or split with one line used for voice
and the others for data. However, interest in ISDN is waning because the technology is
too expensive for the increase in data rates that it provides. The focus is on technologies
that can provide megabit rates into homes, which traditionally has been the weak link.
For comparison, links between telephone switching stations and between cities (trunks)
are almost entirely fiber-optic with gigabit rates.

Technically we can state that ISDN consists of two 64 kbps bearer channels (B channels) plus one 16 kbps data channel (D channel), referred to as the $2B + D$ access. The
maximum combined rate is thus 144 kbps that can be sent over most links between the
customer and the local exchange—links which were originally installed to carry analog
telephone signals over pairs of twisted copper lines. This is made possible by replacing
the slow analog switch at the central station (also referred to as the local exchange or
central office) by a faster and intelligent digital switch which basically is a large, digital
computer with many input/output ports. The availability of these central office digital
switches—which have diverse capabilities, depending on their manufacturer—is grow-
ing rapidly. As shown in Fig. 9.24, ISDN is delivered to the user from a digital switch
through a user interface, called the *Basic Rate Interface* (BRI), which contains the three
separate channels or "pipes." ISDN telephones can, of course, call to and receive calls
from ordinary telephones, since the digital and analog systems are fully interconnected.
To obtain more bandwidth, more complex interfaces which bundle $23B$ channels (the
$23B + D$ access) or more are also available from the telephone companies. The $23B + D$
access gives a transmission rate of 1.544 Mbps and is designed for transmission through
a standard T-1 trunk. Even more complex systems are available at the local exchanges,
with most of these services being narrowband ISDN, with limited bandwidth and data
rate speeds. The next section will consider high-speed, broadband networks.

---

[31]Recall that the fax and modem use analog signals for transmission over copper lines. A process called
modulation is used to take the computer's binary ones and zeros and convert them to a series of analog tones.
On the receiving end, a process called demodulation converts the tones back to their digital equivalents. The
name modem comes from MOdulate/DEModulate.

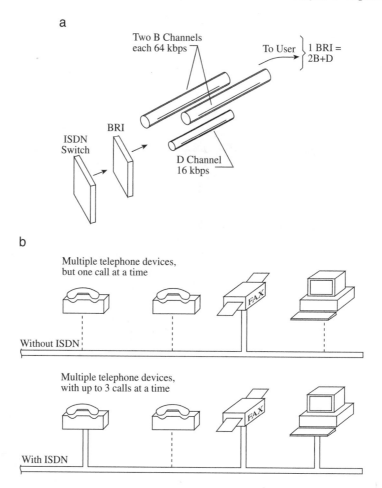

FIGURE 9.24   (a) Principles of ISDN. A BRI delivers three separate channels: two *B* channels for user voice, data, image, and audio, and one *D* channel for packet-switched call setup and signaling. Other functions can be programmed into the digital switch to meet specific needs. (b) With ISDN three separate "conversations" can take place at the same time. Shown are simultaneous conversations with a telephone, with a fax machine, and with a computer over a single line.

9.5.11   Circuit Switching

In ISDN, which is a circuit-switched technology, the signals from all customers are digital and remain digital until the destination. Voice or any other analog sources are first coded by pulse code modulation, that is, an analog waveform is sampled, quantized, and encoded digitally for transmission over the network.

In circuit switching[32] a connection (link) is established between two end points before data transfer is begun and torn down when end-to-end communication is finished. The end points allocate and reserve the connection bandwidth, once in place, for the entire duration, even if no data are transmitted during temporary lulls. Even when the connection uses multiplexed signals, designated bits in the multiplex frame will be given over and remain allocated for that link throughout the life of that connection. The transmission delay from end-to-end is the propagation delay through the network, which typically is very small—a fraction of a second even for satellite links. Small delays can be of great advantage when a message needs to be transmitted in real-time.

Circuit-switched links (also referred to as circuit-oriented service) are ideal for a popular type of multimedia known as streaming media traffic. Streaming media is analogous to broadcast media in that the audio or video material is produced as soon as a computer receives the data over the Internet. By its very nature, streaming media has to flow continuously to the user's computer, so it cannot follow the same traffic rules as conventional data which can be "bursty"[33] and tolerate long delays (e-mail traffic, for example). Typically, we classify data as bursty; however, video can also be bursty but less extreme. Circuit switching, once established, is equivalent to a direct connection. It has a desirably small and constant delay across the circuit and data arrive in the same order that they were sent—and in that sense it is ideally suited for voice, music, and video. On the other hand, circuit switching, even though fast and ideal for streaming, can be very inefficient since most of the links will be idle at any given time.

9.5.12 Broadband ISDN and Asynchronous Transfer Mode (ATM)

The continuing information revolution requires services with ever-increasing data rates. Even the individual consumer, who is limited by dial-up telephone lines using 56 kbps modems or ISDN, is increasingly demanding high-speed broadband access into the home to achieve the best Internet experience possible. "Broadband" is the term used to describe high-speed access to data, voice, and video signals. With this capability, Web pages appear faster, audio and video files arrive quickly, and more than one person in a household can access the Internet at the same time. Broadband technologies such as cable, satellite, and digital subscriber line (DSL) bring Internet information to the home, in any format, with quality results. A service, newer than ISDN, called *broadband ISDN* (B-ISDN), which is based on the asynchronous transfer mode (ATM), supports connectivity at rates up to, and soon exceeding, 155 Mbps. There are major differences between broadband ISDN and ISDN. Specifically, end-to-end communication is by asynchronous transfer mode rather than by synchronous transfer mode (STM)[34]

---

[32]The telephone network is an example in which circuit switching provides a dedicated path or circuit between stations (during a voice conversation, for example).

[33]The bit rate can change abruptly between high bit rates and low or zero bit rates.

[34]STM is a circuit-switched networking mechanism where a connection (link) is established between two end points before data transfer is begun and shut down when end-to-end communication is finished. Data flow and arrive in an orderly fashion (i.e., in the same order that they were sent) but the connection bandwidth stays in place, even if no data are transmitted during temporary lulls.

as in ISDN. Since the early 1990s, the most important technical innovation to come out of the broadband ISDN effort is ATM. It was developed by telephone companies, primarily AT&T. It is a high-data-rate transmission technique—in some ways similar to traditional telephone circuit switching—made possible by fast and reliable computers that can switch and route very small packets of a message, called *cells*, over a network at rates in the hundreds of Mbps.

Generally speaking, *packet switching* is a technique in which a message is first divided into segments of convenient length, called *packets*. Once a particular network software has divided (fragmented) the message, the packets are transmitted individually, and if necessary are stored in a queue at network nodes[35] and orderly transmitted when a time slot on a link is free. Finally the packets are reassembled at the destination as the original message. It is how the packets are fragmented and transmitted that distinguishes the different network technologies. For example, short-range Ethernet networks typically interconnect office computers and use packets of 1500 bytes called *frames*; the Internet's TCP/IP uses packets of up to 64 kbytes called *datagrams*; and ATM uses very short packets (cells) of 53 bytes.

From this point forward, when we speak of packet switching, we understand it to be the TCP/IP connectionless service[36] used by the Internet in which the message transmission delay from end-to-end depends on the number of nodes involved and the level of traffic in the network. It can be a fraction of a second or take hours if network routes are busy. Since packets are sent only when needed (a link is dedicated only for the duration of the packet's transmission), links are frequently available for other connections. Packet switching therefore is more efficient than circuit switching, but at the expense of increased transmission delays. This type of a connectionless communication with variable-length packets (referred to as *IP packets* or *datagrams*) is well suited for any digital data transmission which contains data that are "bursty," i.e., not needing to communicate for an extended period of time and then needing to communicate large quantities of information as fast as possible, as in e-mail and file transfer. But, because of potentially large delays, packet switching is not as suitable for real-time services (streaming media) such as voice or live video where the amount of information flow is more even but very sensitive to delays and to when and in what order the information arrives. Because of these characteristics, separate networks are used for voice and video (STM), and data. This is a contentious issue at present because practically minded corporations prefer a single data network that could also carry voice in addition to data because there is more data than voice traffic.

ATM combines the efficiency of packet switching with the reliability of circuit switching. A connection, called a *virtual circuit* (VC), is established between two end points

---

[35]Communication nodes transport information and are fast computers (also called routers or switches) that act as high-speed network switches. Terminal nodes, such as telephones, printers, and computers, use or generate information. Nodes are interconnected by links of transmission media, such as copper wire and cable, fiberoptic cable, or wireless technology.

[36]Before we continue with the development of ATM we have to understand that the present Internet uses a packet-switching service that is based on TCP/IP protocols. TCP/IP has "powered" the Internet since its beginnings in the 1970s (known then as ARPANET) and continues to be the dominant Internet switching technology with updated and more sophisticated protocols.

before data transfer is begun and torn down when end-to-end communication is finished. ATM can guarantee each virtual circuit the *quality of service* (QoS) it needs. For example, data on a virtual circuit used to transmit video can experience small, almost constant delays. On the other hand, a virtual circuit used for data traffic will experience variable delays which could be large. It is this ability to transmit traffic with such different characteristics that makes ATM an attractive network technology.

The most recent implementation of ATM improves packet switching by putting the data into short, fixed-length packets (cells). These small packets can be switched very quickly (Gbps range) by ATM switches[37] (fast computers), are transported to the destination by the network, and are reassembled there. Many ATM switches can be connected together to build large networks. The fixed length of the ATM cells, which is 53 bytes, allows the information to be conveyed in a predictable manner for a large variety of different traffic types on the same network. Each stage of a link can operate asynchronously (asynchronous here means that the cells are sent at arbitrary times), but as long as all the cells reach their destination in the correct order, it is irrelevant at what precise bit rate the cells were carried. Furthermore, ATM identifies and prioritizes different virtual circuits, thus guaranteeing those connections requiring a continuous bit stream sufficient capacity to ensure against excessive time delays. Because ATM is an extremely flexible technology for routing[38] short, fixed-length packets over a network, ATM coexists with current LAN/WAN (local/wide area network) technology and smoothly integrates numerous existing network technologies such as Ethernet and TCP/IP.

How does ATM identify and prioritize the routing of packets through the network? In the virtual circuit packet-switching technique, which ATM uses, a call-setup phase first sets up a route before any packets are sent. It is a connection-oriented service and in that sense is similar to an old-fashioned circuit-switched telephone network because all packets will now use the established circuit, but it also differs from traditional circuit switching because at each node of the established route, packets are queued and must wait for retransmission. It is the order in which these packets are transmitted that makes ATM different from traditional packet-switching networks. By considering the quality of service required for each virtual circuit, the ATM switches prioritize the incoming cells. In this manner, delays can be kept low for real-time traffic, for example.

In short, ATM technology can smoothly integrate different types of traffic by using virtual circuits to switch small fixed-length packets called *cells*. In addition, ATM can provide quality of service guarantees to each virtual circuit although this requires sophisticated network control mechanisms whose discussion is beyond the scope of this text.

---

[37] A *switch* (once called a *bridge*) is a multi-input, multi-output device whose job is to transfer as many packets as possible from inputs to the appropriate outputs using the header information in each packet. It differs from a *router* mainly in that it typically does not interconnect networks of different types. ATM switches handle small packets of constant length, called *cells*, over large distances, whereas, for example, Ethernet switches handle larger packets of greatly variable length over shorter distances.

[38] In a packet-switching system, "routing" refers to the process of choosing a path over which to send packets, and "router" (a network node connected to two or more networks) refers to a computer making such a choice.

Summarizing, we can say that we have considered three different switching mechanisms: circuit switching used by the telephone network, datagram packet switching used by the Internet, and virtual circuit switching in ATM networks. Even though ATM was developed primarily as a high-speed (Gbps rates over optical fibers), connection-oriented networking technology that uses 53-byte cells, modern ATM networks accept and can deliver much larger packets. Modern ATM is very flexible and smoothly integrates different types of information and different types of networks that run at different data rates. Thus connectionless communication of large file-transfer packets, for example, can coexist with connection-oriented communication of real-time voice and video. When an ATM switch in a network receives an IP packet (up to 64 kbytes) or an Ethernet packet (1500 bytes), it simply fragments the packet into smaller ATM cells and transmits the cells to the next node.

9.5.13 Transmission Control Protocol/Internet Protocol (TCP/IP)

The Internet, which can be defined as a network of networks, requires a common language for exchanging information between many different kinds of computers, such as PCs, workstations, and mainframes, and many different computer networks, such as local area networks (LANs, which are good for interconnecting computers within a few miles) and wide area networks (WANs, which are good for hundreds or thousands of miles). This language, which is a set of protocols called *TCP/IP*, determines how computers[39] connect and send and receive information. Integrating different networks and different computers into a single operational network using TCP/IP is usually referred to as internetworking. The two main protocols of the Internet language are the transmission control protocol (TCP) and the Internet protocol (IP). TCP controls communication between the various computers, while IP specifies how data are routed between the computers on the Internet. For example, if e-mail is to be sent, the application protocol SMTP (Simple Mail Transfer Protocol) formats the message, and if a Web page is to be sent the application protocol HTTP (HyperText Transfer Protocol) formats the page. Then TCP/IP, which is installed on the user's computer, uses TCP to packetize the message (now called TCP packets). Source and destination addresses are then added by IP to form IP packets. IP then routes the IP packets over the Internet. TCP/IP is a connectionless type of packet switching also called *IP datagram packet switching*. In the datagram approach, a message to be sent is first divided into small packets,[40] called *IP packets* or *datagrams*, each with the same destination address. The individual IP packets include information on the packet's origin and destination, as well as error checking information. Each IP packet sent is independent of packets that were

---

[39]"Computers" is used in a broad sense. For example, an internet is composed of multiple networks interconnected by computers called *routers*. Each router is directly connected to many networks. By contrast, a host computer typically connects directly to only one network. A *protocol* is a set of rules or conventions governing the exchange of information between computer systems. *Handshaking* refers to the procedures and standards (protocols) used by two computers or a computer and a peripheral device to establish communication.

[40]IP packets, or datagrams, are the main units of communication over the Internet. They can be as large as 64 kbytes, so they are much larger than ATM cells.

transmitted before. The packets generally follow different routes with different delays and arrive at the destination out of sequence and must be reassembled (the packets must be switched around, hence the name packet switching) at the destination node, using the addressing information that is carried by each packet. It is for the destination node to figure out how to reorder or reassemble them in the original message. The message can experience significant delay from the time that it was sent; even packets can be missing if nodes along the route crashed. However, the destination computer can request that the sender resend missing or corrupted packets. Thus for delivering real-time services such as streaming media with live music and video, the *best effort* of an IP datagram packet-switched network like the Internet can frequently be marginal, especially in congested data traffic. On the other hand, if only a few packets are sent, datagram delivery is quick, can be routed away from congestion, and can find alternate routes to bypass a failed node.

Recall that the most fundamental Internet service is to provide a packet delivery system. Before the Internet, as the number of local networks increased, communication between networks which used different protocols needed to be established. A *gateway*[41] provided such an interface. However, as networks and network traffic increased enormously, the use of gateways to translate the different protocols of the many interconnected networks became inefficient and costly—a standard was needed. In the 1970s ARPANET adopted TCP/IP and forced all connected hosts throughout the world to comply. This technology, which makes the Internet work, continues to function today just as it was envisioned by its founders nearly 30 years ago. We now have a virtual network that encompasses multiple physical networks and offers a connectionless datagram delivery system. The Internet is therefore more than just a collection of networks: it is an interconnected system in which strict conventions allow computers called *IP routers* to connect different networks (ATM, Ethernet, etc.), which in turn allows any computer to communicate with any other computer in the world.

To understand IP routing we first have to look at the architecture of the Internet. It is a multilayer architecture and is simply referred to as the TCP/IP architecture. We can think of the modules of protocol software on each computer as being stacked vertically into layers, with each layer taking the responsibility for handling one part of the communication problem. The TCP/IP model consists of four layers:

- **Application Layer.** It contains protocols designed to implement applications such as ftp (file transfer protocol, used for moving files from one computer to another) and telnet (enables one to interact with a remote computer as if your terminal is directly connected to that computer), and allows these applications on different computers to communicate with each other.

- **Transport Layer.** Also known as the TCP layer or end-to-end protocol, it transfers datagrams across networks and supervises the end-to-end delivery. End-to-end means the two computers at the end of a communication link which imple-

---

[41] A computer that connects two distinctly different communications networks together. Used so that one local area network computer system can communicate and share its data with another local area network computer system; essentially a protocol converter.

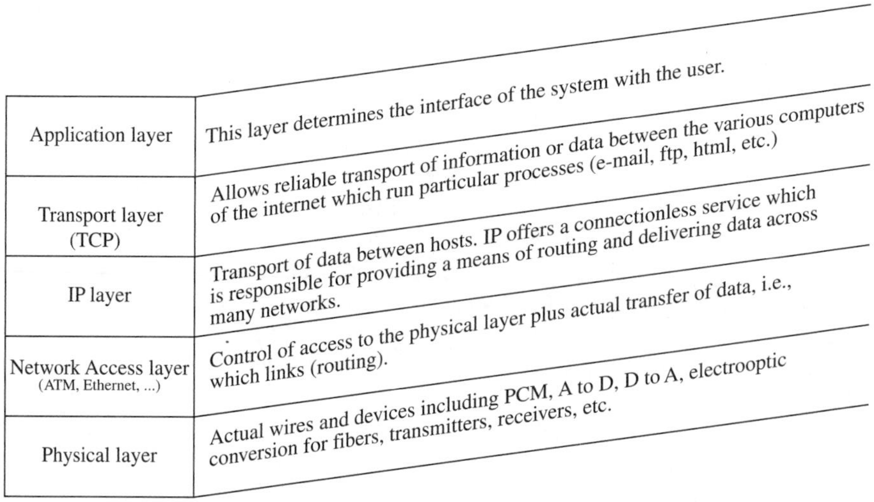

Application layer	This layer determines the interface of the system with the user.
Transport layer (TCP)	Allows reliable transport of information or data between the various computers of the internet which run particular processes (e-mail, ftp, html, etc.)
IP layer	Transport of data between hosts. IP offers a connectionless service which is responsible for providing a means of routing and delivering data across many networks.
Network Access layer (ATM, Ethernet, ...)	Control of access to the physical layer plus actual transfer of data, i.e., which links (routing).
Physical layer	Actual wires and devices including PCM, A to D, D to A, electrooptic conversion for fibers, transmitters, receivers, etc.

FIGURE 9.25   Layered architecture of a network. The physical is the lowest, and the application layer the highest, in the architecture.

ment the transport layer. Computers within the network (which is supervised by the IP layer), such as IP routers and ATM switches, are unaware of the transport layer.

- **IP Layer.** Sometimes referred to as the *network layer*, it supervises the addressing of nodes and the routing of packets across networks; in other words, it implements the end-to-end delivery. It relieves the transport layer of the need to know anything about the underlying technology for data transmission. It is in this layer that the various transmission and switching technologies such as ATM switching and IP routing are implemented. On the other hand, in the simple case of a direct end-to-end link (receiver and sender), there is little need for this layer since the transport layer in combination with the physical layer performs the necessary functions of managing the link.

- **Network Access Layer.** Also known as the *network interface layer* or *data link layer*, it is responsible for accepting IP datagrams and transmitting them over a specific network. The interfacing can be with many different network technologies, ranging from ATM to Ethernet, Point-to-Point, Fiber Distributed Data Interface (FDDI), and so on.

These layers and their role in transferring data are detailed in Fig. 9.25; an additional layer which refers to the physical wiring and cabling is also shown. Note, however, that the Internet architecture does not imply strict layering. This is a system in flux, searching for faster switching and speedier delivery of packets. For example, programmers, when designing applications, are free to bypass the defined transport layer and directly use the IP layer or any underlying layer by writing special code, or they can define new

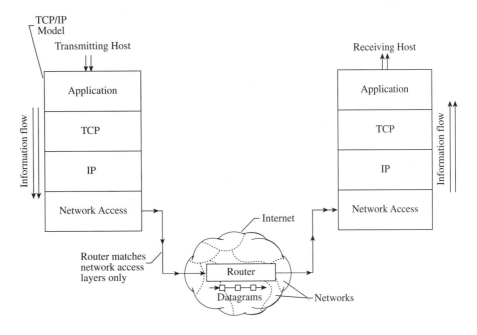

FIGURE 9.26   The standardized TCP/IP architecture of protocol software in layers. The arrows show the direction of a message, initiated by a sender, flowing over the Internet to a receiver.

abstractions that can run on top of any existing protocols as long as these satisfy standards and can be implemented. Intense research in software and hardware continues to increase the speed and the number of packets that can be delivered over existing lines. In this way the Internet architecture has evolved significantly in performance since its beginnings as ARPANET.

Why layering?

It breaks down the problem of designing and building a network into more manageable parts. In a modular design, if a new service needs to be added, the functionality of only a single layer might need to be modified while preserving the functionality of the remaining layers. The layer architecture which defines how traffic is transported through a network allows us to visualize how communication proceeds through the protocol software when traversing a TCP/IP internet, for example, when packets are transmitted using SLIP (Serial Line Internet Protocol) or PPP (Point-to-Point Protocol) between the modem of your home computer and that of your office computer, or how packets are exchanged between office computers that are connected by Ethernet. Figure 9.26 details how a message, originated in the application program of a sender or host computer, is transferred down through the successive layers of protocol software on the sender's machine.[42] The message, now in the form of a stream of bits, is forwarded over the Internet to a receiver where the message is transferred up through the successive layers of protocol software on the receiver's computer. When the message reaches the highest layer in the TCP/IP protocol suite, which is the application protocol layer, it is still not in a format that an application program can use. It is the job of the application layer to transform the message into a form usable by an application program. If the message was a Web page, for example, it will be in HTML, which this layer will convert (using the HTTP application protocol) for use by an application program such as a *browser* like AOL/Netscape Navigator or MS Explorer. Hence all browsers use the same HTTP protocol to communicate with Web servers over the Internet. Similarly, when sending electronic mail using your favorite e-mail program, the SMTP is used to exchange electronic mail. Similarly, RTP (real-time transfer protocol) is designed to transmit audio and video over the Internet in real time. Other applictions are Telnet (allows a user to log on to another computer) and FTP (used for accessing files on a remote computer). Clearly, it is important to distinguish between an application program and the application layer protocol that it uses. As the application protocols sit on top of the transport layer, all use the same TCP connection. This process is reversed when the receiver sends a message and information flows from right to left in Fig. 9.26.

OSI Architecture

To complete the section on Internet communication, we have to include another, later-developed model, the *Open System Interconnection* (OSI) architecture, which is frequently referred to in the media but in practice plays a lesser role than the TCP/IP standard. In general, when reference is made to Internet architecture the TCP/IP model is implied. Nevertheless, OSI protocols find use, especially when new communication protocol standards are developed. Unlike the TCP/IP suite which evolved from research, the OSI protocol suite originated in committee in 1977. Figure 9.27 shows the OSI protocol stack and any relationship between the seven-layer OSI protocols and the four-layer Internet protocols. The relationship between corresponding layers is not exact but only refers to a similarity in function. Of importance is to realize that a layered architecture is designed so that software implementing layer n on the destination computer receives exactly what the software implementing layer n on the source computer sent.

We see that the application layer is divided into three layers in the OSI model. We are not going to detail the OSI model; however, the reader should understand that, for

---

[42]Sender to receiver communication begins when an application on the sender host generates a data message that is then encapsulated by the appropriate application protocol and passed down to the transport layer where a TCP header, containing sender and receiver ports as well as a sequence number, is attached. The TCP segment is now passed down to the IP layer, which incorporates (encapsulates) the segment into an IP datagram with a new header which contains the receiver host address. Finally it passes to the network access layer where a header and a trailer are added which identify the sender's network. We note that the process of encapsulation is repeated at each level of the protocol graph. The appended datagram is now referred to as a frame and is sent over a physical wire, cable, or optic fiber as a series of bits. The encapsulation procedure is reversed when the frames arrive at the receiver host, pass up the protocol layers where headers are removed from the data, and finally are reconstructed as the sent message by the receiver's application program.

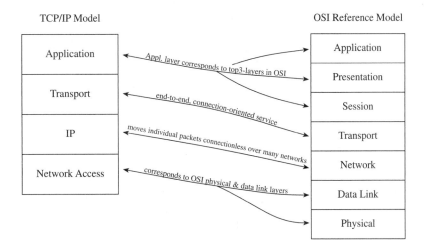

FIGURE 9.27   The OSI architecture model and its relationship to the Internet (TCP/IP) model.

example, if a media release by a software company states that faster layer 5–7 switching was achieved, that "5–7" refers to the session, presentation, and application layers in the OSI architecture. Using Fig. 9.27, we see that the application layer in the TCP/IP model corresponds to the highest three layers in the OSI model. Perhaps the most similarity exists between the transport layers in both models and the IP layer in TCP/IP and the network layer in OSI. As mentioned before, the transport layers provide end-to-end connection-oriented (also called virtual circuit switching) service, including reliability checks, while the destination host communicates with the source host. On the other hand, the next lower layer in both models (IP and network layer) provides connectionless service for packets over many networks. The network access layer, which accepts IP datagrams and transmits them over a specific network such as an Ethernet or Token Ring, etc. The network access layer which is an interface to a specific network technology corresponds to the physical and data link layers in OSI.

9.5.14   ATM versus TCP/IP

At this point it seems appropriate to highlight the relationship between ATM and TCP/IP. It appears that TCP/IP's function is also that of ATM's, namely, the routing of packets over a network. For example, a recent article stated,

> Virtually every major carrier today is in the planning stage for new broadband networks. A core part that is being put in place [is] ATM machines [ATM carries voice, data, and video traffic] as well as optical capability and Internet-protocol capability [IP is another technology that transmits voice, data, and video].

The implication is that there are two different technologies doing the same thing. It is how these packets are transmitted that differentiates the two technologies. A quick comparison between these two technologies follows. Traditional packet switching on the Internet is understood to be connectionless service using TCP/IP protocols in which packets (called IP packets or datagrams) can follow entirely different paths and do not arrive in the same order as sent. On the other hand, ATM switching is a connection-oriented service, also based on packets (called cells) that follow a predefined path (called a virtual circuit) in which packets arrive in the same order as sent. At present, TCP/IP is sufficiently sophisticated that it can integrate many technologies, including ATM and Ethernet (a popular local area network technology that interconnects many computers over short distances such as all office computers in a company). If we define a network as a set of computers, routers, links, and servers, then depending on the protocols used we can classify it as an Ethernet, Internet, ATM net, etc. The interconnection is by routers and switches that can use any of the above technologies or a combination of them. For example, the Internet, which is a network of networks, might use ATM in one network, Ethernet in another, and TCP/IP yet in another, but all networks must seamlessly interconnect according to TCP/IP protocols.

Even though ATM shows great promise as a fast switching technology ideally suited for the Internet, the older and simpler TCP/IP, which "powered" the original Internet (ARPANET) in the 1970s, still continues to be the dominant switching technology for the Internet. This is perhaps because a large installed base of TCP/IP-based applications cannot exploit the benefits of ATM or perhaps IP developers simply do not want the sophistication of ATM in view of advances made in TCP/IP. At this point we can simply state that current TCP/IP protocols are extremely flexible in that almost any underlying technology can be used to transfer TCP/IP traffic; thus ATM can function under TCP/IP—as can Ethernet—to provide fast routing of packets between nodes. Evidently, intense research is improving the older technologies while evolving new ones.

The reader should realize that any vagueness or even ambiguity in the presentation of this material reflects tensions as well as political issues between the TCP/IP and ATM camps. TCP/IP adherents push an "IP-only network" and claim that the "best effort" of an IP packet delivery system is adequate even for streaming media and that ATM as a transport mechanism for IP is not really needed. They also point out that ATM has an undesirable complexity because ATM networks must be aware of all virtual circuit connections by maintaining extensive routing tables, and furthermore, any circuit-switched network, real or virtual, requires considerably more bookkeeping than an IP packet-switched network to supervise the connections, which in turn gives control to companies running ATM. They add that traditional telephone companies do not like the Internet because it is a network they cannot control.

On the other hand, ATM adherents claim that as the Internet grows in complexity, the sophistication of ATM networks will be better suited for future demands such as video conferencing and streaming media that generate constant bit rate traffic and require a small end-to-end delay. Furthermore, ATM is very flexible in allocating quality of service to connections, which IP cannot do. For example, they point out that ATM networks can reserve bandwidth for applications that need small delays whereas IP networking can only guarantee a "best effort" to accomplish this. The reader should come away from this discussion realizing that there is more than one way to design networks and that Ethernets, internets, and ATMs are in a constant state of flux.

Because virtual circuit packet switching requires a call setup, it uses more Internet resources and in that sense is not as efficient as datagram packet switching, which does not require a call setup before any transmission can take place. Also datagram packets can be large, reducing the proportion of overhead information that is needed. On the other hand, connection-oriented service allows for better control of network usage, resulting in better service to the user, that is, delay and packet losses can be controlled, whereas packet switching is a "free-for-all" meaning "first come first serve" or "best effort" service. Clearly, one cannot state that one technique is superior to the other. Furthermore, at this time, intense research is taking place to improve the speed of both techniques.

9.5.15   Digital Subscriber Line (DSL)

As interest in narrow band ISDN is fading, and cable and satellites promise speedier Internet access, telephone companies are turning to broadband services such as DSL to provide high-speed Internet access for the home. Digital subscriber lines and other advanced forms of multiplexing are designed to use as the transport medium the billions of dollars' worth of conventional copper telephone lines which the local telephone companies own, without requiring any new wires into the home. Telephone and other telecommunication companies desire to give their networks a packet-switch orientation and are trying to convert current voice-oriented networks that also carry data into more efficient data-oriented networks that will also carry voice. One such service over a single pair of twisted copper wires is referred to as voice-over DSL. The difference between DSL and traditional telephone service is that DSL converts voice into digital 0 and 1's and sends them in packets over copper wire. Packets from several conversations as well as bits of e-mail and other data travel together in seeming random order. For voice service over DSL, the trick is to deliver the voice packets to the right destination at the right time in the appropriate order, so that the "reassembled" conversations sound natural.

The demand for more network capacity, or bandwidth, closer to the home customer, which is causing telephone companies to deploy DSL, is also causing increased installations of optical fibers across the country, increasing the network's backbone capacity and bringing it closer to the neighborhoods. This is important for DSL which is a copper-based, high-speed but short-distance service in which the customer can be no more than a few miles from a telephone switching station. Figure 9.14 clarifies this restriction and shows that copper lines attenuate a 1 MHz signal by 9 dB in a length of 1 km. At the present, to run broadband fiberoptic lines into homes is too expensive, thus the "last mile" copper wire link between the telephone company's central office and home remains in place.

DSL, which accommodates simultaneous Internet and voice traffic on the same line, can relieve the bottlenecks in the last mile to the home. In DSL, the 1 MHz bandwidth is divided into two greatly unequal parts: the low end of the spectrum, 4 kHz, is used for voice traffic and acts as an ordinary telephone connection[43] (POTS), while the high end, which is practically the entire spectrum, is used for data, typically Internet traffic. The 1 MHz of bandwidth which is available for DSL translates into high-speed data rates of up to 10 Mbps. Of course the high frequencies introduce problems such as high noise and high attenuation that did not exist at 4 kHz; hence sophisticated software and hardware techniques have to be applied at the central office to counter these. To reduce the effects of noise, line-coding techniques are applied to control the frequency band and digital signal processors (DSPs) are applied to restore the original signals from distorted and noisy ones.

There are various flavors of DSL (also referred to as xDSL), for example, *asymmetric* and *very high-speed*. But they have one universal characteristic: the higher the data speed, the shorter the distance between home and switching station must be. In addition all are equipped with modem pairs, with one modem located at a central office and the other at the customer site. Before we give a list of the various types of DSL, let us define a few terms.

- **Symmetrical.** A service in which data travel at the same speed in both directions. Downloads and uploads have the same bandwidth.

- **Asymmetrical.** A service that transmits at different rates in different directions. Downloads move faster than uploads.

- **Downstream.** Traffic is from the network to the customer.

- **Upstream.** Traffic from the customer to the network operating center.

Available types of DSL are:

- **ADSL.** Asymmetric digital subscriber lines deliver traffic at different speeds, depending on its direction, and support a wide range of data services, especially interactive video. ADSL provides three information channels: an ordinary telephone (POTS) channel, an upstream channel, and a higher-capacity downstream channel. These are independent, i.e., voice conversation can exist simultaneously with data traffic. These channels can be separated by frequency-division multiplexing. Downstream speed 1.5–7 Mbps; upstream 16–640 kbps; range 2–3.4 miles.

- **ADSL Lite.** A slower version of ADSL designed to run over digital loop carrier systems and over lengths of more than 3 miles. Dowstream 384 kbps–1.5 Mbps; upstream 384–512 kbps.

---

[43]Recall (Examples 9.5 and 9.17) that copper-based local loops can carry analog voice signals in a bandwidth of only 4 kHz. To convert the analog voice signal to a digital one, we use Nyquist's sampling theorem, which states that an analog signal must be sampled at twice its maximum frequency. Hence the telephone central office samples the received signal at 8 kHz, represents a sample amplitude by 256 levels (or 8 bits, $2^8 = 256$), provided the noise on the analog line is sufficiently low to allow a division into 256 levels, and thus obtains a bit stream of 64 kbps. The 4 kHz limit on the local loop also imposes an upper limit on the speed of analog modems, which now is 56 kbps and most likely will not be exceeded. DSL bypasses this limit by using frequencies up to 1 MHz to carry digital signals over standard copper telephone lines.

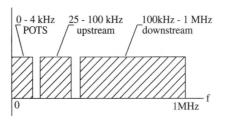

FIGURE 9.28   Frequency ranges in ADSL signals.

- **HDSL.** High-bit-rate digital subscriber lines provide T1 service in both directions for applications that require communications symmetry, such as voice, corporate intranets, and high-volume e-mail. Typical use is between corporate sites. 1 Mbps up- and downstream; range 2–3.4 miles.

- **IDSL.** Uses ISDN-ready local loops. An international communications standard for sending voice, video, and data over digital telephone lines. Up to 144 kbps up- and downstream; range 3.4–4.5 miles.

- **SDSL.** Single-pair symmetric high-bit-rate digital subscriber lines operate on a single copper twisted pair. The advantage is a reduction from two wire pairs to just one. 128 kbps–2 Mbps up- and downstream; range 2 miles.

- **RDSL.** Rate-adaptive digital subscriber lines offer adjustable downstream and upstream rates. This service can adapt its bit rates according to line conditions or customer desires. For example, if a line is noisy, the bit rate can be decreased, making this service more robust. Downstream 40 kbps–7 Mbps; upstream up to 768 kbps; range 2–3.4 miles.

- **VDSL.** Very high-bit-rate asymmetric digital subscriber lines provide very high bandwidth downstream, but have distance limitations and require fiberoptic cable. Originally developed to provide video-on-demand over copper phone lines. Downstream 13–52 Mbps; range 1000 ft.

9.5.16   Cable Modems

Another technology for connection to the home (the "last mile") and an alternative to DSL is the cable modem, which can deliver speeds up to 10 Mbps, or six times faster than a dedicated T-1 line. This technology uses the cable TV (CATV) network which currently reaches 95% of American homes. In this approach, a subset of the available cable channels is made available for transmission of digital signals. As discussed previously, a TV channel provides a 6 MHz bandwidth which can support a 40 Mbps digital stream, with downstream and upstream rates determined by the cable company. Unlike DSL, which is distance-limited, cable's drawback is that its bandwidth must be shared by all subscribers in a neighborhood. Like DSL, cable modems will connect your home to the cable company, with the cable company then defining the traffic on its network.

The implication is that digital traffic can slow down severely if suddenly many customers go on-line in a neighborhood. To reduce such bottlenecks, the cable company can allocate more channels to digital data traffic by taking away TV program channels, a decision most likely to be based on business criteria such as advertising, which might not always be to the liking of the cable modem customer.

9.5.17   Ethernet

A local area network (LAN) is a communication network consisting of many computers such as PCs and workstations that are placed within a local area, such as a single room, a building, or a company. Popular LANs are the Ethernet developed by the Xerox Corporation and the Token Ring developed by IBM. The most popular local area network is the Ethernet, which can connect up to a few hundred computers typically situated in the same building. In an Ethernet, the computers are connected to a common coaxial cable by use of Ethernet interface boards and provide inexpensive, relatively high-speed network access to individual users. We can call this a bus because all connected computers share a single communication channel and a message is received by all computers connected to the network. Ethernet transmission rate is 10 Mbps, but the Fast Ethernet protocol operates at 100 Mbps and gigabit Ethernets are also available. Ethernets make excellent "last mile" alternatives in places such as campus dormitories, for example. The maximum size of Ethernet packets is 1500 bytes. When a computer wants to send a packet to another computer, it puts the source address and the destination address on the packet header and transmits the packet on the cable. All computers read the packet and only the computer with a matching destination address on the packet can receive it. All other computers discard the packet.

9.5.18   The Internet

The Internet is a world-wide network of computers linked by a combination of telephone lines, cable, and fiber optics. It is based on a common addressing system and communication protocol called TCP/IP (transmission control protocol/Internet protocol). TCP/IP converts any type of digital data into smaller packets of data that can be transmitted in quick bursts over any communication line that happens to be available at that time. One packet, which can be a part of computer file, is sent by cable, for example, while the second packet of the same file can be sent by a completely different route and method (satellite, microwave) to its destination where the packets are reassembled into the original file. This makes the Net flexible and very efficient, regardless of geographical distances. The Internet began as ARPANET (Advanced Research Projects Agency Network), established in 1969 by the Department of Defense to provide secure links for research organizations and quickly broadened to academics and the NSF (National Science Foundation), which had implemented a parallel network called NSFNet. The NSF took over TCP/IP technology and established a distributed network of networks capable of handling a large amount of traffic. Currently the World Wide Web (WWW, or Web) is the leading retrieval service for the Internet. It was begun in 1989 at the particle physics laboratory CERN in Geneva, Switzerland. HyperText Transfer Protocol (HTTP) standardized communications between servers and clients using friendly, point-and-click browser software such as Explorer and Netscape in place of arcane Unix-based commands. Servers are network computers that store and transmit information to other computers on the network, while clients are programs that request information from servers in response to a user request. A document on the World Wide Web, commonly called a Web site or a home page, is written in HyperText Markup Language (HTML) and is assigned an on-line address referred to as a Uniform Resource Locator (URL). The ease with which hypertext allows Web users access to other documents has made electronic mail (e-mail), file transfer (via FTP protocol), remote computer access (via telnet), bulletin boards, etc., common among the general public, whereas before the Web appeared the Internet was a province for the technically savvy such as academic, government, and business professionals. Now the Internet allows you to send messages and documents electronically, transfer documents and software from another computer to your own, engage in group discussion, and put your own information out on the Net for everyone else to see. Writing to a colleague or visiting the Louvre is easily done on the Net.

EXAMPLE 9.19   Compare typical access to the Internet from home and from work.

From home, connection is usually via a modem and a telephone line or cable to an internet service provider (ISP). At work, the type of connection depends on whether the company has a computer system with Internet access. If it does, your PC or workstation may be connected via Ethernet cable. If it does not, then your connection is likely via modem and telephone line to an ISP. Typically one will have an Ethernet connection at work and a modem/phone connection at home. If this is the case, different software is needed to match the two different environments. Furthermore, the software for each environment is again divided, depending if one is sending e-mail or browsing the Web, because the resource requirements are very different. Before looking at the four different situations, let us consider the common point: the Internet connection.

Ethernet. Let us begin with a typical company network: a computer system connected in some kind of network with Internet access. The most popular "kind of network" is a switched or shared 10 or 100 MB/s Ethernet—less popular are Token Ring and ATM networks. Your work PC or workstation may then be connected via Ethernet cable to the computer system. An Ethernet cable can either be a twisted copper pair or coaxial cable. Such cables are commonly used to connect computers in a local area network (LAN) which provides the high-speed and wide bandwidth path for the data flow between the computers. The company LAN is then connected to the Internet, also by a high-speed and wide bandwidth connection. A bottleneck can develop when either the speed within the LAN or the connection speed between the LAN and the Internet is inadequate.

Modems. The company computer system normally has a pool of modems so off-site computers can dial in and be connected to the system by use of modem and telephone lines. These connections are much slower (typically 56 kbps) than Ethernet cables because using the analog telephone lines requires a modem at home which must perform a digital-to-analog conversion and another modem at the company which reverses the process and performs the analog-to-digital conversion. If the Ethernet is lightly loaded it can be up to 1000 times faster than the modem connection, but if heavily loaded only about 10 times faster. The speed difference is because Ethernet is digital and telephone lines are analog unless you are using the newer ISDN, SDSL, or ADSL telephone connections which are also digital and much faster than their analog counterparts.[44] When making speed comparisons one should keep in mind that the telephone was made for humans and adapted to use for computers, whereas Ethernet was created for computers from the start. More recently cable connections that use a cable modem have become available. Both cable and DSL support high-speed transmission (up to 10 Mbps) of voice, data, and video—for DSL over conventional copper phone lines and for cable over coaxial copper cable. Comparing the two newer services, we find that both require modems, either a DSL modem or a cable modem. DSL provides dedicated service over a single telephone line; cable modems offer a dedicated service over a shared medium. Although cable modems have greater downstream bandwidth, that bandwidth is shared among all users on a line. DSL users do not share service, so speed will be constant. DSL exists simultaneously with the plain old telephone service (POTS) on the same copper lines. No new lines are required to maintain telephone service, and the new services are unaffected by placing or receiving calls during use. Cable's limitation is that speed will vary with the number of users on a line, whereas DSL's limitation is that it will not work at distances greater than 4 miles from the central phone office, so not all phone customers can use it. In a typical LAN there might be, say, 20 personal computers of virtually every kind, including Windows machines, Macintoshes, and workstations which are connected to several servers within a company department or group. This set of computers is also connected to the rest of the company computer system. The servers may perform one or more of several functions, acting as a common source of software packages for word processing, spreadsheet and data base applications, and file storage and backup, as well as being the host machine for e-mail. The host is usually a computer running Unix, which is the most common operating system that enables several users to use the machine at one time (multiuser) and individual users to run several programs at the same time (multitasking). The Unix OS includes a comprehensive mail-delivery program, sendmail, that implements the Simple Mail Transport Protocol. On systems that provide full Internet access, all tasks[45] use TCP/IP. This use of standard protocols is the key reason why users can interact with each other using a wide variety of hardware and software over the Internet. The use of proprietary software by some commercial service providers restricts what users can do.

E-mail. To use electronic mail requires an e-mail package either on the host or on the client (your PC). *PINE* and *ELM* are packages installed on the host whereas *PCmail*, *Messenger*, *Outlook*, and *Eudora* are on PCs. All of these packages interact with the sendmail program that connects with the network but the amount of information that is passed between your PC and the host is relatively small. The host is always connected

---

[44]Ethernet can have a speed of 10 Mbps or even 100 Mbps. The best modems are only 56 kbps. ISDN comes as 2–64 kbps "channels" which equal 128 kbps uncompressed. Both DSL (distance-limited) and cable (number of users limited) can have speeds of up to 10 Mbps.

[45]Tasks include e-mail, telnet (a mechanism to establish an interactive session from your machine to a remote host machine when both machines are connected through the Internet—invoking the telnet program on your machine will initiate the telnet connection), FTP (file transfer protocol), etc.

to the network and stores the mail it receives until you decide to open it on your PC and act on it. To use e-mail, you need a terminal emulation software package on your PC that connects with the host, for example, *ProComm* and *Hyperterminal*. Alternatively, one will use SLIP (Serial Line IP) or PPP (Point to Point Protocol) with a modem to simulate being connected directly to a LAN. These packages either help you send commands to the e-mail package on the host (PINE, ELM, etc.) or help your PC-based e-mail package (Eudora, etc.) interact with the SMTP server. Web-based e-mail such as Hotmail is also popular. It is a free e-mail system (www.hotmail.com)—but you need Internet access to use it. The advantage with Web-based mail (in addition to Hotmail there is *Yahoo*, and many others) is that you can receive and send e-mail anywhere you can find access to the Web, for example, an Internet cafe anywhere in the world.

If you are connected via a fast Ethernet cable and you do not pay for connect time, you may not be concerned about the time for message composing, editing, reading, and so on. On the other hand, if you are connected via a relatively slower telephone line and modem combination, or if you pay for connect time, you may prefer to use a PC-based e-mail package that enables you to prepare and read messages off-line at your leisure and connect briefly to transmit messages between PC and server. In either case, the faster modems with data compression and error correction are much preferred. Data compression alone gives an effective speed increase of at most four times.

Web Browsers. Installed on PCs, these are based on Windows, Mac OS, or X-Windows (Unix). The most common examples are AOL/Netscape Navigator (originally Mosaic) and MS Internet Explorer. Their use requires your PC to be connected as a node on the Internet, which means that your PC must be assigned an IP address by your system manager. The system manager controls your access to the Internet and authorizes your log-on ID.

If you are connected via Ethernet to a company LAN, you will have a permanent IP address. This is a fast connection that will not cause delays in receiving data, especially when the Web pages contain a high density of graphics. The major causes of delay are most likely the network bandwidth limitations and network traffic congestion. Some Web server sites are overloaded because many users access their pages. Some sites have not installed a connection to the Internet that is adequate for the traffic they are generating.

If you are connected via a modem, you need either a permanent IP address or a temporary one that is assigned each time you log on. You also need a SLIP or PPP connection. PPP is a newer and more general protocol than SLIP. A SLIP or PPP connection requires two conditions: one, you have a communications software package on your PC that supports a SLIP or PPP connection, and two, the network that provides your Internet connection has a modem pool that supports SLIP or PPP. A third and obvious condition is that the network must have a full Internet access connection to the Internet. Delays have two main causes: traffic congestion (either because of an insufficient bandwidth connection or because of servers that are too small) and modems. At times it pays to turn off the graphics display, an option in Web browsers. This alone may speed up your display response time if text is the primary message and the graphics are merely window dressing. On the other hand, if graphics are content, one may lose important information. ▪

9.6   SUMMARY

In this concluding chapter we have pulled together concepts in digital electronics and fundamental concepts in information and communication theory and showed that these seemingly diverse concepts have led to a new form of communicating, embodied by the Internet. Underlying these notions is the trend toward communication that is completely digital. We showed that information can be quantified in terms of bits, that the flow of information can be expressed as a rate in bits per second (bps), and that analog messages can be changed to digital ones by using the principles of the sampling theorem. Bandwidth and channel capacity of a transmission link or trunk were explored, and we found that bandwidth can be traded for noise in a transmission system.

As popular as AM and FM are for analog transmission, so is *pulse code modulation* (PCM) for digital signaling. PCM uses discrete time samples of the analog signal, but instead of simply transmitting the amplitudes of the discrete analog samples at discrete times, it first quantizes each amplitude of the sampled analog signal into a number of discrete levels. A binary code is then used to designate each level at each sample time. Finally, the binary code is transmitted as a sequence of on–off pulses.

The fact that several pulses are needed to code each sample increases the bandwidth. Suppose that the effective analog signal bandwidth is W; then a minimum bandwidth B required to transmit a series of pulses that are sufficiently short so that n of them will fit into a sampling period, in accordance with (9.20), is given by $B = Wn = W \log_2 S$, where S is the number of quantizing levels. Thus the bandwidth is increased by the number n of bits used in the binary code. The advantage of binary PCM is that a decision between two states (in a received signal) is easier and therefore less prone to error than one between several levels. For this reason binary PCM is commonly used.

PCM allows repeated noise-free regeneration of the original digital signal, which implies that a digital signal can be sent over large distances without deterioration. The technique of multiplexing, workable only for digital signals, permits many digital signals originating from diverse sources such as voice, video, or data to be sent at the same time over a single transmission cable, which is a very efficient way of sending messages and has no counterpart in analog signaling.

Access to the Internet over standard telephone lines using modems, even at the 56 kbps speed, can be painfully slow. A technique to allow digital signals to be carried on analog telephone lines with a bandwidth much larger than the original 4 kHz for which they were originally designed is ISDN, which even in its most basic form can carry digital signals at a rate of 128 kbps. The ever-increasing data rates required by the continuing technological revolution has led to the introduction of broadband ISDN or B-ISDN, with data rates in the gigabit per second (Gbps) range to be transmitted over fiberoptic and coaxial cables. B-ISDN is based on the asynchronous transfer mode (ATM), which is a connection-oriented, high-speed, packet-switched digital communication process that can convey many different types of information, including real-time audio and video, which ordinary packet switching could not handle before ATM because packets can be delayed in transit while waiting for access to a link, and this would corrupt the received audio or video signal. Packet switching is ideal for file transfer in connectionless communication with large packets, while real-time voice and video communication work best with a connection-oriented scheme and small packets. The design of ATM is such that it can handle both of these tasks as it is designed to prioritize different transmissions and thus can guarantee voice and video and an acceptably small time delay. It was originally designed for a speed of 156 Mbps over fiberoptic cables, but now the speed is in the gigabit range, which is desirable for use on the Internet.

Office machines connected together in a single room or building form a local area network which is usually governed by Ethernet or by Token Ring protocols. Ethernet is a standard interconnection method with a basic data rate of 10 Mbps over a cable or bus. It is a bus because all connected hosts share it and all receive every transmission. A host interface (an Ethernet card) on each computer chooses packets the host should receive and filters out all others.

The Internet was initially created as a communications network to link university, government, and military researchers. It is a network connecting many computer networks and based on a common addressing system and communications protocol called TCP/IP. A message is transmitted through this network, which sends it by any of a seemingly infinite number of routes to the organization through which the receiver has access to the Internet. It is a connectionless, packet-switched technology originally designed for "bursty" data such as e-mail and file transfer, but TCP/IP has undergone intense research and the "best effort" of the latest TCP/IP protocols can also deliver streaming media with acceptable results. The primary uses of the Internet are electronic mail (e-mail), file transfer (ftp), remote computer access (telnet), bulletin boards, business applications, and newsgroups. The World Wide Web is the leading information retrieval service of the Internet. Browser software such as Netscape Navigator and Internet Explorer allows users to view retrieved documents and Web pages from the Internet.

Problems

1. After asking what sex the newborn baby is, you are told, "It's a boy." How much information did you receive?
   Ans: 1 bit.

2. You are told that you are on square 51 of a numbered 64-square checkerboard. How much information did you receive?

3. A 64-square checkerboard is used as display board. Each square can be either lit or dark. What is the information content of such a display?
   Ans: 64 bits per display.

4. A computer uses 32 bits for addressing. How many addresses can be stored?

5. A computer needs to address 2000 locations. What is the minimum number of bits that are needed?
   Ans: 11 bits.

6. What is the resolution available with 12-bit sound?

7. If the sound intensity level changes by a 100:1 ratio, what is the dynamic range in dB of the sound?
   Ans: 20 dB.

8. If a TV set is to have a horizontal resolution of 600 lines (or 600 pixels), that is,

be able to display 600 alternating dark and light spots along a single horizontal line, what is the

(a) information in bits per pixel if each pixel can have nine gradations of light intensity?
(b) information in bits per line?
(c) bandwidth that the TV circuitry must have?

9. What is the information rate of a light display with four on–off lights which can change once a second?
   Ans: 4 bps.
10. What is the information rate of HDTV (high-definition television) if each pixel has 16 brightness levels with 700 lines of horizontal resolution, it has 1200 lines per frame, and it refreshes at 30 frames per second?
11. The term *baud* gives the signaling rate in symbols per second. For binary systems signaling rate in bauds is equivalent to data rate in bits per second because in a binary system the signaling element takes on one of two different amplitudes: either a one or a zero can be encoded, according to the amplitude value. Suppose each signaling element takes on one of four amplitudes. Each amplitude then corresponds to a specific 2-bit pair (00, 01, 10, 11). In this 2-bit case, bit rate = 2· baud rate. Find the data rate in bps (bits per second) for a digital system using 16 different signaling states and transmitting at 1200 baud.
   Ans: 4800 bps.
12. The highest frequency in a telemetry signal representing the rudder position of a ship is 2 Hz. What is the shortest time for the rudder to change to a distinctly new position? *Hint*: refer to Fig. 9.3*a*, which shows that a signal change from a min to a max takes place in a time of a half-cosine.
13. What is the approximate bandwidth of speech if the highest frequency in speech is 3 kHz?
   Ans: 3 kHz.
14. What is the exact bandwidth of speech if the highest frequency in speech is 3 kHz and the lowest is 100 Hz?
15. If a piece of music in which the highest frequency is 10 kHz is to be converted to digital form, what is the minimum sampling frequency?
16. If an amplifier which has a bandwidth of 100 kHz can amplify 10 s pulses satisfactorily, what bandwidth is needed if the pulses are shortened to 5 s ?
   Ans: 200 kHz.
17. What is the frequency content of a DC signal, of a sinusoid of frequency 1 MHz, and of an extremely short pulse?
18. When converting an analog signal to digital form, explain why the analog signal must first be passed through a low-pass filter which has a cutoff frequency that is half of the sampling frequency.
19. Explain aliasing.
20. Aliasing is a major source of distortion in the reconstruction of signals from their samples. Give two ways to avoid aliasing. *Hint*: one relates to the speed of sampling, and the other to filtering.

21. What is oversampling?
22. Assume the highest frequency present in a speech signal is 3.5 kHz. If this signal is to be oversampled by a factor of 2, find the minimum sampling frequency.
   Ans: 14 kHz.
23. A time-domain signal $v(t)$ has the spectrum $V(f)$ shown in Fig. 9.29 with $f_h = 25$ kHz and $W = 10$ kHz. Sketch $V_s(f)$ for sampling frequencies $f_s = 60$ kHz and 45 kHz and comment in each case on the reconstruction of $v(t)$ from $v_s(t)$.

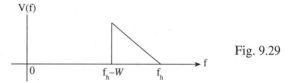

Fig. 9.29

24. What is the maximum number of levels that can be encoded by using 10 bits?
25. It is desired to digitize a speech signal by sampling and quantizing the continuous time signal. Assume the highest frequency present is 4 kHz and quantization to 256 levels (8 bits) is adequate. Find the number of bits required to store 2 minutes of speech.
   Ans: $7.68 \cdot 10^6$ or 7,680,000.
26. In pulse amplitude modulation (PAM) where an analog signal is sampled and represented by varying-height pulses, a minimum bandwidth on the order of $1/(2T)$ is required (see (9.20)) to transmit a pulse train. In pulse code modulation (PCM) the varying amplitude of each pulse is further represented by a code consisting of several pulses (the amplitude of these pulses is constant in binary coding which uses simple on–off pulses). This increases the required bandwidth in proportion to the number of pulses in the code. For example, if 8 quantizing levels are used in a binary PCM system, 3 pulse positions (that is, 3 bits) are used, and thus the PCM signal requires 3 times more bandwidth than if 1 pulse were transmitted (the reason is that 3 pulses are now sent in the same time that it took to send 1 pulse). Similarly, the bandwidth is increased by a factor of 4 for 16 levels in a binary system. Extending this to a binary system with n quantization levels, we see that the required bandwidth is increased by a factor of at least $\log_2 n$.

   (a) In binary PCM coding, find the increase of the transmission bandwidth if 4, 64, and 256 quantization levels are used.
      Ans: 2, 6, 8.
   (b) Find the increase in bandwidth required for transmission of a PCM signal that has been quantized to 64 levels if each pulse is allowed to take on the following number of levels: 2, 3, and y.
      Ans: 6, 4(3.79), and $\log_y 64$.

27. A message has a bandwidth of 4 kHz. If the signal is to be sampled for PCM,

   (a) what is the minimum sampling frequency to avoid aliasing?
   (b) what is the required sampling frequency if there is to be a guard band of 2.5 kHz?

28. The highest frequency component of high-fidelity sound is 20 kHz. If 16 bits are used to encode the quantized signal, what bandwidth is needed as a minimum for the digital signal?
Ans: 320 kHz.

29. A music signal needs to be digitized by sampling and quantizing the continuous time signal. Assume the highest frequency present is 10 kHz and quantization by 8 bits is adequate. Find the number of bits required to store 3 minutes of music.

30. High-fidelity audio has a maximum frequency of 22 kHz. If 20 bits are used to encode the quantized signal, what bandwidth is needed as a minimum for the digital signal?

31. The maximum frequency of a television signal is 4 MHz.

    (a) What is the minimum sampling rate for this signal?
    (b) If 10 bits are used to encode the sampled signal, what is the final bit rate?

32. Determine the number of quantizing levels that can be used if the number of bits in a binary code is 6, 16, and y.
Ans: 64, 65,536, 2^y.

33. When digitizing an analog signal,

    (a) how are guard bands introduced?
    (b) what is the purpose of guard bands?

34. The maximum frequency in an analog signal is 4 kHz. After sampling, the signal is to be reconstructed by passing the samples through a single-section, low-pass RC filter of the type shown in Fig. 9.7*b*.

    (a) Specify the appropriate sampling rate and the filter bandwidth B.
    (b) Relate the sampling rate and filter bandwidth to the distortion of the reconstructed signal.

35. An altitude meter converts altitude to an analog electrical signal. A specific meter has a time constant of 2 s.

    (a) Find the signal bandwidth that can be associated with such a meter.
    (b) If the analog signal is to be converted to a digital signal by sampling, how frequently must the meter output be sampled?
    Ans: 0.08, 0.16.

36. The video bandwidth of a TV channel is given as 4.2 MHz. Using Eq. (9.17), verify that the calculated horizontal resolution for a TV set is 449 lines.

37. Given the number of horizontal raster lines per frame and the number of frames per second in the NTSC standard, calculate the time in seconds allowed for the electron beam to trace one horizontal scanning line.
Ans: 63.5 μsec/line.

38. What is the horizontal line frequency of a typical TV set? (The horizontal section of the circuitry in a TV set which produces an electrical sawtooth signal that is used to sweep the electron beam from side to side must therefore operate at this frequency.)

39. If there are 16 shades of gray per pixel, 500 pixels per line, 525 lines per frame, and 30 frames per second in a TV set, calculate the maximum number of pixels per frame and calculate the maximum information rate in Mbits per second in the TV display.

Ans: 262,500; 31 Mbps.

40. What is the highest frequency present in a signal if a measurement of the shortest half-cosine rise time in that signal is found to be 0.1 s?

41. What is the bandwidth of a 5 V, 0.2 s pulse? Assume its shape approximates that of a square pulse.

Ans: 5 MHz.

42. (a) If a single 5 V, 0.2 s pulse or a series of such pulses (like a square wave) is to be transmitted, what must the transmission bandwidth be for recognition of the pulses at the output of the transmission system?

Ans: 2.5 MHz.

(b) What is the minimum bandwidth required to transmit a pulse train with pulse length of 1 ns?

43. Explain the difference in the bandwidth of Problems 41 and 42(a).

44. Find the minimum channel bandwidth required for pulse detection and resolution of a sequence of 10 s pulses which are randomly spaced. The spacing between pulses varies between 4 and 15 s.

45. Determine the bandwidth required for a data rate of 10 Mbps. Assume simple pulses, equally spaced, such as in the square wave shown in Fig. 9.12, to represent the information bits.

Ans: 5 MHz.

46. A 300 Ω transmission line is connected directly to the 50 Ω input of a television set. Find the percentage of power traveling on the transmission line that is available to the TV set.

Ans: 49%.

47. Assume that an infinitely long 300 Ω transmission line is suddenly connected to a 6 V battery. Calculate the power drain on the battery.

48. A mismatched transmission line is a line that is terminated in a load impedance that is different from the characteristic impedance of the transmission line.

(a) Find the reflection coefficient of a matched line.
(b) Find the reflection coefficient of a shorted line.
(c) Find the reflection coefficient of an open-circuited line.
(d) Find the reflection coefficient of a line in which the load absorbs half the power that is incident on the load.
(e) Find the load impedance for the case of (d).
   Ans: 0, -1, 1, ± 0.707, $Z_L = Z_o \cdot 5.83\ \Omega$ or $Z_o \cdot 0.17\ \Omega$.

49. A 100 m-long RG 58 coax cable is terminated with 150 Ω impedance.

(a) Find the reflection coefficient.
(b) If a very short, 10 V pulse is launched at the input, calculate the time it will take for the pulse to return to the input and voltage be.

50. Using Fig. 9.14, estimate the factor by which the bandwidth of a fiberoptic cable is larger than that of a twisted-wire copper cable.
    Ans: $5 \cdot 10^4$.

51. Define pulse dispersion.

52. An antenna dish needs to concentrate 5 GHz electromagnetic energy into a circle of 1 km in diameter at a distance of 500 km. Find the diameter of the dish.
    Ans: 73 m.

53. A cellular network has 200 antennae. If a switch to PCS (personal communication service) is to be made, how many more antennae would be needed?

54. If a signal-to-noise ratio is given as 20 dB, what is the voltage ratio?
    Ans: 10-to-1.

55. A speech signal has a bandwidth of 3 kHz and an rms voltage of 1 V. This signal exists in background noise which has a level of 5 mV rms. Determine the signal-to-noise ratio and the information rate of the digitized signal.

56. Determine the channel capacity needed for transmission of the digitized signal of the previous problem.
    Ans: 48 kbps.

57. If stereo music requires a bandwidth of 16 kHz per channel and 8 bits of PCM accuracy, how many channels of a T-1 carrier system would be required to transmit this information?
    Ans: 8.

58. If 56 dB of digitization noise is acceptable in a system, determine the maximum number of bits that can be used in this system when encoding an analog signal.

59. In the design of a PCM system, an analog signal for which the highest frequency is 10 kHz is to be digitized and encoded with 6 bits. Determine the rate of the resulting bit stream and the signal-to-noise ratio.
    Ans: 120 kbps, 36 dB.

60. A PCM system uses a quantizer followed by a 7-bit binary encoder. The bit rate of the system is 60 Mbps. What is the maximum message bandwidth for which the system operates satisfactorily?

61. How many megabytes does a 70-min, stereo music compact disc store if the maximum signal frequency is 20 kHz and the signal-to-noise ratio is 96 dB?
    Ans: 672 MB.

62. In a binary PCM transmission of a video signal with a sampling rate of $f_s = 10$ MHz, calculate the signaling rate needed for a SNR ≈ 50 dB.

63. Frequency-division multiplexing, as shown in Fig. 9.19, is used to separate messages in the frequency domain and assign a distinct frequency slot to each message. Four messages, each with a bandwidth B of 4 kHz, are to be multiplexed by amplitude modulation (AM). For proper demultiplexing a guard band of 3 kHz is needed.

    (a) If a low-pass filter with steeper cutoff characteristics could be used in demultiplexing, would the guard band have to be smaller or larger?

    (b) Determine the lowest carrier frequencies f_{c1}, f_{c2}, f_{c3}, and f_{c4} to accomplish frequency-division multiplexing.

(c) Determine the highest frequency component that the transmission system must be capable of propagating.

Ans: (a) smaller; (b) 0, 11 kHz, 22 kHz, 33 kHz; (c) 37 kHz.

64. Ten digital signals are to be time-division multiplexed and transmitted over a single channel. What must the capacity of the channel be if each signal is sampled at 10 kHz and encoded with 10 bits?

65. Common telephone systems have a bandwidth of 3 kHz and a signal-to-noise ratio of 48 dB. Determine the Shannon capacity of such a system.

Ans: 48 kbps.

66. In PCM the sampled value is quantized and the quantized value transmitted as a series of numbers. If, for example, there are 50 such levels, quantizing will produce a set of integers between 0 and 49. If a particular message sample has a quantized level of 22, what number string would be transmitted if the numbers are coded into binary pulse strings?

Ans: 010110.

67. What is the sampling frequency for a telephone signal that has a message bandwidth of 3 kHz if there is to be a guard band of 2 kHz?

68. Twenty-five input signals, each band-limited to 3 kHz, are each sampled at an 8 kHz rate and then time-multiplexed. Calculate the minimum bandwidth required to transmit this multiplexed signal in the presence of noise if the pulse modulation is

(a) PAM (pulse amplitude modulation).

(b) binary PCM (pulse code modulation) with a required level resolution of 0.5%.

Ans: (a) 100 kHz; (b) 800 kHz.

69. The T-1 telephone system multiplexes together 24 voice channels. Each voice channel is 4 kHz in bandwidth and is sampled every 1/8000 of a second, or every 125 μs. If the 24 channels are to share the system by time-division multiplexing,

(a) what is the minimum system bandwidth if the voice signal is sampled for PAM?

(b) what is the minimum system bandwidth if the voice signal is sampled for 8-bit binary PCM?

INDEX

$$t_1 = \ln2 \, (R_1 + R_2) \, C_1$$

$$t_2 = \ln2 \; R_2 C_1$$

$$f = \frac{1}{t_1 + t_2}$$

$$f = \frac{1}{\ln_2 \left[R_2 C_1 + (R_1 + R_2) \, C_1 \right]}$$

$$f = \frac{1}{\ln_2 \left[R_2 C_1 + R_1 C_1 + R_2 C_1 \right]}$$

$$f = \frac{1}{\ln2 \left[R_1 + 2R_2 \right] C_1}$$

$$f = \frac{1.44}{(R_1 + 2R_2) \, C_1}$$